C000220267

BIOCHEMICAL SOCIETY SYMPOSIA

No. 70

PROTEASES AND THE REGULATION OF BIOLOGICAL PROCESSES

BIOCHEMICAL SOCIETY SYMPOSIUM No. 70

held at Imperial College London, December 2002

Proteases and the Regulation of Biological Processes

ORGANIZED AND EDITED BY

J. SAKLATVALA, H. NAGASE AND G. SALVESEN

PORTLAND PRESS

Published by Portland Press,
59 Portland Place, London W1B 1QW, U.K.
on behalf of The Biochemical Society
Tel: (+44) 20 7580 5530; e-mail: editorial@portlandpress.com
http://www.portlandpress.com

ISBN 1 85578 155 7   ISSN 0067-8964

British Library Cataloguing in Publication Data
A catalogue record for this book is available from the British Library

Typeset by Portland Press Ltd
Printed in Great Britain by Bell and Bain Ltd, Glasgow

Contents

Preface

The Biochemical Society's Annual Symposium for 2002, Proteases and the Regulation of Biological Processes, was held at Imperial College, London on 16–18 December. It was dedicated to Dr Alan J. Barrett, on the occasion of his 65th birthday, to honour his research achievements and contributions to the field of proteolytic enzymes. The meeting brought together international leaders in the field to discuss recent advances in protease research. This volume contains contributions from the speakers at the Symposium and their colleagues.

Proteases were originally identified as protein-degrading enzymes in digestive juice or tissue homogenates, which led to the concept that they completely 'destroy' proteins. The current view, however, sees proteases much more as conducting specific, limited proteolysis and thereby, playing pivotal roles in many biological processes such as fertilization, development, morphogenesis, cell signalling, transcriptional control, blood coagulation, cell division, migration, growth, apoptosis and the activation of cytokines and growth factors. In fact, approximately 2% of human genes encode proteolytic enzymes, which emphasizes the importance of proteolysis. Under physiological conditions, proteolytic processes are precisely regulated. Dysregulation of proteolysis in organisms is deleterious: abnormal development and diseases are attributed to aberrant activities of proteases. Thus, the field has attracted researchers from many different scientific disciplines including biology, biochemistry, biophysics, medicinal chemistry, medicine, dentistry, veterinary medicine and agriculture.

The Symposium's scientific session was opened by Wolfram Bode's lecture on structure–function relationships of matrix metalloproteinases (MMPs) and tissue inhibitors of metalloproteinases (TIMPs). Dusan Turk then discussed new structures of lysosmal cysteine proteases and challenges for the design of synthetic inhibitors. In the session on antigen presentation, Jennifer Rivett gave a presentation on the role of proteasomes in the MHC class I pathway and Colin Watts discussed the role of legumain (asparaginyl endopeptidase) in the MHC class II pathway. Pericellular proteolysis by membrane-bound proteases has made considerable advancement in recent years; Roy Black discussed the biochemistry of ADAM-17 [a disintegrin and metalloproteinase (tumour necrosis factor α-converting enzyme)], and Judith Bond gave a presentation on the regulation of meprin assembly and activation. Gillian Murphy then explained how membrane-bound MMPs and ADAMs are regulated by TIMPs, and Richard Farndale discussed structural interactions between collagen and cell surface proteins.

The session on inflammation and tissue destruction began with Jeremy Saklatvala's presentation on inflammation and intracellular signalling pathways. Tim Cawston described the regulation of collagenolysis by cytokines, and John Mort discussed the role of MMPs and aggrecanases in the destruction of the cartilage matrix. Roger Dean then described protein oxidation and proteolysis in diseased tissues. Robin Poole, who was scheduled to talk on collagenases and osteoarthritis, was unable to attend the meeting, but we are very pleased that he has contributed to this volume.

In the session devoted to proteinase inhibitors and potential therapy, Andy Docherty reviewed the successes and failures of therapeutic intervention with protease inhibitors and newer approaches to therapies. Robin Carrell presented elegant video depictions of the conformational changes in serpins, showing how they form irreversible complexes with target proteases. Magnus Abrahamson reviewed the structure and function of cystatins, and Hideaki Nagase presented mutagenesis studies that were aimed at generating selective TIMPs. A lot of progress has been made towards our understanding of how β-amyloid is generated from the membrane-bound precursor protein. Jordan Tang described memapsin 2 (β-secretase or BACE) as a drug target for Alzheimer's disease. Robert Rawson discussed the intricate regulation of the release of sterol regulatory element binding protein as an example of regulated intramembranous proteolysis.

As part of the intracellular proteolysis session, Guy Salvesen presented a hypothesis for the activation of apical caspases, a key step in apoptosis, and Frank Uhlmann presented separase, a new cysteine protease involved in chromosomal segregation, uncovered by bioinformatic tools pioneered by Alan Barrett. The final area to be covered at the Symposium was proteases and cancer. Bonnie Sloane discussed the role of cathepsin B in cancer progression, and Motoharu Seiki described the dynamics of membrane-type 1 MMP in cell migration. The Symposium ended with Peter Friedl's fascinating demonstration of the distinctions between protease-dependent and protease-independent tumour cell migration.

The field of proteases is vast and is still growing rapidly, with continued discoveries of new enzymes and delineation of the biological roles of known ones. The topics selected for this 3-day meeting could not cover every aspect of the field, but it is our hope that the reader will find this volume a concise overview of recent progress in the field. We are most grateful to the authors for their contributions.

We also would like to thank Rhonda Oliver, of Portland Press, for suggesting the meeting, and Mike Cunningham, also of Portland Press, for his assistance in compiling this volume.

Jeremy Saklatvala
Hideaki Nagase
Guy Salvesen

A tribute to Alan J. Barrett

Alan Barrett is widely known for his seminal contributions to the study of proteolytic enzymes. He read natural sciences at the University of Cambridge and then worked in the University's Biochemistry Department with Professor D.H. Northcote for his PhD. He was awarded this degree in 1964 for his thesis entitled 'Plant pectic polysaccharides'. He then went to work with Dame Honor Fell at the Strangeways Research Laboratory in Cambridge, where he remained for 30 years until he moved to the Babraham Institute, Cambridge. Honor Fell and her colleagues were investigating the effects of vitamin A on the skeleton. She found that when embryonic limb-bone rudiments in organ culture were treated with vitamin A, there was a dramatic loss of protein–polysaccharide (today called proteoglycan) from the extracellular matrix of cartilage. The changes were of great interest as they were similar to those observed in cartilage in rheumatoid arthritis or osteoarthritis. Although he was recruited for his knowledge of polysaccharides, Alan quickly realized that protein–polysaccharide was being lost as a result of proteolysis. He therefore set about isolating proteolytic enzymes, cathepsins, from tissues. Little was known about these at the time, in contrast with the better under-stood digestive enzymes, trypsin, chymotrypsin and pepsin.

He developed rigorous methods to assay and purify cathepsin D and cathepsin B, and, later, many other tissue proteinases. His pioneering work in obtaining genuinely pure proteins opened the door for mechanistic studies, for the generation of selective antibodies to investigate tissue and cellular localization, and for the development of specific substrates and inhibitors. His work with John Dingle (and Honor Fell) made a major contribution to our understanding of the lysosomal system and the role of proteases in both the destruction and normal turnover of connective-tissue proteins.

In the 1970s, he turned his attention to protein inhibitors of proteases, such as α_2-macroglobulin, the major general plasma proteinase inhibitor, and to smaller inhibitors, such as the cystatins. The α_2-macroglobulin project demonstrated that this protein bound to endopeptidases of all catalytic classes, thereby inhibiting their activity against protein, but not peptide, substrates. These studies led to the famous 'trap hypothesis', which proposed that proteases cleave a sensitive region of the α_2-macroglobulin peptide chain, causing a major conformational change that results in physical entrapment of the attacking enzyme. Around this time, Alan also formalized the classification of peptidases into the four main catalytic classes — metalloproteases, serine, cysteine and aspartic proteases — that are so well known and used today.

As more and more proteases and inhibitors were discovered and information on them burgeoned, Alan realized there was a need for a systematic compilation of information on these molecules. He wrote, with Ken McDonald, the two-volume *Mammalian Proteases: A Glossary and Bibliography* in 1980 and 1986. He also edited *Proteinase Inhibitors* with Guy Salvesen in 1986, two volumes of *Methods in Enzymology, Proteolytic Enzymes: Serine and Cysteine Peptidases* (volume 244) in 1994 and *Proteolytic Enzymes: Aspartic and Metallo Peptidases* (volume 248) in 1995, and the *Handbook of Proteolytic Enzymes*, with Neil Rawlings and Fred Woessner, in 1998. These books proved invaluable works of reference to those in the field. Since 1996, he has worked with Neil Rawlings to create the MEROPS database. This is, and will continue to be, among the most important resources for anyone seeking information on proteolytic enzymes.

When Alan set out to purify tissue proteases in the 1960s, there was little concept of the importance of proteolysis as a regulatory mechanism in biology. Proteases were thought of as purely degradative molecules, whether in the digestive tract or the lysosome. The understanding of zymogens, and the elucidation of the blood-clotting cascades in the 1960s and 1970s were the first inklings that limited, specific proteolysis was going to be an important general biological mechanism. Today, we know that a host of processes, ranging from fertilization to cell death, are controlled by proteolysis. We hope that this volume, with contributions from the speakers at the Biochemical Society's Annual Symposium in 2002, which was organized to celebrate Alan's 65th birthday with his many friends, colleagues and collaborators, reflects this diversity.

Jeremy Saklatvala
Hideaki Nagase
Guy Salvesen

Abbreviations

Aβ	β-amyloid
ACE	angiotensin-converting enzyme
AD	Alzheimer's disease
ADAM	a disintegrin and metalloproteinase
ADAMTS	ADAM with thrombospondin motifs
AEP	asparagine endopeptidase
AOPP	advanced oxidation products of proteins
AP	activator protein
Apaf	apoptotic protease activating factor
APC	anaphase-promoting complex
APP	amyloid precursor protein
ARE	AU-rich element
AIIt	annexin II heterotetramer
ATF	activating transcription factor
BCR	B cell antigen receptor
BSL	back-side loop
BTC	betacellulin
CARD	caspase recruitment domain
C/EBP	CAAT/enhancer-binding protein
CEW cystatin	chicken egg white cystatin
COX	cyclo-oxygenase
CRP	collagen-related peptide
2D	two-dimensional
3D	three-dimensional
DED	death effector domain
DISC	death-inducing signalling complex
DNPH	2,4-dinitrophenylhydrazine
DUSP	dual-specificity phosphatase
ECM	extracellular matrix
EGF	epidermal growth factor
EGFR	EGF receptor
ER	endoplasmic reticulum
ERK	extracellular-signal-regulated kinase
EST	expressed sequence tag
FADD	Fas-associated protein with death domain
c-FLIP$_L$	flice-like inhibitory protein
FN domain	fibronectin domain
GpVI	glycoprotein VI

GRE	glucocorticoid response element
GRP	gastrin-releasing peptide
HB-EGF	heparin-binding epidermal growth factor
HCCAA	hereditary cystatin C amyloid angiopathy
HCII	heparin cofactor II
ICE	IL-1β-converting enzyme
I-domain	inserted domain
IL-1	interleukin-1
IL-1R	IL-1 receptor
IL-1ra	IL-1 receptor antagonist protein
IRAK	IL-1-receptor-associated kinase
JNK	c-Jun N-terminal kinase
LDL	low-density lipoprotein
L-domain	left domain
LPS	lipopolysaccharide
$\alpha_2 M$	α_2-macroglobulin
MAPK	mitogen-activated protein kinase
MAPKAPK	MAPK-activated protein kinase
MAP kinase	mitogen-activated protein kinase
MBP	myelin basic protein
MIDAS	metal ion-dependent adhesion site
MKP	MAPK phosphatase
MMP	matrix metalloproteinase
MPR	mannose 6-phosphate receptor
MT-MMP	membrane-type MMP
NFκB	nuclear factor κB
NGF-R	nerve growth factor receptor
N-TIMP	TIMP N-terminal domain
OA	osteoarthritis
OB fold	oligosaccharide/oligonucleotide-binding fold
OSM	oncostatin M
PAI	plasminogen activator inhibitor
PEX domain	haemopexin-like domain
RA	rheumatoid arthritis
R-domain	right domain
Rip	regulated intramembrane proteolysis
SCAP	SREBP-cleavage-activating protein
SLAM	selected lymphocyte antibody method
S1P/S2P	site-1/site-2 protease
SPP	signal peptide peptidase
SREBP	sterol regulatory element binding protein
TACE	TNFα-converting enzyme
TGF	transforming growth factor
TIMP	tissue inhibitor of metalloproteinases
TIR	Toll and IL-1 receptor
TLR	Toll-like receptor
TM	transmembrane
TNF	tumour necrosis factor

TNFR	TNF receptor
tPA	tissue plasminogen activator
TRAF	TNF receptor-associated factor
TTCF	tetanus toxin C fragment
TTP	tristetraprolin
uPA	urokinase-type plasminogen activator
UTR	untranslated region
ΔZn TACE	TACE having a deletion in its Zn-binding domain

Biochem. Soc. Symp. **70**, 1–14
(Printed in Great Britain)
© 2003 Biochemical Society

1

Structural basis of matrix metalloproteinase function

Wolfram Bode[1]

Max-Planck-Institut für Biochemie, D-82152 Martinsried, Germany

Abstract

The matrix metalloproteinases (MMPs) constitute a family of multidomain zinc endopeptidases which contain a catalytic domain with a common metzincin-like topology. The MMPs are involved not only in extracellular matrix degradation, but also in a number of other biological processes. Normally, their proteolytic activity is regulated precisely by their main endogenous protein inhibitors, in particular the tissue inhibitors of metalloproteinases (TIMPs). Disruption of this balance results in serious diseases, such as arthritis, tumour growth and metastasis, rendering the MMPs attractive targets for inhibition therapy. Knowledge of their tertiary structures is crucial for a full understanding of their functional properties. Since the first publication of atomic MMP structures in 1994, much more structural information has become available on details of the catalytic domain, on its interaction with synthetic and protein inhibitors, on domain organization and on the formation of complexes with other proteins. This review will outline our current knowledge of MMP structure.

Introduction

The matrix metalloproteinases (MMPs; matrixins; EC 3.4.24.–) form a large family of structurally and functionally related zinc endopeptidases that are found in species from plants via hydra to humans [1,2]. Collectively, the MMPs are capable of degrading, both *in vitro* and *in vivo*, all kinds of extracellular matrix protein components, such as interstitial and basement membrane collagens, proteoglycans, fibronectin and laminin, and are thus implicated in connective tissue remodelling processes associated with embryonic development, pregnancy, growth, wound healing, etc. Many of these MMPs, however, are also involved in the shedding and release of latent growth factors, cytokines and cell surface receptors, in the activation of proMMPs, and in the inactivation of proteinase and angiogenesis inhibitors, thus participating in diverse physio-

[1]e-mail bode@biochem.mpg.de

logical processes (for a recent review, see [3]). Normally, the degradative potential of the MMPs is held in check by the endogenous specific tissue inhibitors of metalloproteinases (TIMPs). Disruption of this MMP–TIMP balance can result in diseases such as rheumatoid arthritis, osteoarthritis, atherosclerosis, heart failure, fibrosis, pulmonary emphysema, tumour growth, and cell invasion and metastasis. The therapeutic inhibition of MMPs is a promising approach for the treatment of some of these diseases, and the MMP structures are therefore attractive targets for rational inhibitor design (for recent references, see [4–7]).

Due to their identical catalytic zinc environment, a characteristic methionine-containing tight turn below this catalytic zinc ion and the unique topology of their catalytic domain, the MMPs have been grouped in the 'metzincin' zinc endopeptidase superfamily [8,9]. Vertebrate MMPs are synthesized as latent multidomain proteinases which are transported to the cell surface, where they either remain membrane-bound or are secreted into the extracellular space. Individual vertebrate MMPs have been named according to their presumed substrates or classified by sequential (mainly chronological) numbers, which run from MMP-1 to MMP-28 (omitting numbers 4–6). The MMP gene family in humans encodes, besides three pseudogenes, 25 related proteinases, of which 22 have been well characterized (for details, see a recent review [10]). These MMPs have counterparts in other mammals (such as mouse, rat, rabbit, pig, cattle and horse), in other vertebrates, in invertebrates and in plants (see [3,11]), which together form the MMP or matrixin subfamily A of the metalloproteinase M10 family in MEROPS clan MB [12].

All of these MMPs are synthesized with an N-terminal signal sequence, which is removed upon insertion into the endoplasmic reticulum, yielding the latent proenzymes. All human proMMPs have in common an N-terminal prodomain of approx. 80 amino acids and an adjacent characteristic catalytic domain which consists of approx. 175 amino acid residues (see Figure 1) [except in MMP-2 (gelatinase A) and MMP-9 (gelatinase B), which have additional 175-residue fibronectin type II domain (FN) inserts]. In MMP-23 [13,14], the N-terminal part of the pro-domain seems to be modified into a transmembrane (TM) domain [15]. All other human MMP pro-domains contain a 'cysteine-switch' PRCXXPD consensus sequence [16], whose unpaired cysteine blocks the catalytic zinc in the latent pro- form. In all human MMPs, the linker between the pro-domain and the catalytic domain is susceptible to proteolytic activation cleavage; one-third of the classified MMPs exhibit an alkaline RX(R/K)R consensus cleavage site, rendering them activatable by proconvertases/furin [17, 58].

Except for MMP-23 (which has a different C-terminal domain) and the two matrilysins MMP-7 and MMP-26, all human MMPs contain an approx. 195-residue C-terminal haemopexin-like (PEX) domain connected covalently to the catalytic domain through a linker of up to 70 residues. In the membrane-type MMPs (MT-MMPs), the polypeptide chains have additional C-terminal tails, which either include a TM helix and terminate with a short cytoplasmic domain [17], or whose termini act as a glycosylphosphatidylinositol membrane-anchoring signal, which becomes replaced by the glycosylphosphatidylinositol

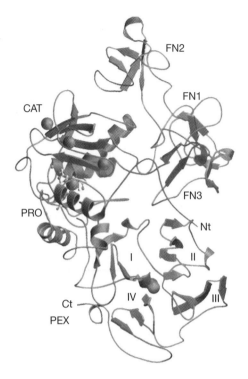

Figure 1 Ribbon plot of proMMP-2 [44]. The proMMP-2 stucture comprises [from the N-terminus (Nt) to the C-terminus (Ct)] the pro-domain (PRO; in front); the conserved part of the catalytic domain (CAT; upper left ribbon) enveloping the catalytic and the structural zinc ions (smaller spheres) and two calcium ions (larger spheres); three FN domains (FN1–FN3; upper right side) inserted in the sV–hB loop (see Figure 2 for nomenclature); and the PEX domain (lower ribbon, comprising blades I–IV and containing up to three ions in its central channel), connected via a hinge segment. The molecule is shown such that its catalytic domain is almost in standard orientation (see Figure 2).

membrane anchor [18]. Most MMPs are glycosylated to different extents and at different sites.

The TIMP family contains four members (TIMPs-1 to -4), which after optimal topological superposition exhibit 41–52% sequence identity (see Figure 3 below; for references, see [7]). In addition to their inhibitory role, these TIMPs seem to have other functions, such as growth factor-like and anti-angiogenic activity (see, e.g., [19,20]). The TIMPs, each of which contains between 184 and 194 amino acid residues, form tight 1:1 complexes with MMPs, with K_i values down to sub-femtomolar levels [21]. Except for the rather weak interaction between TIMP-1 and most MT-MMPs, the TIMPs do not seem to differentiate much between the various MMPs. In addition, TIMP-1 is capable of binding to proMMP-9 (progelatinase B), TIMPs-2 and -4 can bind to proMMP-2 (progelatinase A), while TIMP-3 binds to both [22]. The complex between the MMP-14 (MT1-MMP) and TIMP-2 acts as a cell-surface-bound 'receptor' for proMMP-2 activation *in vivo* [23–26] by presenting the bound proMMP-2 such that an adjacent uninhibited MMP-14 molecule sets a first cut in the pro-domain of MMP-2 (see below), followed by the final activation cleavage by another MMP-2 molecule [27]. MMP-14 is a particularly efficient activator, with its specific 'MT-loop' (Figure 2) presumably serving as a specific exosite for proMMP-2 (see below), while MMP-15 (MT2-MMP) seems to act via an alternative, TIMP-2-independent pathway [28].

Since early 1994, a large number of MMP and TIMP structures have been published, determined mostly by X-ray techniques. The first X-ray crystal structures, determined with the help of 'classical' heavy metal derivatives, were of the catalytic domains (blocked by various synthetic inhibitors) of human fibroblast collagenase (collagenase 1/MMP-1) [29–32] and human neutrophil collagenase (collagenase 2/MMP-8) [32–34] (Figure 2). At the same time, an NMR structure of the catalytic domain of stromelysin 1/MMP-3 [35] became available. These structures were later complemented by additional X-ray and NMR structures of the catalytic domains of MMP-1, matrilysin/MMP-7 [36], MMP-3, MMP-8, MMP-14 [37], collagenase 3/MMP-13 [38], mouse stromelysin 3/MMP-11 [39] and macrophage metalloelastase/MMP-12 [40,41]. In 1995 the first X-ray crystal structure of an MMP pro-form (C-terminally truncated prostromelysin 1) was

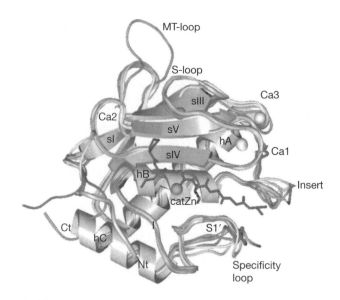

Figure 2 Ribbon structure of the MMP catalytic domain shown in standard orientation. Superposition of the catalytic (CAT) domains of superactivated MMP-8 [34] (ribbon) and MMPs-1, -2 (representing only the conserved part; see Figure 1), -3, -7, -13 and -14 (thin ropes), shown together with a heptapeptide substrate bound productively in the active-site cleft [55]. Strands (sI–sV) and helices (hA–hC) are labelled, and some notable loops are indicated. The catalytic (catZn) and structural (centre and top) zinc ions and the calcium ions (Ca1–Ca3; flanking) are shown as dark grey and light grey spheres respectively. Ct, C-terminus; Nt, N-terminus.

published [42], and the first structure of a mature full-length MMP (pig fibroblast collagenase 1/MMP-1) [43] was described. In 1999 the structure of full-length proMMP-2 (progelatinase A) [44] was reported (Figure 1), as was the FN domain-deleted catalytic domain of MMP-2 [45]. Quite recently, the FN domain-deleted catalytic domain [46] and the full-length pro-catalytic domain [47] of MMP-9 were published, and added as new members to the MMP structure zoo (see [48]).

With regard to the TIMPs, a first preliminary NMR model of human N-TIMP-2 (N-terminal domain of TIMP-2) was presented in 1994 [49], revealing that the polypeptide framework of the N-terminal part of the TIMPs resembles the so-called OB (oligosaccharide/oligonucleotide-binding)-fold proteins. In 1997, the first structure of a complete TIMP, i.e. human deglycosylated TIMP-1 in complex with the catalytic domain of human MMP-3, was published [50], and this was followed by the X-ray structure of TIMP-2 in complex with the catalytic domain of MMP-14 [37]. Quite recently, these TIMP complexes were complemented by the proMMP-2–TIMP-2 complex [51].

In the following, the polypeptide fold, the domains and the structural determinants of the MMP and TIMP structures and their detailed interaction will be described. The MMP nomenclature used is based on the cDNA sequence of (human) fibroblast collagenase-1/MMP-1 as the reference MMP (see [48,52]). For the assignment of peptide substrate residues and substrate recognition sites on the proteinase, the nomenclature of Schechter and Berger [53] will be used: P1, P2 etc. and P1′, P2′ etc. indicate the residues in the N- and C-terminal directions respectively from the scissile peptide bond of a bound peptide substrate (analogue); and S1, S2 etc. and S1′, S2′ etc. the equivalent binding sites on the enzyme. TIMP residues will be numbered according to TIMP-1 or TIMP-2.

Structures and mechanisms

The MMP catalytic domain

The catalytic domains of the MMPs exhibit the shape of an oblate ellipsoid. In the standard orientation, which in this chapter and in most other MMP papers is preferred for the display of the catalytic domains, a small active-site cleft notched into the flat ellipsoid surface extends horizontally across the domain surface to bind peptide substrates from left to right (Figure 2). This cleft, harbouring the catalytic zinc, separates the smaller lower subdomain from the larger upper subdomain.

The upper subdomain (Figure 2), formed by the first three-quarters of the polypeptide chain, consists of a five-stranded β-pleated sheet, which is flanked by three surface loops on its convex side and by two long regular α-helices on its concave side embracing a large hydrophobic core. The polypeptide chain starts on the molecular surface of the lower subdomain, where its N-terminal Phe^{100} ammonium group, if correctly tailored, can become engaged in a surface-located salt bridge with Asp^{250} (see [34,54]). The chain turns to the upper domain, passes β-strand sI, the amphipatic α-helix hA and β-strands sII, sIII, sIV and sV, and enters the active-site helix hB (for nomenclature, see Figure 2).

In the classical MMPs, strands sII and sIII are connected by a relatively short loop bridging strand, sI; in the MT-MMPs (MMP-14, -15, -16 and -24), however, this loop is expanded into the spur-like, solvent-exposed MT-specific loop [37]. In all MMPs, strands sIII and sIV are linked via an S-shaped double loop, which is connected via the structural zinc and the first of two to three bound calcium ions to the β-sheet. This S-loop extends into the cleft-sided bulge, continuing in the antiparallel edge-strand sIV; this bulge–edge segment is of prime importance for binding of peptide substrates and inhibitors (Figure 2). The sIV–sV connecting loop, together with the sII–sIII bridge, sandwich the second bound calcium. After strand sV, the chain (except in both gelatinases, where it turns towards the 185-residue FN domains, which form the large cloverleaf-like FN domain [44] consisting of three tandem copies of FN modules; see Figure 1) passes the large open sV–hB loop, before entering the active-site helix hB. This helix provides the first (His[218]) and the second (His[222]) His residues which, together with His[228], ligand the catalytic zinc and the catalytic Glu[219] residue between them, all of which represent the N-terminal part of the zinc-binding consensus sequence HEXXHXXGXXH that is characteristic of the metzincin superfamily [8,9].

This active-site helix stops abruptly at Gly[225], where the peptide chain bends downwards and descends to present the third zinc-liganding histidine, His[228], and then runs through a wide right-handed spiral terminating in the 1,4-tight 'Met-turn' (Figure 2). The chain then turns back to the molecular surface to an almost invariant Pro[238], forms with a conserved Pro[238]-Xaa-Tyr[240] segment (the S1′ wall-forming segment) the outer wall of the S1′ pocket and then runs through the wide specificity loop of slightly variable length and conformation, before it passes the C-terminal α-helix hC, which ends with the conserved Tyr[260]-Gly[261] pair.

Specificity determinants

Bounded at the upper rim by the bulge–edge segment and the second part of the S-loop, and at the lower side by the third zinc-liganding imidazole and the S1′ wall-forming segment, the active-site cleft of all MMPs is relatively flat on the left ('non-primed') side, but carves into the molecular surface at the catalytic zinc and to the right ('primed') side, levelling off again to the surface further to the right (Figure 2). In unliganded MMPs, the catalytic zinc residing at its centre is co-ordinated by the three imidazole Nε2 atoms of the three histidines (His[218], His[222] and His[228]) and by a fixed water molecule, which simultaneously is within hydrogen-bonding distance of the carboxylate group of the catalytic Glu[219]. In complexes of MMPs with hydroxamate inhibitors, this water molecule is replaced by two oxygen atoms, which, together with the three imidazole nitrogens, ligand the catalytic zinc in a trigonal-bipyrimidal (penta-co-ordinate) manner [33]. As in all other metzincins, the zinc-imidazole ensemble of the MMPs is placed above the distal ε-methylsulphur moiety of the strictly conserved Met[236] residue of the Met-turn, which forms a hydrophobic base.

Immediately to the right of the catalytic zinc invaginates the S1′ specificity pocket, which differs considerably in size and shape among the various MMPs. Of the synthetic inhibitors that have been determined in complex with an MMP, only the Pro-Leu-Gly-hydroxamic acid inhibitor [33,34] binds to the left-hand subsites (the non-primed subsites S3–S1) alone, antiparallel to the edge strand ('left-side inhibitor'). A few synthetic inhibitors bind across the whole of the active site, while for the majority of synthetic inhibitors studied so far the extended peptide moiety interacts with the primed right-hand subsites ('right-side inhibitors'), inserting between the (antiparallel) bulge–edge segment and the (parallel) S1' wall-forming segment of the cognate MMP, causing formation of a three-stranded mixed β-sheet (Figure 2). An L-configured P1'-like side chain can extend into the hydrophobic bottleneck of the S1' pocket [55]. This P1'–S1' interaction is the main determinant of the affinity of inhibitors and the cleavage position of peptide substrates.

With particular dependence on the length and character of residue 214 in the N-terminal part of the active-site helix hB, the size of the S1' pocket differs considerably among the MMPs (Figure 2). In MMP-1 and MMP-7, for example, the side chains of Arg^{214} and Tyr^{214} respectively extend into the S1' opening, limiting it to a size and shape still compatible with the accommodation of medium-sized P1' residues, rendering it less accessible to very large (synthetic) side chains, in agreement with peptide cleavage studies on model peptides. More recently determined MMP-1 structures with synthetic inhibitors show, however, that the Arg^{214} side chain can swing out of its normal site, thus also allowing binding of synthetic inhibitors with larger P1' side chains [38]. The smaller Leu^{214} residue in MMPs-2, -3, -9, -12, -13 and -14 does not block the internal S1' 'pore', which extends right through the molecule to the lower surface, i.e. it is more like a long, solvent-filled tube. In spite of its small Leu^{214} residue, however, the S1' pocket of MMP-8 is of medium size and is closed at the bottom, due to the Arg^{243} side chain extending into the S1' space from the specificity loop [33]. In MMP-9, the side chain of Arg^{424} (corresponding to residue 241 in MMP-1) can also close the end of the pocket [46].

By replacing the zinc-chelating groups of such peptide left- and right-side inhibitors with a normal peptide bond, a contiguous peptide substrate was constructed to mimic the probable binding geometry of a normal substrate–MMP encounter complex [55]. Accordingly, the peptide substrate chain is aligned in an extended manner with the continuous bulge–edge segment, causing formation of an antiparallel two-stranded β-pleated sheet, which on the right-hand side is expanded into a three-stranded mixed parallel–antiparallel β-sheet, due to additional alignment with the S1' wall-forming segment. A bound peptide substrate (such as the hexapeptide shown in Figure 2) can form five and two inter-main-chain hydrogen bonds respectively with the two crossing-over MMP segments. Similar to the reaction mechanism suggested previously for the more distantly related zinc endopeptidase thermolysin [56], the MMP-catalysed cleavage of the scissile peptide bond will probably proceed via a general-base mechanism [55].

The MMP pro-domain

Currently, the only proMMP structures known are those of C-terminally truncated proMMP-3 [42] and proMMP-9 [47], and of full-length proMMP-2 (Figure 1) [44]. The pro-domain peptide has an egg-like shape, attached with its rounded-off side to the active site of the catalytic domain. It consists essentially of three mutually perpendicularly packed α-helices and an almost invariant Pro[90]-Arg-Cys-Gly-Val-Pro-Asp[96] 'cysteine switch' loop running through the active-site cleft in the opposite direction to bound substrates, extending into the catalytic domain. From Pro[107] onwards, the polypeptide chains of the three proMMPs are in register with those of the mature (i.e. activated) MMPs.

In MMP-11, MMP-28 and the six MT-MMPs, the connecting segment between the switch loop of the pro-chain and the N-terminus of the mature MMP is elongated further, by up to 22 residues (MMP-28), with the pro-domain-chain terminating in an Arg-Xaa-Arg/Lys-Arg sequence typical of cleavage by furin-like convertases [17]. The intracellular cleavage of these MMPs in the *trans*-Golgi network results in expression of the active enzymes at the cell surface [18,57].

The proteolytic activation of these 'classical' proMMPs (see [54]) seems to proceed via a stepwise mechanism. Some early cleavages occurring in the flexible, exposed helix 1–helix 2 loop (the 'bait') might not only expose the hydrophobic core of the pro-domain, but also destabilize this domain, thus exposing other (downstream) cleavage sites and weakening and finally disrupting the Cys–catalytic-zinc interaction. This will lead eventually to liberation and increased flexibility of the Xaa[99]-Phe/Tyr[100] activation cleavage peptide bond (as originally predicted by the 'cysteine switch hypothesis' [16]).

The PEX domain

Except for MMPs-7, -23 and -26, all vertebrate/human MMPs are expressed with a C-terminal PEX domain. Some of these PEX domains have been shown to be involved in substrate recognition and to confer substrate specificity, most dramatically in the collagenase subfamily, where the ability to cleave native triple-helical collagen is associated with the covalently bound PEX domain (for references, see [10,25]). The PEX domains exhibit the structure of a four-bladed β-propeller of pseudo four-fold symmetry [43] (see Figure 1). The polypeptide chain is organized essentially in four β-sheets (blades), I–IV, which are arranged almost symmetrically around a central axis in consecutive order. Each propeller blade is twisted and consists of four antiparallel β-strands connected in a W-like topology. The first, innermost strands of all four blades enter the propeller at one site and run almost parallel to one another along the propeller axis, forming a central funnel-shaped tunnel, which opens slightly towards the exit and often accommodates some ions. The fourth strands of blades II and III are interrupted by characteristic β-bulges, which allow these strands to remain in-phase with the antiparallel third strands, in spite of the overall sheet curvature. In all four blades, the outer segments loop around the periphery of the disc and end in short helical segments. The C-terminus of the blade IV helix is connected to the entering strand of blade I via a single disulphide bridge, rigidifying the whole domain.

In the two full-length MMP structures available [43,44], the catalytic domain and the PEX domain make non-covalent contacts only along small domain edges, with the outermost strand of the first blade of the PEX domain propeller contacting the C-terminal helix hC of the catalytic domain (see Figure

1). The Pro-rich linker between the two domains bulges backwards and runs in a loose manner, antiparallel to helix hC, before it joins the PEX moiety. The linkers of the four collagenases are shortest of all MMPs, which might add to the rigid juxtaposition of their catalytic and PEX domains. The interdomain linkers are considerably longer, however, in the MT-MMPs, possibly giving more freedom to the relative placement of the catalytic and PEX domains. MMP-9 exhibits the longest interdomain linker, which is heavily O-glycosylated and has been predicted to be type-V-collagen-like (for references, see [59]).

Chimaeric constructs made in order to elucidate the structural features defining the triple-helicase specificity of collagenases have shown that the catalytic domain, the hinge and the PEX domain possess important determinants, and that all three must be arranged correctly to act in concert to confer helicase specificity (for references, see [48]). The Glu(Asn)^{209}-Tyr-Asn-Leu segments of MMPs-1 and -8 respectively, which precede the active-site helix and are located in the corridor connecting the catalytic domain and the PEX domain, seem to form a critical exosite for triple-helical collagen substrates [60,61]. Intriguingly, Tyr^{209} in MMPs-1 and -8 is preceded by a *cis*-bond, distinguishing these more powerful type I and II collagenases from other collagen-degrading MMPs, such as MMP-2, MMP-13 and MMP-14 (for a more detailed discussion, see [61]). Triple-helical collagen substrates thus may bind into the active-site cleft of the cognate MMP with partial unwinding from P3 to P3′, bend at the 209 exosite, and interact with blade II of the PEX domain [62]. Other potential triple-helicase mechanisms have recently been discussed in detail by Overall [10].

TIMP complexes and the proMMP activation complex

The wedge-shaped TIMP molecules consist of an N-terminal segment, an all-β-structure N-terminal part, an all-helical centre, and a C-terminal β-turn structure [50]. The N-terminal half, consisting essentially of a closed five-stranded β-barrel, and the C-terminal half of the polypeptide chain form two opposing subdomains (Figure 3). The TIMP edge is formed by five sequential, seperate chain segments, namely the extended N-terminal segment Cys^1–Pro^5 and two flanking loops on either side provided by the N- and C-terminal parts. Particularly remarkable features of TIMP-2 are the quite elongated sA–sB β-hairpin loop and the much longer negatively charged flexible C-terminal tail [37].

In complexes with MMPs, the wedge-shaped TIMPs bind with their edge into the entire length of the active-site cleft of their cognate MMPs [37,50], with some rigidification of the participating loops [63–65]. The first five TIMP residues, Cys^1–Pro^5, bind to the MMP active-site cleft in a substrate- or product-like manner, i.e. similar to the way in which the P1, P1′, P2′, P3′ and P4′ residues of peptide substrates insert between the bulge and the wall-forming segments, forming five intermolecular inter-main-chain hydrogen bonds. Cys^1 is located directly above the catalytic zinc, with its N-terminal α-amino nitrogen and its carbonyl oxygen atoms co-ordinating the catalytic zinc together with the three imidazole rings from the cognate MMP. The Thr/Ser side chain of the second TIMP residue extends, similar to the side chain of a peptide substrate P1′ residue, into the S1′ pocket of the cognate MMP, without filling this pocket properly [7].

Figure 3 ProMMP-2 activation complex. Shown is a suggested model of the quarternary complex formed by one membrane-bound full-length 'receptor' MMP-14 molecule (yellow), inhibited by a TIMP-2 molecule (blue) [37], whose C-terminal part binds to the PEX2 domain proMMP-2 ligand (red) [51]. This complex in turn presents its scissile peptide bond to another membrane-bound full-length MMP-14 activator (green). See the text for further details.

It has been known for some time that MMP-14, via a bound TIMP-2 molecule, fixes a proMMP-2 molecule through its PEX domain, presenting the Asn[37]–Leu[38] scissile peptide bond to a second, non-inhibited MMP-14 molecule [23,24,26]. Participation of the negatively charged C-terminal tail of TIMP-2 had been demonstrated [66,67], and the binding site for TIMP-2 on proMMP-2 could be narrowed down to the junction of blades III and IV on the peripheral rim of the MMP-2 PEX domain by site-directed mutagenesis of surface-located alkaline residues [68]. The recent crystal structure of the proMMP-2–TIMP-2 complex [51] shows these TIMP-2–PEX2 interactions at atomic resolution. The two components interact via mixed hydrophobic/polar interfaces, essentially through two distinct but adjacent binding regions, formed by (i) strands 3 and 4 of blade IV with the central area of the C-terminal domain of TIMP-2, and (ii) strands 2–4 of blade III with the C-terminal tail of TIMP-2.

On the basis of the experimental MMP-14–TIMP-2 [37] and proMMP-2–TIMP-2 [51] structures, and using other information, modelling of the functional quarternary proMMP-2 activation complex can be attempted (Figure 3). The MMP-14 molecule contains, on the basis of its sequence, a 35-residue catalytic domain–PEX domain linker, a modelled PEX14 domain (i.e. PEX domain of MMP-14), a 30-residue linker between the PEX and TM domains (which, on the basis of secondary structure predictions should, besides a short β-strand, comprise mainly non-classifiable secondary structure elements), a 20-residue α-helical TM14 domain perforating the membrane, and a 26-residue cytoplasmic domain (Figure 3). The orientation and position of the PEX14 domain relative to the catalytic domain must (due to a considerable clash with the PEX2 domain of MMP-2 in the ternary/tertiary complex; see below) differ substantially from that observed in the full-length MMPs-1 and -2, which would be facilitated by the much longer catalytic domain–PEX domain linker of MMP-14. TIMP-2 docks to this 'receptor' MMP-14, as observed in the MMP-14–TIMP-2 structure, with its C-terminal negatively charged tail stretched, as observed in the proMMP-2–TIMP-2 structure. The proMMP-2 ligand will dock with this TIMP-2, as observed in the latter complex. The primary cleavage site around Asn^{37}–Leu^{38} in proMMP-2 is not susceptible to MMPs in its usual conformation. Thus it must (facilitated by the extremely high temperature factors of the primed-side residues) unfold and adapt to allow another MMP-14 molecule to attack the Asn^{37}–Leu^{38} bond, while the relatively rigid Cys^{31}–Cys^{36} non-primed-side loop must rearrange. Modelling trials show that the catalytic domain of this second MMP-14 could embrace the (quite flexible and plastic) pro-domain of proMMP-2 by virtue of its double S-loop, its exposed sIV–sV loop and its MT-specific loop, thus constraining the position and orientation of the catalytic domain of MMP-14. The PEX14 domain plus the adjacent TM domain and the cytoplasmic domain are arranged such that the two PEX14 domains (i.e. one from each MMP-14 molecule) form a homodimer, similar to that observed for the crystalline but non-covalent PEX9 dimer [69], so that the two TM domains and two cytoplasmic domains of the MMP-14 molecules will also stay together. Such homophilic complex formation of membrane-bound MMP-14 has been suggested by the homodimerization of PEX14 and the functional dependence of activation of MMP-14 on possession of the correct PEX14 domain [70], and by the importance of oligomerization through the cytoplasmic domains [71,72].

The valuable contributions of Dr H. Nagase, Dr H. Tschesche, Dr K. Brew, Dr R. Huber and Dr K. Maskos are gratefully acknowledged.

References

1. Woessner, J.F. and Nagase, H. (2000) Matrix Metalloproteinases and TIMPs, Oxford University Press, New York
2. Brinckerhoff, C.E. and Matrisian, L.M. (2002) Nat. Rev. Mol. Biol. **3**, 207–214
3. Sternlicht, M.D. and Werb, Z. (2001) Annu. Rev. Cell Dev. Biol. **17**, 463–516
4. Giavazzi, R. and Taraboletti, G. (2001) Crit. Rev. Oncol. Hematol. **37**, 53–60
5. Dove, A. (2002) Nat. Med. (N.Y.) **8**, 95
6. Hidalgo, M. and Eckhardt, S.G. (2001) J. Natl. Cancer Inst. **93**, 178–193
7. Brew, K., Dinakarpandian, D. and Nagase, H. (2000) Biochim. Biophys. Acta **1477**, 267–283
8. Bode, W., Gomis-Rüth, F.-X. and Stöcker, W. (1993) FEBS Lett. **331**, 134–140
9. Stöcker, W., Grams, F., Baumann, U., Reinemer, P., Gomis-Rüth, F.X., McKay, D.B. and Bode, W. (1995) Protein Sci. **4**, 823–840
10. Overall, C.M. (2002) Mol. Biotechnol. **22**, 51–86
11. Massova, I., Kotra, L.P, Fridman, R. and Mobashery, S. (1998) FASEB J. **12**, 1075–1095

12. Barrett, A.J., Rawlings, N.D. and Woessner, Jr, J.F. (1998) Handbook of Proteolytic Enzymes, Academic Press, London

13. Pei, D. (1999) FEBS Lett. **457**, 262–270

14. Velasco, G., Pendas, A.M., Fueyo, A., Knauper, V., Murphy, G. and Lopez-Otin, C. (1999) J. Biol. Chem. **274**, 4570–4576

15. Pei, D., Kang, T. and Qi, H. (2000) J. Biol. Chem. **275**, 33988–33997

16. Van Wart, H.E. and Birkedal-Hansen, H. (1990) Proc. Natl. Acad. Sci. U.S.A. **87**, 5578–5582

17. Sato, H., Takino, T., Okada, Y., Cao, J., Shinagawa, A., Yamamoto, E. and Seiki, M. (1994) Nature (London) **370**, 61–65

18. Itoh, Y., Kajita, M., Kinoh, H., Mori, H., Okada, A. and Seiki, M. (1999) J. Biol. Chem. **274**, 34260–34266

19. Gomez, D.E., Alonso, D.F., Yoshiji, H. and Thorgeirsson, U.P. (1997) Eur. J. Cell Biol. **74**, 111–122

20. Murphy, G. and Willenbrock, F. (1995) Methods Enzymol. **248**, 496–510

21. Murphy, G., Houbrechts, A., Cockett, M.I., Williamson, R.A., O'Shea, M. and Docherty, A.J.P. (1991) Biochemistry **30**, 8097–8102

22. Butler, G.S., Apte, S.S., Willenbrock, F. and Murphy, G. (1999) J. Biol. Chem. **274**, 10846–10851

23. Strongin, A.Y., Collier, I.E., Bannikov, U., Marmer, B.L., Grant, G.A. and Goldberg, G.I. (1995) J. Biol. Chem. **270**, 5331–5338

24. Kinoshita, T., Sato, H., Takino, T., Itoh, M., Akizawa, T. and Seiki, M. (1996) Cancer Res. **56**, 2535–2538

25. Murphy, G. and Knäuper, V. (1997) Matrix Biol. **15**, 511–518

26. Butler, G.S., Will, H., Atkinson, S.J. and Murphy, G. (1997) Eur. J. Biochem. **244**, 653–657

27. Deryugina, E.I., Ratnikov, B., Monosov, E., Postnova, T.I., DiScipio, R., Smith, J.W. and Strongin, A.Y. (2001) Exp. Cell Res. **263**, 209–223

28. Morrison, C.J., Butler, G.S., Bigg, H.F., Roberts, C.R., Soloway, P.D. and Overall, C.M. (2001) J. Biol. Chem. **276**, 47402–47410

29. Lovejoy, B., Cleasby, A., Hassell, A.M., Longley, K., Luther, M.A., Weigl, D., McGeehan, G., McElroy, A.B., Drewry, D., Lambert, M.H. and Jordan, S.R. (1994) Science **263**, 375–377

30. Borkakoti, N., Winkler, F.K., Williams, D.H., D'Arcy, A., Broadhurst, M.J., Brown, P.A., Johnson, W.H. and Murray, E.J. (1994) Nat. Struct. Biol. **1**, 106–110

31. Stams, T., Spurlino, J.C., Smith, D.L., Wahl, R.C., Ho, T.F., Qoronfleh, M.W., Banks, T.M. and Rubin, B. (1994) Nat. Struct. Biol. **1**, 119–123

32. Spurlino, J.C., Smallwood, A.M., Carlton, D.D., Banks, T.M., Vavra, K.J., Johnson, J.S., Cook, E.R., Falvo, J., Wahl, R.C., Pulvino, T.A. et al. (1994) Proteins Struct. Funct. Genet. **19**, 98–109

33. Bode, W., Reinemer, P., Huber, R., Kleine, T., Schnierer, S. and Tschesche, H. (1994) EMBO J. **13**, 1263–1269

34. Reinemer, P., Grams, F., Huber, R., Kleine, T., Schnierer, S., Pieper, M., Tschesche, H. and Bode, W. (1994) FEBS Lett. **338**, 227–233

35. Gooley, P.R., O'Connell, J.F., Marcy, A.I., Cuca, G.C., Salowe, S.P., Bush, B.L., Hermes, J.D., Esser, C.K., Hagmann, W.K., Springer, J.P. and Johnson, B.A. (1994) Nat. Struct. Biol. **1**, 111–118

36. Browner, M.F., Smith, W.W. and Castelhano, A.L. (1995) Biochemistry **34**, 6602–6610

37. Fernandez-Catalan, C., Bode, W., Huber, R., Turk, D., Calvete, J.J., Lichte, A., Tschesche, H. and Maskos, K. (1998) EMBO J. **17**, 5238–5248

38. Lovejoy, B., Welch, A.R., Carr, S., Luong, C., Broka, C., Hendricks, R.T., Campbell, J.A., Walker, K.A.M., Martin, R., Van Wart, H. and Browner, M.F. (1999) Nat. Struct. Biol. **6**, 217–221

39. Gall, A.L., Ruff, M., Kannan, R., Cuniasse, P., Yiotakis, A., Dive, V., Rio, M.C., Basset, P. and Moras, D. (2001) J. Mol. Biol. **307**, 577–586

40. Lang, R., Kocourek, A., Braun, M., Tschesche, H., Huber, R., Bode, W. and Maskos, K. (2001) J. Mol. Biol. **312**, 731–742

41. Nar, H., Werle, K., Bauer, M.M., Dollinger, H. and Jung, B. (2001) J. Mol. Biol. **312**, 743–751

42. Becker, J.W., Marcy, A.I., Rokosz, L.L., Axel, M.G., Burbaum, J.J., Fitzgerald, P.M.D., Cameron, P.M., Esser, C.K., Hagmann, W.K., Hermes, J.D. and Springer, J.P. (1995) Protein Sci. **4**, 1966–1976

43. Li, J.-Y., Brick, P., O'Hare, M.C., Skarzynski, T., Lloyd, L.F., Curry, V.A., Clark, I.M., Bigg, H.F., Hazleman, B.L., Cawston, T.E. and Blow, D.M. (1995) Structure **3**, 541–549

44. Morgunova, E., Tuuttila, A., Bergmann, U., Isupov, M., Lindqvist, Y., Schneider, G. and Tryggvason, K. (1999) Science **284**, 1667–1670

45. Dhanaraj, V., Williams, M.G., Ye, Q.-Z., Molina, F., Johnson, L.L., Ortwine, D.F., Pavlovshy, A., Rubin, R., Skean, R.W., White, A.D. et al. (1999) Croat. Chem. Acta **72**, 575–591

46. Rowsell, S., Hawtin, P., Minshull, C.A., Jepson, H., Brockbank, S.M., Barratt, D.G., Slater, A.M., McPheat, W.L., Waterson, D., Henney, A.M. and Pauptit, R.A. (2002) J. Mol. Biol. **319**, 173–181

47. Elkins, P.A., Ho, Y.S., Smith, W.W., Janson, C.A., D'Alessio, K.J., McQueney, M.S., Cummings, M.D. and Romanic, A.M. (2002) Acta Crystallogr. D Biol. Crystallogr. **58**, 1182–1192

48. Bode, W. and Maskos, K. (2000) Matrix Metalloproteinase Protocols **151**, 45–77

49. Williamson, R.A., Martorell, G., Carr, M.D., Murphy, G., Docherty, A.J., Freedman, R.B. and Feeney, J. (1994) Biochemistry **33**, 11745–11759

50. Gomis-Rüth, F.X., Maskos, K., Betz, M., Bergner, A., Huber, R., Suzuki, K., Yoshida, N., Nagase, H., Brew, K., Bourenkov, G.P. et al. (1997) Nature (London) **389**, 77–81

51. Morgunova, E., Tuuttila, A., Bergmann, U. and Tryggvason, K. (2002) Proc. Nat. Acad. Sci. U.S.A. **99**, 7414–7419

52. Bode, W., Fernandez-Catalan, C., Tschesche, H., Grams, F., Nagase, H. and Maskos, K. (1999) Cell. Mol. Life Sci. **55**, 639–652

53. Schechter, I. and Berger, A. (1967) Biochem. Biophys. Res. Commun. **27**, 157–162

54. Nagase, H. (1997) Biol. Chem. **378**, 151–160

55. Grams, F., Reinemer, P., Powers, J.C., Kleine, T., Pieper, M., Tschesche, H., Huber, R. and Bode, W. (1995) Eur. J. Biochem. **228**, 830–841

56. Matthews, B.W. (1988) Acc. Chem. Res. **21**, 333–340

57. Kang, T., Nagase, H. and Pei, D. (2002) Cancer Res. **62**, 675–681

58. Pei, D. and Weiss, S.J. (1995) Nature (London) **375**, 244–247

59. Opdenakker, G., Van den Steen, P.E. and Van Damme, J. (2001) Trends Immunol. **22**, 571–579

60. Chung, L., Shimokawa, K., Dinakarpandian, D., Grams, F., Fields, G.B. and Nagase, H. (2000) J. Biol. Chem. **275**, 29610–29617

61. Brandstetter, H., Grams, F., Glitz, D., Lang, A., Huber, R., Bode, W., Krell, H.-W. and Engh, R.A. (2001) J. Biol. Chem. **276**, 17405–17412

62. Ottl, J., Gabriel, D., Murphy, G., Knauper, V., Tominaga, Y., Nagase, H., Kroger, M., Tschesche, H., Bode, W. and Moroder, L. (2000) Chem Biol. **7**, 119–132

63. Muskett, F.W., Frenkiel, T.A., Feeney, J., Freedman, R.B., Carr, M.D. and Williamson, R. (1998) J. Biol. Chem. **273**, 21736–21743

64. Tuuttila, A., Morgunova, E., Bergmann, U., Lindquist, Y., Maskos, K., Fernandez-Catalan, C., Bode, W., Tryggvason, K. and Schneider, G. (1998) J. Mol. Biol. **284**, 1133–1140

65. Wu, B., Arumugam, S., Gao, G., Le, G., Semenchenko, V., Huang, W., Brew, K. and VanDoren, S.R. (2000) J. Mol. Biol. **295**, 257–268

66. Cao, J., Sato, H., Takino, T. and Seiki, M. (1995) J. Biol. Chem. **270**, 801–805

67. Butler, G.S., Butler, M.J., Atkinson, S.J., Will, H., Tamura, T., van Westrum, S.S., Crabbe, T., Clements, J., d'Ortho, M.P. and Murphy, G. (1998) J. Biol. Chem. **273**, 871–880

68.   Overall, C.M., King, A.E., Bigg, H.F., McQuibban, A., Atherstone, J., Sam, D.K., Ong, A.D., Lau, T.T., Wallon, U.M., DeClerck, Y.A. and Tam, E. (1999) Ann. N.Y. Acad. Sci. **878**, 747–753

69.   Cha, H., Kopetzki, E., Huber, R., Lanzendorfer, M. and Brandstetter, H. (2002) J. Mol. Biol. **320**, 1065–1079

70.   Itoh, Y., Takamura, A., Ito, N., Maru, Y., Sato, H., Suenaga, N., Aoki, T. and Seiki, M. (2001) EMBO J. **20**, 4782–4793

71.   Lehti, K., Lohi, J., Juntunen, M.M., Pei, D. and Keski-Oja, J. (2002) J. Biol. Chem. **277**, 8440–8448

72.   Rozanov, D.V., Deryugina, E.I., Ratnikov, B.I., Monosov, E.Z., Marchenko, G.N., Quigley, J.P. and Strongin, A.Y. (2001) J. Biol. Chem. **276**, 25705–25714

Biochem. Soc. Symp. **70**, 15–30
(Printed in Great Britain)
© 2003 Biochemical Society

2

Papain-like lysosomal cysteine proteases and their inhibitors: drug discovery targets?

Dušan Turk[1], Boris Turk and Vito Turk

Department of Biochemistry and Molecular Biology, Jozef Stefan Institute, Jamova 39, Ljubljana, 1000 Slovenia

Abstract

Papain-like lysosomal cysteine proteases are processive and digestive enzymes that are expressed in organisms from bacteria to humans. Increasing knowledge about the physiological and pathological roles of cysteine proteases is bringing them into the focus of drug discovery research. These proteases have rather short active-site clefts, comprising three well defined substrate-binding subsites (S2, S1 and S1′) and additional broad binding areas (S4, S3, S2′ and S3′). The geometry of the active site distinguishes cysteine proteases from other protease classes, such as serine and aspartic proteases, which have six and eight substrate-binding sites respectively. Exopeptidases (cathepsins B, C, H and X), in contrast with endopeptidases (such as cathepsins L, S, V and F), possess structural features that facilitate the binding of N- and C-terminal groups of substrates into the active-site cleft. Other than a clear preference for free chain termini in the case of exopeptidases, the substrate-binding sites exhibit no strict specificities. Instead, their subsite preferences arise more from the specific exclusion of substrate types. This presents a challenge for the design of inhibitors to target a specific cathepsin: only the cumulative effect of an assembly of inhibitor fragments will bring the desired result.

Introduction

The physiological roles of lysosomal cysteine proteases now emerging from research are bringing these proteases increasingly into focus as drug targets for a wide range of diseases, such as cancer, rheumatoid arthritis and osteoarthritis, multiple sclerosis and muscular dystrophy (reviewed in [1]). For many diseases resulting from excess proteolysis, no inhibitors have yet been

---

[1]To whom correspondence should be addressed (e-mail Dusan.Turk@ijs.si).

identified with the necessary profile for use as therapy. Thus research into the physiological roles of proteases and into the identification of substances able to modulate them will remain a priority of both science and the pharmaceutical industry for the foreseeable future. For now, drug design proceeds hand in hand with the discovery of biological roles for these enzymes. The inhibitor that unravels a specific role can immediately become a drug candidate.

It is estimated that there are around 500–600 proteases in the human genome, of which about 400 have already been identified [2]. In lysosomes there are around 60 hydrolases [3], which include a group of about a dozen papain-like lysosomal cysteine proteases (papain-like cathepsins). Their relatively small size, their uniquely reactive cysteine thiol group (pK_a between 2.5 and 3.5 [4]) and their unique reactive mechanism make these enzymes attractive targets for drug design. There are 11 human enzymes currently known (cathepsins B, C, F, H, L, K, O, S, V, X and W) [1,5], and it is quite likely that the list has already been completed. Human gene databank searches have not indicated any new members of the family (B. Turk and A. Sali, unpublished work).

The use of biochemical techniques led to the discovery of the classical cathepsins (B, C, H, L and S), cathepsin C being the first [6], whereas the other cathepsins (F, K, O, V, X and W) were discovered in the 1990s by means of DNA manipulation techniques. Only cathepsins O and W have not yet been characterized biochemically. The papain-like fold was revealed in the early days of crystallography [7], whereas structural characterization of cathepsins began in the early 1990s with the cathepsin B structure [8]. Currently, crystal structures of all human representatives or their mammalian analogues, except for cathepsins O and W, have been determined and are available in the Protein Databank (PDB; Table 1). The relevance of cathepsins as potential drug targets is best indicated by the fact that four (cathepsins K, S, V and recently F) out of nine structures of cathepsins were published by industrial research groups. The structures of cathepsins L [9] and S [10] have also been reported by industrial groups (but not in the public domain). An additional publication describing complexes of

Table 1 Primary citations and PDB codes of cathepsin structures.
The parallel entries indicate that some of the structures were determined simultaneously by several groups.

Cathepsin	PDB code	Reference
B	1HUC	[8]
K	1ATK	[70]
1MEM	[47]	
H	8PCH	[60]
L	1ICF	[69]
V	1FH0	[48]
X	1EF7	[59]
C	1JQP	[67]
1K3B	[66]	
F	1M6D	[49]
S	1GLO	[71]

cathepsin S with synthetic inhibitors is in preparation (M. Cygler and V. Rath, personal communication).

Cathepsin physiology

Protein digestion in lysosomes was long believed to be the major physiological role of the papain-like lysosomal cysteine proteases [11]. Analyses of gene knockouts suggested that this function is not dependent on any single cathepsin [12–17]. Moreover, analyses of gene knockouts and of locations of mutations on genes of lysosomal cysteine proteases responsible for some hereditary diseases revealed several specific biological functions. These functions are a result of limited proteolysis of their target substrates, and rely additionally on co-localization and timing. For example, cathepsin K has been found to be crucial in bone remodelling [12,18].

Cathepsin S is the major processing enzyme of the MHC class II-associated invariant chain, and is thus essential for the normal functioning of MHC class II-associated antigen processing and presentation [15,16]. Cathepsins L and F have been shown to participate in the same process, primarily in tissues or cells not expressing cathepsin S [15,19], although the role of the former has probably been taken over by cathepsin V in humans [20].

Cathepsin L-deficient mice developed periodic hair loss and epidermal hyperplasia, indicating that cathepsin L is involved in epidermal homoeostasis and regular hair follicle morphogenesis and cycling [17]. At 1 year old, cathepsin L-deficient mice [21] exhibited histomorphological and functional alterations of the heart, resulting in dilated cardiomyopathy, which is a frequent cause of heart failure.

Cells derived from cathepsin C-deficient mice failed to show activation of groups of serine proteases from granules of immune (cytotoxic T lymphocytes, natural killer cells) and inflammatory (neutrophils, mast cells) cells involved primarily in the defence of the organism, demonstrating that cathepsin C is involved in this activation [13,22]. The current list of unprocessed zymogens of proteases in cathepsin C knockout mice includes granzymes A, B and C, cathepsin G, neutrophil elastase and chymase. More information about the processes in which the lysosomal papain-like cysteine proteases participate can be found elsewhere [1,5,11,18,23,24].

Cathepsin pathology

Lysosomal cysteine proteases have been found to be associated with a number of pathologies, including cancer, inflammation, rheumatoid arthritis and osteoarthritis, multiple sclerosis, muscular dystrophy, pancreatitis, liver disorders, lung disorders, lysosomal disorders, Batten's disease, diabetes and myocardial disorders. In many of these diseases, the lysosomal enzymes have been found in the extracellular and extralysosomal environment in their (zymogenic) 'pro' forms, which are substantially more stable than the mature enzymes (reviewed in [5,11,18,23]).

A few genetic disorders have been traced to the genes encoding lysosomal cysteine proteases. Pycnodysostosis, which is characterized by severe bone abnormalities, is associated with a loss-of-function mutation of cathepsin K [25], while a loss-of-function mutation in the cathepsin C gene leads to Papillon–Lefevre syndrome, characterized in humans by palmoplantar keratosis and severe, early-onset periodontitis [26,27]. It has been suggested that these effects are likely to be the result of incomplete processing of some as yet unidentified proteases presumably involved in establishing or maintaining the structural organization of the epidermis at the extremities and the integrity of tissues surrounding the teeth, as well as in the processing of proteins such as keratins [28]. In addition, cathepsin C may be involved in chronic airway diseases such as asthma [29].

Interestingly, the down-regulation of natural inhibitors, as demonstrated by a mutation in the gene for stefin B, predisposes affected individuals to a hereditary form of myoclonal epilepsy [30,31].

Cathepsin structure

Papain-like lysosomal cysteine proteases share the common fold of a papain-like structure. Cathepsin L, as a typical endopeptidase, has been chosen as the representative of the family (Figure 1). A papain-like fold consists of two domains, reminiscent of a closed book with the spine to the front. The domains separate on the top in a 'V'-shaped active-site cleft, in the middle of which are found residues Cys^{25} and His^{159}, one from each domain, which form the catalytic site of the enzyme. The most prominent feature of the left (L-) domain is the central α-helix, which is approx. 30 residues long; the right (R-) domain forms a kind of a β-barrel, which includes a shorter α-helical motif. (The terms left and right domains refer to the standard view shown in Figure 1.) The enzymes are monomeric proteins, with molecular masses between 22 and

Figure 1 Fold of cathepsin L. Cathepsin L (PDB code 1ICF) is shown as a ribbon in its standard orientation – a view along the two-domain interface with the central α-helix in vertical orientation and the active site on the top. The side chains of catalytic residues Cys^{25} and His^{159} are shown as atom balls. The figure was prepared using the program Ribbons [74].

28 kDa. The only exception is cathepsin C, which is a tetrameric molecule with a molecular mass of 200 kDa [32].

Cathepsin substrate-binding sites

The nomenclature of the assignment of sites of interaction between a polypeptide substrate and a protease originates from the work of Schechter and Berger [33] (Figure 2). They showed that the kinetics of substrate hydrolysis are influenced by polypeptide chain length up to seven amino acids, and concluded that there are seven substrate-binding sites on the papain molecule. Three decades later, when a sufficient number of protease–inhibitor structures had become available, the definition of Schechter and Berger of substrate-binding sites on the papain-like enzymes was revisited and redefined [34]. The inspection of structures revealed that the base and walls of the substrate-binding sites are formed by four chain segments comprising two shorter loops on the L-domain (residues 19–25 and 61–69) and two longer loops on the R-domain (residues 136–162 and 182–213) (Figures 1 and 3b). A third loop from the L-domain might be involved as well if the disulphide Cys^{22}–Cys^{65}, which connects the two L-domain loops at the top, is considered an additional loop closure.

The superimposed structures of complexes of substrate-analogue inhibitors with cathepsins and papain-like enzymes were divided into five groups. Figures 4(a) and 4(c) show substrate-analogue inhibitors, presumably revealing the geometry of binding of substrate residues. Substrate residues bind along the active-site cleft in an extended conformation, with the side chains oriented alternately left and right – towards the L- and R-domains. Residues P2, P1 and P1′ bind into well defined binding sites. Positioning of these residues is governed by interactions that involve both main-chain as well as side-chain atoms. The S2 binding site is actually the only deep binding pocket, whereas the S1 and S1′ sites provide a binding surface. The positioning of the P3 residue is mediated only by side-chain interactions. For this reason, the binding geometries of the latter are scattered over a broad area and are unique for each substrate–protease pair. On the prime side of the binding cleft, the P2′ residue binding site appears to be quite well defined. However, current knowledge is based on specific interactions between the inhibitor CA030 and the parts of cathepsin B structure responsible for its carboxydipeptidase activity (Figure 4c) [35]. It is likely that interactions within the S2′ site of an endopeptidase may be different.

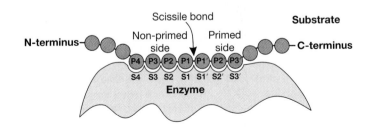

Figure 2 Illustration of the Schechter and Berger [33] definition of substrate-binding sites. The Figure was prepared using the program Ribbons [74].

(a)

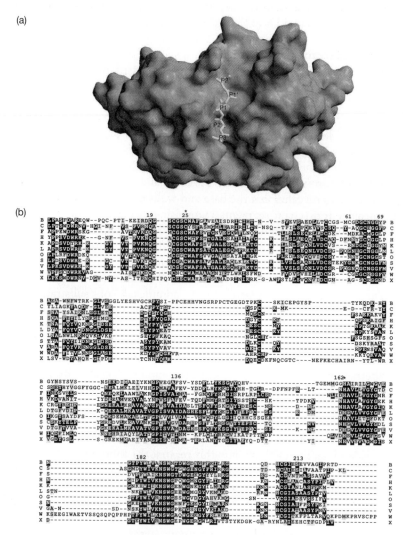

Figure 3 Substrate-binding sites of cathepsins. (a) View from the top:
model of a polyalanine substrate bound into the active-site cleft of cathepsin L.
Modelling of the binding geometry of a substrate is based on information gained
from crystal structures of substrate-analogue inhibitors and their interactions
with a papain-like protease active site (see Figures 4a–4c). Substrate residues are
shown in stick representation and are denoted using the Schechter and Berger
nomenclature [33]. Cathepsin L is shown in dark grey surface representation.
The surface of the catalytic cysteine side chain is highlighted. The cathepsin L surface was generated with GRASP [77]. The Figure was prepared using MAIN [75]
and rendered with Raster3D [76]. (b) Structure-based amino acid alignment of
sequences of the papain-like domains of all known human cathepsins. The structural alignment was made using the program Modeller [72]. The sequences of the
remaining cathepsins F, O and W were aligned to the template with the ClustalW
program [73]. The structures were taken from the PDB and sequences from the
SwissProt or GenBank databases.

Main-chain interactions do not constrain the positioning of the substrate into binding sites beyond S3 and S2′. These substrate residues dock on the surfaces of the underlying enzymes in their own way (Figure 4a). In particular, for the non-primed binding sites, the structures suggest that a common S4 binding site, and also an S3′ site, do not exist. Therefore it was suggested that the substrate residue binding regions beyond S2 and S2′ should not be called sites, but areas [34]. The papain-like proteases thus represent a special class of proteolytic enzymes with the smallest number of substrate-binding sites, as opposed to chymotrypsin-like serine proteases with six [36] and aspartic proteases with eight [37] binding sites.

Low-molecular-mass inhibitors

The presence of a rather small number of substrate-binding sites seems to facilitate covalent interactions with low-molecular-mass inhibitors. Covalent interactions, however, impose tight constraints on the geometry of binding. It is thus not surprising that natural low-molecular-mass inhibitors such as E64 [trans-epoxysuccinyl-L-leucylamido-(4-guanidino)butane] and leupeptin interact covalently with the reactive site and bind only into the non-primed binding side of the active-site cleft, whereas inhibitors that bind into both sides (primed as well as non-primed) have been designed only recently.

The first inhibitor constructs, the structures of which were determined in complex with papain, were based on the chloromethyl reactive group [38]. These studies revealed binding of substrate residues in the non-primed binding sites of the enzyme (Figure 4a). Around the same time, a natural cysteine protease inhibitor, named E64, was discovered [39,40]. E64 utilizes an epoxysuccinyl group to interact covalently with the reactive-site cysteine (Figure 5). The structures of E64 [41] and its analogues [42] revealed that they bind into the non-primed region of the active site; however, they bind in the direction of propeptide binding, opposite to that of substrate binding (Figure 4b).

Replacement of the terminal agmatine residue with proline was essential for the conversion of E64, a general inhibitor of papain-like cysteine proteases, into the cathepsin B-specific inhibitor CA030 (Figure 6). The crystal structure of CA030 in complex with human cathepsin B [35] revealed the essence of the change. CA030 binds, in contrast with E64, into the primed binding side in the direction of substrate binding. Switching the sides of binding was made possible by specific interactions. The carboxylic group of the C-terminal residue of CA030 mimics the C-terminus of a substrate and docks against the occluding loop residues His[110] and His[111] (Figure 4c; also see Figure 7). Alignment of the binding geometries of E64 and CA030 (Figure 6) [35] shows that the epoxysuccinyl group possesses internal symmetry with two carboxylic heads, mimicking a polypeptide C-terminus to which amino acid residues can be attached. The synthesis of double-head inhibitors such as NS134 and CLIK066 followed [43–45] (Figure 6). The binding geometry of the double-head inhibitor design has been recently confirmed by the crystal structures of complexes of cathepsin L and cathepsin B with these inhibitors ([46]; I. Stern, N. Schaschuke, L. Moroder and D. Turk, unpublished work).

(a)

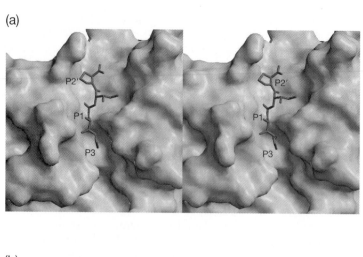

(b)

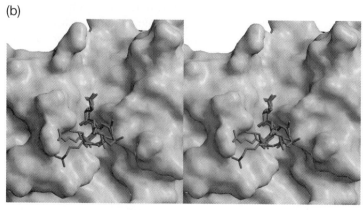

(c)

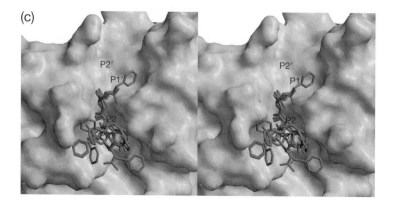

(d)

(e)

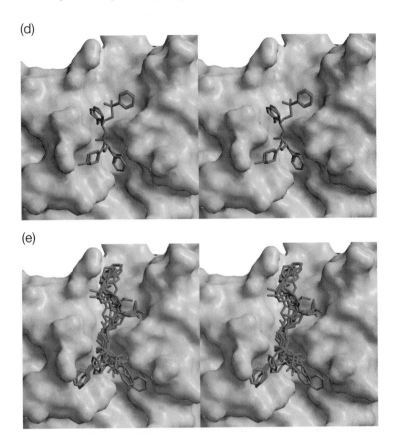

Figure 4 Binding geometry of low-molecular-mass inhibitors. The inhibitors, shown in stick representation, from structures of complexes with papain-like cysteine proteases (PDB codes given below in parentheses) are superimposed on the top of the cathepsin L surface. Complexes with plant enzymes are also included. See Figure 6 for the structures of representative inhibitors. (**a**) Substrate-analogue inhibitors: fluoro- and chloro-methylketone-based inhibitors and leupeptin are shown as dark sticks. Inhibitors are taken from the structures of complexes with the following enzymes: cruzipain (1AIM, 2AIM), papain (1PAD, 1POP, 5PAD, 6PAD), glycyl endopeptidase (1GEC) and cathepsin B (1THE, 1CTE). (**b**) E64 and a derivative are shown as dark sticks. Inhibitors are taken from structures of complexes with the following enzymes: actinidin (1AEC), caricain (1MEG), cathepsin K (1ATK) and papain (1PE6, 1PPP). (**c**) Inhibitor CA030 from the complex with cathepsin B (1CSB). (**d**) Vinyl sulphone-based inhibitors taken from structure of complexes with cathepsin K (1MEM) {N-[1S-(2-phenylethyl)-3-phenylsulphonylallys]-4-methyl-2R-piperazinyl carbonylaminovaleramide} and cathepsin V (1FH0) {4-morpholin-4-yl-piperidine-1-carboxylic acid [1-(3-benzensulphonyl-1-propyl-allylcarbamoyl)-2-phenylethyl]-amide}. (**e**) Group of cathepsin K inhibitors manufactured by GlaxoSmithKline (1AYU, 1AYV, 1AYW, 1AU0, 1BGO, 1AU2, 1AU3, 1AU4); a representative example is N-{2-[1-(N-benzyloxy-carbonylamino)-3-methylbutyl] thiazol-4-ylcarbonyl}-N′-(benzyloxycarbonyl-L-leucinyl)hydrazide (1AYV). The Figures were prepared with MAIN [75] and rendered with Raster3D [76].

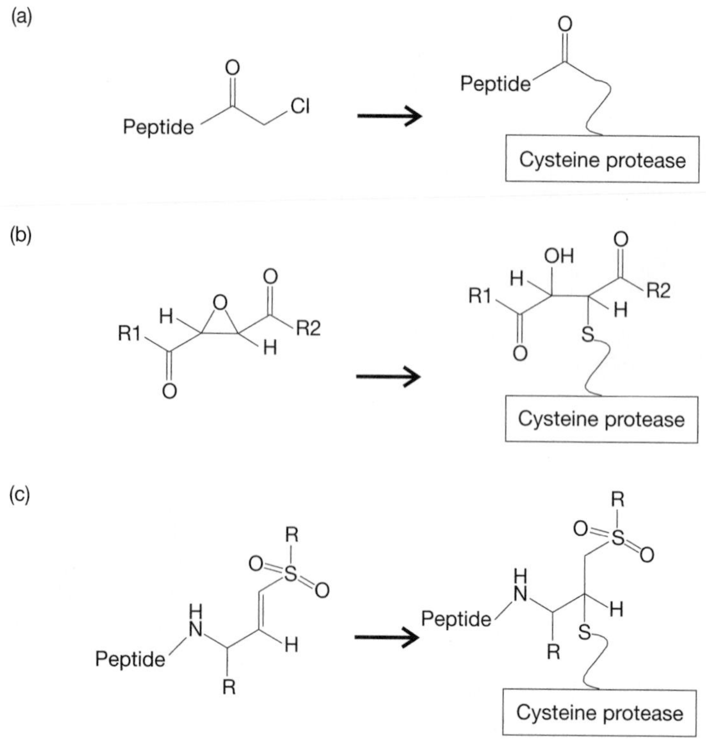

(a)

(b)

(c)

Figure 5 Three of the most frequent reactive groups before and after binding irreversibly to the reactive-site cysteine. (a) Chloromethylketone; **(b)** epoxysuccinyl; **(c)** vinyl sulphone.

Inhibitor constructs using the vinyl sulphone reactive group (Figure 4d) [47–49], and exceptionally even a chloromethyl inhibitor with a long side chain of a P1-mimicking residue [50] (Figure 4a), can reach into the S1' binding site. A series of compounds synthetized at GlaxoSmithKline revealed that the irreversible covalent interaction with the reactive-site cysteine is not mandatory (Figure 4e). These compounds span both sides of the active-site cleft and block the reactive site with fragments that resemble the peptide bond, but which cannot be hydrolysed [51]. Additional information regarding inhibitors and their chemistry can be found elsewhere [24,52,53].

Exopeptidases

The 'V'-shaped active-site cleft of endopeptidases (cathepsins F, L, K, O, S and V) extends along the whole length of the two-domain interface on both sides, whereas the exopeptidases (cathepsins B, C, H and X) possess additional features that block access to parts of the active-site cleft (Figure 7). The role of these features is dual: they prevent binding of longer peptidyl substrates, and they dock with charged N- or C-termini of substrates by utilizing selective electrostatic interactions.

E64

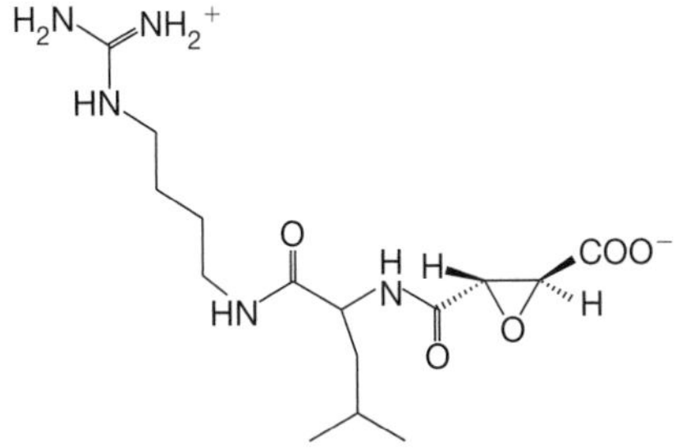

CA030

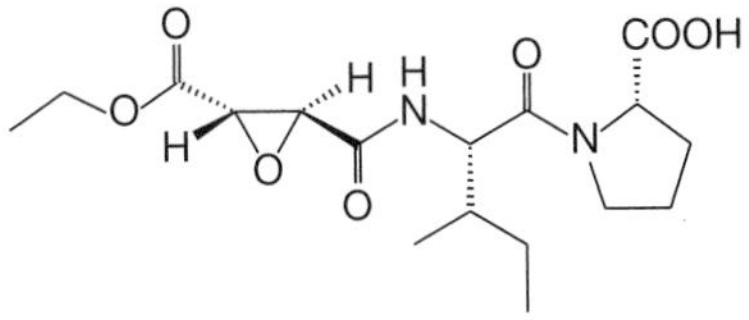

NS134

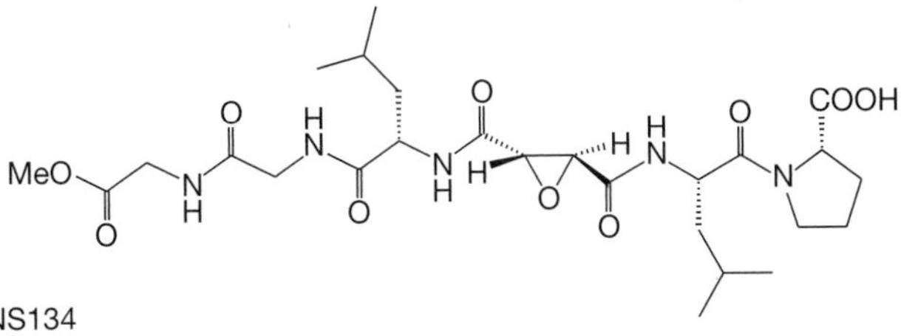

CLIK066

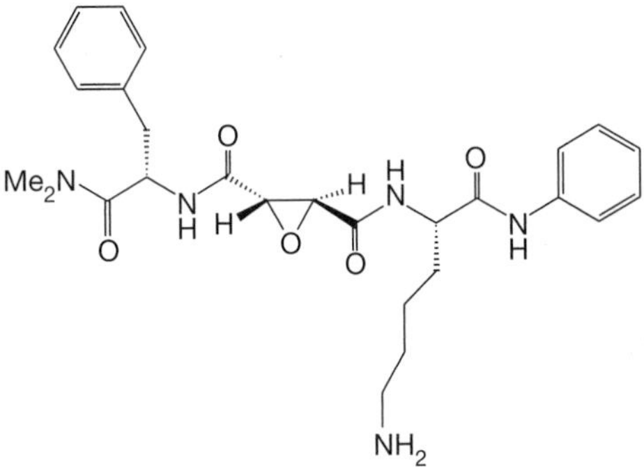

Figure 6 Epoxysuccinyl derivatives aligned on epoxysuccinyl fragment as they bind along the active-site cleft.

The carboxydipeptidase cathepsin B [8] has an approx. 20-residue insertion, termed the occluding loop, that blocks the active-site cleft on the primed binding side beyond S2′. Its two histidine residues, His^{110} and His^{111}, are cruical for docking with the C-terminal carboxylic group of a substrate residue. Interestingly, cathepsin B also exhibits an endopeptidase activity that is made possible by the flexible occluding loop, which can occupy positions outside the active-site cleft [54–56].

Cathepsin X is primarily a carboxymonopeptidase [57] that can also exhibit carboxydipeptidase activity [58]. The crystal structure showed that a histidine residue, His^{23}, positioned within a short loop termed a mini-loop [57], is the anchor for the carboxylic group of the substrate C-terminal residue [59].

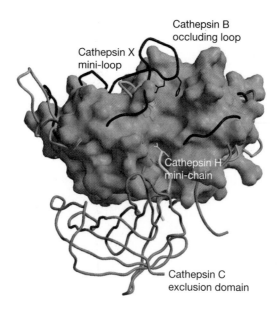

Figure 7 Features of exopeptidases. Chain traces of cathepsin H (8PCH), cathepsin C (1K3B), cathepsin B (1HUC) and cathepsin X (1EF7) are shown superimposed on the cathepsin L structure in a view from the top. The surface of cathepsin L is shown in grey. Dark grey denotes chain traces of cathepsins B and C, whereas the traces of cathepsins H and X are brighter. Structural elements that facilitate the exopeptidase activities are labelled. Residues that have a crucial role in exopeptidase specificity are shown in stick representation.

In the free enzyme structure, the histidine ring occupies the position that in related cathepsins is the S2' substrate-binding site. A simple modelling study (manual rotation about the side-chain bonds) suggested that the histidine ring can adapt a position equivalent to His[110] of cathepsin B, thereby allowing the carboxydipeptidase mode.

Cathepsin H is an amino-monopeptidase. The crystal structure of the pig enzyme [60] revealed the binding of an eight-residue segment of the propeptide, called the mini-chain. The mini-chain is covalently fastened to the body of the enzyme via a disulphide bond. It binds into the active-site cleft of the enzyme in the direction of a bound substrate. The negatively charged carboxylic group of its C-terminal residue, Thr[83P], attracts the positively charged N-terminus of a substrate and thereby facilitates the aminopeptidase activity of cathepsin H. Thr[83P] actually mimics a substrate P2 residue by occupying the position that in related enzymes is the S2 binding site. The mini-chain is fastened additionally to the enzyme surface by a four-residue insertion (Lys[155A] to Asp[155D]) and a carbohydrate chain attached to Asn[112]. Interestingly, the positioning of the cathepsin H mini-chain closely resembles the positioning of the C-terminus of a distant homologue, bleomycin hydrolase [61].

Cathepsin C (also termed dipeptidyl peptidase I) is an amino-dipeptidase. Cathepsin C has four independent active sites located on the external surface of the tetrahedral molecule. The exposure of the active sites makes cathepsin C unique when compared with other oligomeric proteolytic enzymes. Thus the 20 S proteasome [62,63], bleomycin hydrolase [61], tryptase [64] and tricorn protease [65] have their active sites on the inside surface. Proteasomes are barrel-like structures composed of four rings of α- and β-subunits, which cleave unfolded proteins captured in the central cavity into short peptides. Tryptases are flat tetramers with a central pore in which the active sites reside. The pore restricts the size of accessible substrates and inhibitors. The active sites of bleomycin hydrolase and tricorn protease are located similarly within the hexameric structure. Its exposed active sites enable cathepsin C to hydrolyse protein substrates in their native state, regardless of their size. Its design, supported by the oligomeric structure, confines the activity of cathepsin C to an amino-dipeptidase and thereby makes it suitable for use in many different environments, where it can selectively activate a group of chymotrypsin-like proteases and presumably also other proteins.

The massive body of the exclusion domain blocks the active site of cathepsin C beyond the S2 binding site [66,67]. An exposed β-hairpin, the first N-terminal residues of the exclusion domain and the carbohydrate ring attached to Asn^5 block undesired access, while Asp^1 with its carboxylic group side chain controls entry into the S2 binding pocket by docking with the N-terminal amino group of a substrate. Asp^1 simultaneously prevents the positively charged side chains of arginine and lysine residues from binding into the S2 binding pocket. An additional special feature of cathepsin C is the dependence of its activity on chloride ions, one of which is located at the bottom of the very long S2 binding pocket.

Conclusions

The papain-like cathepsins are rather non-specific enzymes with no clear substrate recognition site. The major substrate-binding determinants are main-chain atoms, which interact with the conserved structural features of the S2, S1 and S1′ substrate-binding sites. The vast majority of low-molecular-mass inhibitors additionally utilize a covalent interaction with the reactive-site cysteine. This does not imply that specific inhibitors cannot be designed. It does, however, suggest that specificity is not an issue of a single binding site, but rather a sum of all interactions. This suggests that inhibitor constructs interacting with regions on both sides of the reactive site may be advantageous compared with those that bind only into one side. The specificity of exopeptidases is more a matter of exclusive interactions of free-chain termini than of side-chain recognition. The design of exopeptidase inhibitors therefore seems easier. Structural observations raise the question of whether the substrate-binding sites are indeed the only potential determinants for the design of specific inhibitors. The crystal structures of complexes with protein inhibitors [68,69] have revealed that stefins and cystatins, rather non-selective inhibitors of papain-like proteases, interact with the non-specific features located at the bottom of the active-site cleft. However, thyropin, the inhibitory fragment of the p41 form of the MHC class II-associated invariant chain, forms contacts along the active-site cleft as well as with surface loops on the top of the L-domain.

These contacts with the variable loops on the top of the L-domain endow this inhibitor with the ability to differentiate between two otherwise very similar endopeptidases, i.e. cathepsins L and S.

This chapter is dedicated to Professor A. Barrett on the occasion of his 65th birthday. The Slovenian Ministry of Education, Science and Sport and ICGEB are gratefully acknowledged for their financial support.

References

1.   Turk, B., Turk, D. and Turk, V. (2000) Biochim. Biophys. Acta **1477**, 98–111
2.   López-Otín, C. and Overall, C.M. (2002) Nat. Rev. Mol. Cell Biol. **7**, 509–519
3.   Mason, R.W. (1995) Subcell. Biochem. **27**, 159–190
4.   Pinitglang, S., Watts, A.B., Patel, M., Reid, J.D., Noble, M.A., Gul, S., Bokth, A., Naeem, A., Patel, H., Thomas, E.W. et al. (1997) Biochemistry **36**, 9968–9982
5.   Turk, V., Turk, B. and Turk, D. (2001) EMBO J. **20**, 4629–4633
6.   Gutman, H.R. and Fruton, J. (1948) J. Biol. Chem. **174**, 851–858
7.   Drenth, J., Jansonius, J.N., Koekoek, R., Swen, H.M. and Wolthers, B.G. (1968) Nature (London) **218**, 929–932
8.   Musil, D., Zuciæ, D., Turk, D., Engh, R.A., Mayr, I., Huber, R., Popoviè, T., Turk, V., Towatari, T., Katunuma, N. and Bode, W. (1991) EMBO J. **10**, 2321–2330
9.   Fujishima, A., Imau, Y., Nomura, T., Fujisawa, Y., Yamamoto, Y. and Sugawara, T. (1997) FEBS Lett. **407**, 47–50
10. McGrath, M.E., Palmer, J.T., Brömme, D. and Somoza, J.R. (1998) Protein Sci. **7**, 1294–1302
11. Kirschke, H., Barrett, A.J. and Rawlings, N.D. (1995) in Protein Profile, vol. 2 (Sheterline, P., ed.), pp. 1587–1643, Academic Press, London
12. Saftig, P., Hunzicker, E., Wehmeyer, O., Jones, S., Boyde, A., Rommerskirch, W., Moritz, J.D., Schu, P. and von Figura, K. (1998) Proc. Natl. Acad. Sci. U.S.A. **95**, 13453–13458
13. Pham, C.T. and Ley, T.J. (1999) Proc. Natl. Acad. Sci. U.S.A. **96**, 8627–8632
14. Deussing, J., Roth, W., Saftig, P., Peters, C., Ploegh, H.L. and Villadangos, J.A. (1998) Proc. Natl. Acad. Sci. U.S.A. **95**, 4516–4521
15. Nakagawa, T., Roth, W., Wong, P., Nelson, A., Farr, A., Deussing, J., Villadangos, J.A., Ploegh, H., Peters, C. and Rudensky, A.Y. (1998) Science **280**, 450–453
16. Nakagawa, T.Y., Brissette, W.H., Lira, P.D., Griffiths, R.J., Petrushova, N., Stock, J., McNeish, J.D., Eastman, S.E., Howard, E.D., Clarke, S.R.M. et al. (1999) Immunity **10**, 207–217
17. Roth, V., Deussing, J., Botchkarev, V.A., Pauly-Evers, M., Saftig, P., Hafner, A., Schmidt, P., Schmahl, W., Scherer, J., Anton-Lamprecht, I. et al. (2000) FASEB J. **14**, 2075–2086
18. Chapman, H.A., Riese, J.P. and Shi, G.-P. (1997) Annu. Rev. Physiol. **59**, 63–88
19. Shi, G.-P., Bryant, R.A., Riese, R., Verhelst, S., Driessen, C., Li, Z., Brömme, D., Ploegh, H.L. and Chapman, H.A. (2000) J. Exp. Med. **191**, 1177–1186
20. Brömme, D., Li, Z., Barnes, M. and Mehler, E. (1999) Biochemistry **38**, 2377–2385
21. Stypmann, J., Glaser, K., Roth, W., Tobin, D.J., Petermann, I., Matthias, R., Monnig, G., Haverkamp, W., Breithardt, G., Schmahl, W. et al. (2002) Proc. Natl. Acad. Sci. U.S.A. **99**, 6234–6239
22. Wolters, P.J., Pham, C.T.N., Muilenburg, D.J., Ley, T.J. and Caughey, G.H. (2001) J. Biol. Chem. **276**, 18551–18556
23. Barrett, A.J., Rawlings, N.D. and Woessner, Jr, J.F. (eds) (1998) Handbook of Proteolytic Enzymes, Academic Press, London
24. Brömme, D. and Kaleta, J. (2002) Curr. Pharm. Des. **8**, 1639–1658
25. Gelb, B.D., Shi, G.-P., Chapman, H.A. and Desnick, R.J. (1996) Science **273**, 1236–1238

26. Toomes, C., James, J., Wood, A.J., Wu, C.L., McCormick, D., Lench, N., Hewitt, C., Moynihan, L., Roberts, E., Woods, C.G. et al. (1999) Nat. Genet. **23**, 421–424

27. Hart, T.C., Hart, P.S., Bowden, D.W., Michalec, M.D., Callison, S.A., Walker, S.J., Zhang, Y. and Firatli, E. (1999) J. Med. Genet. **36**, 881–887

28. Nuckolls, G.H. and Slavkin, H.C. (1999) Nat. Genet. **23**, 378–380

29. Wolters, P.J., Laig-Webster, M. and Caughey, G.H. (2000) Am. J. Respir. Cell Mol. Biol. **22**, 183–190

30. Pennacchio, L.A., Lehesjoki, A.E., Stone, N.E., Willour, V.L., Virtaneva, K., Miao, J., D'Amato, E., Ramirez, L., Faham, M., Koskiniemi, M. et al. (1996) Science **271**, 1731–1734

31. Lalioti, M.D., Scott, H.S., Buresi, C., Rossier, C., Bottani, A., Morris, M.A., Malafosse, A. and Antonarakis, S.E. (1997) Nature (London) **386**, 847–851

32. Dolenc, I., Turk B., Pungerèiè, G., Ritonja, A. and Turk, V. (1995) J. Biol. Chem. **270**, 21626–21631

33. Schechter, I. and Berger, A. (1967) Biochem. Biophys. Res. Commun. **27**, 157–162

34. Turk, D., Gunèar, G., Podobnik, M. and Turk, B. (1998) Biol. Chem. **379**, 137–147

35. Turk, D., Podobnik, M., Popoviè, T., Katunuma, N., Bode, W., Huber, R. and Turk, V. (1995) Biochemistry **34**, 4791–4797

36. Bode, W. and Huber, R. (1992) Eur. J. Biochem. **204**, 433–451

37. Wlodawer, A. and Gustchina, A. (2000) Biochim. Biophys. Acta **1477**, 16–34

38. Drenth, J., Kalk, K.H. and Swen, H.M. (1976) Biochemistry **15**, 3731–3738

39. Aoyagi, T. and Umezawa, H (1975) in Proteases in Biological Control (Reich, E., Rifkin, D.B. and Shaw, E., eds), pp. 429–454, Cold Spring Harbor Laboratory Press, Cold Spring Harbor, NY

40. Hanada, K., Tamai, M., Yamagishi, M., Ohmura, S., Sawada, J. and Tanaka, I. (1978) Agric. Biol. Chem. **42**, 523–528

41. Varughese, K.I., Su, Y., Cromwell, D., Hasnain, S. and Xuong, N.H. (1992) Biochemistry **22**, 5172–5176

42. Yamamoto, D., Matsumoto, K., Ohishi, H., Ishida, T., Inoue, M., Kitamura, K. and Mizuno, H. (1991) J. Biol. Chem. **266**, 14771–14777

43. Schaschke, N., Assfalg-Machleidt, I., Machleidt, W., Turk, D. and Moroder, L. (1997) Bioorg. Med. Chem. **5**, 1789–1797

44. Katunuma, N., Murata, E., Kakegawa, H., Matsui, A., Tsuzuki, H., Tsuge, H., Turk, D., Turk, V., Fukushima, M., Tada, Y. and Asao, T. (1999) FEBS Lett. **458**, 6–10

45. Schaschke, N., Assfalg-Machleidt, I., Lassleben, T., Sommerhoff, C.P., Moroder, L. and Machleidt, W. (2000) FEBS Lett. **482**, 91–96

46. Tsuge, H., Nishimura, T., Tada, Y., Asao, T., Turk, D., Turk, V. and Katunuma, N. (1999) Biochem. Biophys. Res. Commun. **266**, 411–416

47. McGrath, M.E., Klaus, J.L., Barnes, M.G. and Brömme, D. (1997) Nat. Struct. Biol. **4**, 105–109

48. Somoza, J.R., Zhan, H., Bowman, K.K., Yu, L., Mortara, K.D., Palmer, J.T., Clark, J.M., Mcgrath, M.E. (2000) Biochemistry **39**, 12543–12551

49. Somoza, J.R., Palmer, J.T. and Ho, D.H. (2002) J. Mol. Biol. **322**, 559–568

50. Jia, Z., Hasnain, S., Hirama, T., Lee, X., Mort, J.S., To, R. and Huber, C.P. (1995) J. Biol.Chem. **270**, 5527–5533

51. Thompson, S.K., Halbert, S.M., Bossard, M.J., Tomaszek, T.A., Levy, M.A., Zhao B., Smith, W.W., Abdel-Meguid, S.S., Janson, C.A., D'Alessio, K.J. et al. (1997) Proc. Natl. Acad. Sci. U.S.A. **94**, 14249–14254

52. Shaw, E. (1990) Adv. Enzymol. **63**, 271–297

53. Otto, H.H. and Schirmeister, T. (1997) Chem. Rev. **97**, 133–172

54. Illy, C., Quraishi, O., Wang, J., Purisima, E., Vernet, T. and Mort, J.S. (1997) J. Biol. Chem. **272**, 1197–1202

55. Podobnik, M., Kuhelj, R., Turk, V. and Turk, D. (1997) J. Mol. Biol. **271**, 774–788

56. Nägler, D.K., Storer, A.C., Portaro, F.C., Carmona, E., Juliano, L. and Menard, R. (1997) Biochemistry 36, 12608–12615
57. Nägler, D., Tam, W., Storer, A.C., Krupa, J.C., Mort, J.S. and Menard, R. (1999) Biochemistry 38, 4868–4874
58. Klemenèiè, I., Carmona, A.K., Cezari, M.H., Juliano, M.A., Juliano, L., Gunèar, G., Turk, D., Križaj, I., Turk, V. and Turk, B. (2000) Eur. J. Biochem. 267, 5404–5412
59. Gunèar, G., Klemenièiè, I., Turk, B., Turk, V., Karaoglanovic-Carmona, A., Juliano, L. and Turk D. (2000) Structure 29, 305–313
60. Gunèar, G., Podobnik, M., Pungerèar, J., Štrukelj, B., Turk, V. and Turk, D. (1998) Structure 6, 51–61
61. Joshua-Tor, L., Xu, H.E., Johnston, S.A. and Rees, D.C. (1995) Science 269, 945–950
62. Lowe, J., Stock, D., Jap, B., Zwickl, P., Baumeister, W. and Huber, R. (1995) Science 268, 533–539
63. Groll, M., Ditzel, L., Lowe, J., Stock, D., Bochtler, M., Bartunik, H.D. and Huber, R. (1997) Nature (London) 386, 463–471
64. Pereira, P.J., Bergner, A., Macedo-Ribeiro, S., Huber, R., Matschiner, G., Fritz, H., Sommerhoff, C.P. and Bode, W. (1998) Nature (London) 392, 306–311
65. Brandstetter, H., Kim, J.S., Groll, M. and Huber, R. (2001) Nature (London) 414, 466–470
66. Turk, D., Janjiè, V., Stern, I., Podobnik, M., Lamba, D., Dahl, S.W., Lauritzen, C., Pedersen, J., Turk, V. and Turk, B. (2001) EMBO J. 20, 6570–6582
67. Olsen, J.G., Kadziola, A., Lauritzen, C., Pedersen, J., Larsen, S. and Dahl, S.W. (2001) FEBS Lett. 506, 201–206
68. Stubbs, M.T., Laber, B., Bode, W., Huber, R., Jerala, R., Lenarèiè, B. and Turk, V. (1990) EMBO J. 9, 1939–1947
69. Gunèar, G., Pungerèiè, G., Klemenèiè, I., Turk, V. and Turk, D. (1999) EMBO J. 18, 793–803
70. Zhao, B., Janson, C.A., Amegadzie, B., D'Alessio, K., Griffin, C., Hanning, C.R., Jones, C., Kurdyla, J., McQueney, M., Qiu, X. et al. (1997) Nat. Struct. Biol. 4, 109–111
71. Turkenburg, J.P., Lamers, M.B.A.C., Brzozowski, A.M., Wright, L.M., Hubbard, R.E., Sturt, S.L. and Williams, D.H. (2002) Acta Crystallogr. D Biol. Crystallogr. 58, 451–455
72. Šali, A. and Blundell, T.L. (1993) J. Mol. Biol. 234, 779–815
73. Higgins, D.G., Thompson, J.D. and Gibson, T.J. (1996) Methods Enzymol. 266, 383–402
74. Carson, M. (1997) Methods Enzymol. 277, 493–505
75. Turk, D. (1992) Ph.D. Thesis, Technische Universität, München
76. Merritt, E.A. and Bacon, D.J. (1997) Methods Enzymol. 277, 505–524
77. Nicholls, A., Sharp, K.A. and Honig, B. (1991) Proteins 11, 281–376

Biochem. Soc. Symp. **70**, 31–38
(Printed in Great Britain)
© 2003 Biochemical Society

3

Roles for asparagine endopeptidase in class II MHC-restricted antigen processing

Colin Watts[1], Daniela Mazzeo, Michelle A. West,

Stephen P. Matthews, Doreen Keane, Garth Hamilton,

Linda V. Persson, Jennifer M. Lawson, Bénédicte Manoury

and Catherine X. Moss

Division of Cell Biology & Immunology, School of Life Sciences, University of
Dundee, Dundee DD1 5EH, U.K.

Abstract

The adaptive immune response depends on the creation of suitable peptides from foreign antigens for display on MHC molecules to T lymphocytes.
Similarly, MHC-restricted display of peptides derived from self proteins results
in the elimination of many potentially autoreactive T cells. Different proteolytic
systems are used to generate the peptides that are displayed as T cell epitopes on
class I compared with class II MHC molecules. In the case of class II MHC molecules, the proteases that reside within the endosome/lysosome system of
antigen-presenting cells are responsible; surprisingly, however, there are relatively few data on which enzymes are involved. Recently we have asked whether
proteolysis is required simply in a generic sense, or whether the action of particular enzymes is needed to generate specific class II MHC-associated T cell
epitopes. Using the recently identified mammalian asparagine endopeptidase as
an example, we review recent evidence that individual enzymes can make clear
and non-redundant contributions to MHC-restricted peptide display.

Introduction

Proteases perform two key roles in the class II MHC antigen-processing
pathway: they initiate removal of the invariant chain chaperone for class II
MHC molecules, and they generate peptides from foreign and self proteins for

---

[1]To whom correspondence should be addressed (e-mail c.watts@dundee.ac.uk).

eventual capture and display to T cells [1,2]. Unlike the class I MHC loading pathway, where peptide generation is compartmentally segregated from the loading of nascent class I MHC molecules, the generation of peptides and the loading of newly synthesized class II MHC molecules occur within the same compartmental system. Moreover, most of the proteolytic enzymes that reside within endosomes and lysosomes are rather non-specific [3]. Thus a potential conflict exists between the need for proteolysis to create suitable peptides and the risk that these peptides will be destroyed before capture by class II MHC molecules can be achieved. A related question concerns the identity of the enzymes that mediate endosomal/lysosomal antigen processing. Earlier studies that undertook the large-scale sequencing of class II MHC-associated peptides found that many peptides shared a common core sequence of about 12 residues, but were extended in both the C- and N-terminal directions to different extents [4–6]. In other words they comprised 'nested sets' of related peptides. Different processes can be envisaged to explain this. One would be that proteolysis is required simply in a generalized sense, and that the 'ragged' termini of the MHC-associated peptides reflect the redundant action of many different enzymes. An alternative idea would be that much longer peptides are initially captured by class II MHC molecules, and that these are then trimmed to different extents. In this scenario, the primary cleavages that would create a substrate for MHC capture may or may not be redundant. In any case, they are obliterated by subsequent trimming. Here we review recent data that identify the apparently non-redundant contributions that asparaginyl endopeptidase makes in both the creation and the destruction of T cell epitopes in the class II MHC pathway.

Unchaperoned peptides are rapidly destroyed in the endosome/lysosome compartment

It seems likely that there is limited time available for nascent class II MHC molecules to capture peptides before the latter are destroyed in the endosomes and lysosomes of antigen-presenting cells. Most lysosomal proteases are rather non-specific and work optimally at the acidic pH found in these compartments. Indeed, our early studies demonstrated that, following endocytosis of radiolabelled antigen bound to the B cell antigen receptor (BCR), the only labelled fragments recovered were those that were captured by class II MHC molecules or those protected by continued association with the BCR, a phenomenon we termed 'footprinting' [7,8]. The 'footprinted' antigen fragments, which varied in size between 2 and 17 kDa, arose due to the persistence of a stable association between the BCR and the antigen, i.e. those regions of antigen distant from the binding site to the BCR were more susceptible to digestion than those that were more proximal. Consequently, when different clonal B cells recognizing different epitopes in tetanus toxin were used, a distinct pattern of 'footprinted' fragments was observed in each clone. Later studies have demonstrated that this modulation of processing by antibody (BCR or soluble immunoglobulin) can have striking consequences, both positive and negative, for the spectrum of T cell epitopes displayed on MHC class II molecules [9,10]. Thus antibodies may stabilize processed antigen, increasing the time available for MHC capture, or may interfere by sterically blocking the loading of T cell epitopes.

Processing of the tetanus toxin antigen by asparagine endopeptidase (AEP)

As noted above, we have used the tetanus toxin antigen as a model system with which to analyse processing in the class II MHC pathway. To learn more about the proteases required for the processing and presentation of different antigens, we adopted a simple experimental strategy. Lysosomes and endosomes were purified from human or mouse antigen-presenting cells and, following disruption, were used to digest different antigen substrates. Protease inhibitors were then tested in an attempt to identify the dominant processing enzymes. In several cases a discrete pattern of products was observed in spite of the presumed complexity of the enzyme mixture, and the broad class of protease responsible could be identified. In the case of the C-terminal domain of the tetanus toxin antigen [tetanus toxin C fragment (TTCF)], none of the common inhibitors tested blocked digestion. It turned out that one processing enzyme dominated the digestion pattern [11,12]. This enzyme was a novel mammalian cysteine protease with unusually strict specificity for cleavage after asparagine residues [11]. We refer to this enzyme simply as AEP (for asparagine endopeptidase). AEP is unrelated to the lysosomal cathepsins, and is insensitive to inhibitors of those enzymes such as leupeptin and E64 [11]. It is analogous to plant enzymes called legumains and to asparaginyl endopeptidases described in parasites. There is some identity (~30%) with the transamidase GPI8 that is involved in glycosylphosphatidylinositol anchor biosynthesis, but the significance of this is not clear. AEP is inhibited by type II cystatins such as cystatin C but, interestingly, via a distinct domain distant from that which interacts with the cathepsins ([13]; see also Chapter 15 by Abrahamson et al. in the current volume). This is also true for parasite-encoded homologues of mammalian type II cystatins. In collaboration with the group of R.M. Maizels [14], we showed that a cystatin homologue called Bm-CPI-2 from the nematode parasite *Brugia malayi* was able to block both cathepsins and AEP and, when added exogenously to antigen-presenting cells, suppressed the presentation of some T cell epitopes. However, whether Bm-CPI-2 perturbs immune responses against this pathogen *in vivo* remains to be shown.

Chen and colleagues [15] have provided evidence that mammalian AEP is related to other cysteine proteases with defined specificity for the P1 side chain. These include the caspases, separase and gingipain, which are grouped together in clan CD. AEP/legumain is widely expressed in mammalian tissues, and its physiological substrates are for the most part not known. In addition to the work reviewed here on its role in antigen processing and presentation, AEP has also been implicated in activation of gelatinase A [16] and in osteoclast differentiation [17]. While specific for asparagine in the P1 position, different forms of the enzyme from plants, parasites and mammals appear to have less rigorous, but nonetheless distinct, preferences for residues in flanking positions (e.g. [18]). However, this information comes mostly from cleavage analysis of positional scanning peptide libraries. When tested using native substrates, other factors may obscure these subsite preferences. Dando et al. [19] showed that only approx. 10% of Asn bonds in native proteins were cleaved by the enzyme purified from pig kidney. We have found similar highly selective targeting of Asn residues, with clear examples of proteins that contain surface-disposed Asn residues but which are not substrates. A selection of AEP cleavage sites in native proteins is shown in Table 1, collated from our own data and that of Dando et al. [19]. The preferences seen using peptide libraries, for example for proline in position P3 [18], are not so evident here, with only three out of 37 cleavage sites having this amino acid at P3. However, rates of cleavage are not taken into account. From the limited data available, no very clear preferences are observed. Residues in the P1' and P2' positions are frequently charged or hydrophobic, but there are several exceptions. Proline has not been observed in P1', as also reported by Dando et al. [19]. Interestingly, we have identified some instances where cleavage after aspartic acid occurs, for example in ovalbumin and in lactate dehydrogenase. Another very clear instance of this is in AEP itself, which is autoactivated by cleavages after both Asn and Asp residues ([20]; D.N. Li, S.P. Matthews, A.N. Antoniou, D. Mazzeo and C. Watts, unpublished work).

Table 1 AEP cleavage sites in native protein substrates. Processing sites for AEP/legumain in native proteins are shown. Diverse protein substrates, including several putative human autoantigens, were digested either with AEP/legumain purified from pig kidney or with recombinant autoactivated human enzyme. Cleavage sites were identified by N-terminal sequence analysis of products, radiosequencing of ^{35}S-labelled material (invariant chain) or MS [insulinoma antigen-2 (IA-2)]. Residues shown denote (from left to right) P5–P1 (where P1 is Asn or Asp; shown in bold) and P1'–P5'. Data from Dando et al. [19] are also included, marked by *.

Substrate	Cleavage after Asn	Cleavage after Asp
TTCF	HMLDN↓EEDID	
	DIEYN↓DMFNN	
	QYGTN↓EYSII	
	YTPNN↓EIDSF	
	GNAFN↓NLDRI	
	RVGYN↓APGIP	
		SAVPD↓AAGPT
Fetuin	VDYIN↓KHLPR	
Transferrin*	LAPNN↓LKPVV	
	CEPNN↓KEGYY	
	AKNLN↓EKDYE	
	ARAPN↓HAVVT	
Lactate dehydrogenase	WSGVN↓VAGVS	
	GSGCN↓LDSAR	
		ELGTD↓ADKEH
		GTDAD↓KEHWK
Ovalbumin	GTSVN↓VHSSL	
	ITKPN↓DVYSF	
		EAGVD↓AASVS

Table 1 (contd.)

Substrate	Cleavage after Asn	Cleavage after Asp
IA-2	KSLFN↓RAEGP	
	PAQAN↓MDIST	
	DHLRN↓RDRLA	
	QAEPN↓TCATA	
	QGEGN↓IKKNR	
	SDYIN↓ASPII	
Casein*	EIVPN↓SVEQK	
Concanavalin A, A-chain*	TIDFN↓AAYNA	
	NAAYN↓ADTIV	
Rat α_1-macroglobulin*	YRSSN↓IRTSS	
	ELCGN↓KVAEV	
Invariant chain	MSMDN↓MLLGP	
MBP	HFFKN↓IVTPR	
Invariant chain (human)	LISNN↓EQLPM†	
	THLKN↓TMETI	
	LQLEN↓LRMKL	
Gelatinase A	PDVAN↓YNFFP	
	VANYN↓FFPRK	
	TYTKN↓FRLSQ	
AEP	LMNTN↓DLEES	
		AVPID↓DPED
		DDPED↓GGKH
GAD 65	VERAN↓SVTWN	
Human serum albumin	PKEFN↓AETFT	

†This cleavage site in the invariant chain is in the cytoplasmic tail, and so would not be recognized in living cells.

Constructive versus destructive processing

Two of the substrates listed in Table 1 have been investigated as antigens in the class II MHC system. These are TTCF and myelin basic protein (MBP). Several lines of evidence support the notion that AEP not only dominates TTCF processing *in vitro*, but also mediates the rate-limiting step in the processing of this antigen *in vivo* [12]. To measure antigen presentation, we use human or murine T cell clones and hybridomas that are triggered by the appearance of specific peptides on the surface of appropriate antigen-presenting cells. Following endocytosis of native antigen, there is usually a delay of 1–2 h before a sufficient number of peptide–MHC complexes appear. The efficiency of processing can therefore be measured kinetically or, alternatively, in dose–response titrations where the total T cell response after 48 h is measured. In the case of the TTCF antigen, mutagenesis of the AEP cleavage sites at residues 1219 and 1184 profoundly slowed the presentation of a number of T cell epitopes and shifted the dose–response titration curve such that 10–100 times more antigen was required to elicit the same response [21]. Treatment of cells with inhibitors of AEP, including newly developed acyloxymethylketone inhibitors, also slows the presentation of T cell epitopes in TTCF [21a]. Thus, at least for this antigen, processing at discrete sites by a single enzyme is required for optimal presentation. Other enzymes may well be involved, and can apparently take over when AEP processing sites or the enzyme itself are missing, although presentation is less efficient. In physiological situations, for example in the germinal centre where B cells need to recruit T cell 'help' to avoid apoptotic death, speed of presentation could be crucial, demanding optimal antigen processing [22].

AEP processing is therefore a creative force for T cell epitopes in tetanus toxin. In the case of MBP, however, the situation is reversed. MBP is believed to be an autoantigen in the disease multiple sclerosis. An adaptive T cell response is initiated to this self protein, which is a component of the myelin sheath in the central nervous system. In both humans and mice, this response is frequently directed to a peptide in MBP, the core of which can be mapped to residues 85–99. We can reasonably ask why this T cell response exists; in other words, why is tolerance not established to this region of MBP as it is to other self pro-

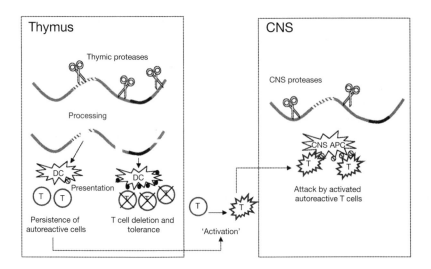

Figure 1 Hypothetical mechanism for the escape of autoreactive T cells from tolerance induction. In the thymus, presentation of 'self' peptides on MHC molecules on thymic dendritic cells (DC) and other antigen-presenting cells (APC) results in the deletion of autoreactive T cells from the repertoire. However, only peptides displayed above a certain threshold level can trigger efficient T cell deletion (e.g. denoted by the black peptide). Those destroyed by processing (denoted by hatching) fail to trigger the deletion of T cells that recognize them, and so these cells escape into the peripheral T cell repertoire. Subsequently, these may become activated, for example by recognition of viral or bacterial peptides that 'mimic' the self peptide. The now activated T cell can cross the blood–brain barrier and potentially recognize the self peptide presented on antigen-presenting cells in the central nervous system (CNS). A key assumption here is that, in the CNS, destructive processing of this peptide is decreased, resulting in an elevated threshold of presentation.

teins and, indeed, to other regions of MBP? [23,24]. The answer may be that this peptide is destroyed by AEP in the class II MHC-positive cells in the thymus that are believed to mediate so-called 'negative selection' of autoreactive T cells (Figure 1). Together with the groups of D. Wraith and L. Fugger, we have recently shown that AEP targets a single Asn residue in the 18.5 kDa form of MBP at Asn[94], thus destroying the T cell epitope comprising residues 85–99 [25]. Therefore antigen presentation in this case is inversely proportional to the level of AEP activity. The activation of MBP-(85–99)-specific T cells is believed to be associated with the pathology of multipe sclerosis; in other words, these T cells apparently recognize this peptide in association with class II MHC in the central nervous system [26]. This clearly implies that the epitope is not always destroyed by AEP, and raises interesting questions concerning the level of expression of the enzyme in different tissues (see Figure 1).

Prospects

As outlined above, differential processing of self proteins in different types of professional antigen-presenting cells at different anatomical locations could contribute to the progression of autoimmune reactions. Manipulation of the levels of individual enzymes, either genetically or chemically, in murine models of disease might allow this to be tested. On the question of redundancy, studies on AEP suggest that the peptides appearing on class II MHC molecules can be modulated by the inhibition or overexpression of a single processing enzyme. Other recent reports on the effects of chemical or genetic ablation of other proteolytic enzymes in the class II MHC pathway, such as cathepsins B, S and L [27–29], also support the idea that redundancy in this system might be less than anticipated. This raises the possibility that it may be possible to manipulate the outcome of both antigen and autoantigen processing using inhibitors of these enzymes, for example in allergy or autoimmunity. However, we require more information before the prospects for modulating the immune system via processing enzymes can be properly assessed.

Work in the authors' laboratory is supported by The Wellcome Trust, the European Union and Medivir UK Ltd. We thank our collaborators T. Tree, M. Peakman, R. Maizels, D. Wraith and L. Fugger for reagents and stimulating discussions.

References
1. Watts, C. (2001) Curr. Opin. Immunol. 13, 26–31
2. Lennon-Dumenil, A.M., Bakker, A.H., Wolf-Bryant, P., Ploegh, H.L. and Lagaudriere-Gesbert, C. (2002) Curr. Opin. Immunol. 14, 15–21
3. Turk, B., Turk, D. and Turk, V. (2000) Biochim. Biophys. Acta 1477, 98–111
4. Chicz, R.M., Urban, R.G., Lane, W.S., Gorga, J.C., Stern, L.J., Vignali, D.A. and Strominger, J.L. (1992) Nature (London) 358, 764–768
5. Hunt, D.F., Michel, H., Dickinson, T.A., Shabanowitz, J., Cox, A.L., Sakaguchi, K., Appella, E., Grey, H.M. and Sette, A. (1992) Science 256, 1817–1820
6. Rudensky, A.Y., Preston-Hurlbert, P., Hong, S.-C., Barlow, A. and Janeway, Jr, C.A. (1991) Nature (London) 353, 622–627

7. Davidson, H.W. and Watts, C. (1989) J. Cell Biol. **109**, 85–92
8. Davidson, H.W., Reid, P.A., Lanzavecchia, A. and Watts, C. (1991) Cell **67**, 105–116
9. Watts, C., Antoniou, A., Manoury, B., Hewitt, E.W., McKay, L.M., Grayson, L., Fairweather, N.F., Emsley, P., Isaacs, N. and Simitsek, P.D. (1998) Immunol. Rev. **164**, 11–16
10. Jaume, J.C., Parry, S.L., Madec, A.M., Sonderstrup, G. and Baekkeskov, S. (2002) J. Immunol. **169**, 665–672
11. Chen, J.M., Dando, P.M., Rawlings, N.D., Brown, M.A., Young, N.E., Stevens, R.A., Hewitt, E., Watts, C. and Barrett, A.J. (1997) J. Biol. Chem. **272**, 8090–8098
12. Manoury, B., Hewitt, E.W., Morrice, N., Dando, P.M., Barrett, A.J. and Watts, C. (1998) Nature (London) **396**, 695–699
13. Alvarez-Fernandez, M., Barrett, A.J., Gerhartz, B., Dando, P.M., Ni, J. and Abrahamson, M. (1999) J. Biol. Chem. **272**, 19195–19203
14. Manoury, B., Gregory, W.F., Maizels, R.M. and Watts, C. (2001) Curr. Biol. **11**, 447–451
15. Chen, J.-M., Rawlings, N.D., Stevens, R.A.E. and Barrett, A.J. (1998) FEBS Lett. **441**, 361–365
16. Chen, J.M., Fortunato, M., Stevens, R.A. and Barrett, A.J. (2001) Biol. Chem. **382**, 777–783
17. Choi, S.J., Reddy, S.V., Devlin, R.D., Menaa, C., Chung, H., Boyce, B.F. and Roodman, G.D. (1999) J. Biol. Chem. **274**, 27747–27753
18. Mathieu, M.A., Bogyo, M., Caffrey, C.R., Choe, Y., Lee, J., Chapman, H., Sajid, M., Craik, C.S. and McKerrow, J.H. (2002) Mol. Biochem. Parasitol. **121**, 99–105
19. Dando, P.M., Fortunato, M., Smith, L., Knight, C.G., McKendrick, J.E. and Barrett, A.J. (1999) Biochem. J. **339**, 743–749
20. Chen, J.M., Fortunato, M. and Barrett, A.J. (2000) Biochem. J. **352**, 327–334
21. Antoniou, A.N., Blackwood, S.L., Mazzeo, D. and Watts, C. (2000) Immunity **12**, 391–398
21a. Loak, K., Li, D.N., Manoury, B., Billson, J., Murton, F., Hewitt, E. and Watts, C. (2003) Biol. Chem., in the press
22. Aluvihare, V.R., Khamlichi, A.A., Williams, G.T., Adorini, L. and Neuberger, M.S. (1997) EMBO J. **16**, 3553–3562
23. Harrington, C.J., Paez, A., Hunkapiller, T., Mannikko, V., Brabb, T., Ahearn, M., Beeson, C. and Goverman, J. (1998) Immunity **8**, 571–580
24. Targoni, O.S. and Lehmann, P.V. (1998) J. Exp. Med. **187**, 2055–2063
25. Manoury, B., Mazzeo, D., Fugger, L., Viner, N., Ponsford, M., Streeter, H., Mazza, G., Wraith, D.C. and Watts, C. (2002) Nat. Immunol. **3**, 169–174
26. Krogsgaard, M., Wucherpfennig, K.W., Canella, B., Hansen, B.E., Svejgaard, A., Pyrdol, J., Ditzel, H., Raine, C., Engberg, J. and Fugger, L. (2000) J. Exp. Med. **191**, 1395–1412
27. Driessen, C., Lennon-Dumenil, A.M. and Ploegh, H.L. (2001) Eur. J. Immunol. **31**, 1592–1601
28. Pluger, E.B., Boes, M., Alfonso, C., Schroter, C.J., Kalbacher, H., Ploegh, H.L. and Driessen, C. (2002) Eur. J. Immunol. **32**, 467–476
29. Hsieh, C.S., deRoos, P., Honey, K., Beers, C. and Rudensky, A.Y. (2002) J. Immunol. **168**, 2618–2625

© 2003 Biochemical Society
Biochem. Soc. Symp. **70**, 39–52
(Printed in Great Britain)

4

Substrate specificity and inducibility of TACE (tumour necrosis factor α-converting enzyme) revisited: the Ala-Val preference, and induced intrinsic activity

Roy A. Black*[1], John R. Doedens*, Rajeev Mahimkar*,

Richard Johnson*, Lin Guo*, Alison Wallace*, Duke Virca*,

June Eisenman*, Jennifer Slack*, Beverly Castner*,

Susan W. Sunnarborg†, David C. Lee†, Rebecca Cowling‡,

Guixian Jin‡, Keith Charrier*, Jacques J. Peschon*

and Ray Paxton*

*Amgen Inc., 51 University Street, Seattle, WA 98101, U.S.A., †Department of Biochemistry and Biophysics, University of North Carolina School of Medicine, Chapel Hill, NC 27599, U.S.A., and ‡Department of Biological Chemistry, Wyeth-Ayerst Research, Pearl River, NY 10965, U.S.A.

Abstract

Tumour necrosis factor α (TNFα)-converting enzyme (TACE/ADAM-17, where ADAM stands for a disintegrin and metalloproteinase) releases from the cell surface the extracellular domains of TNF and several other proteins. Previous studies have found that, while purified TACE preferentially cleaves peptides representing the processing sites in TNF and transforming growth factor α, the cellular enzyme nonetheless also sheds proteins with divergent cleavage sites very efficiently. More recent work, identifying the cleavage site in the p75 TNF receptor, quantifying the susceptibility of additional peptides to cleavage by TACE and identifying additional protein substrates, underlines the complexity of TACE–substrate interactions. In addition to substrate speci-

[1]To whom correspondence should be addressed (e-mail blackra@amgen.com).

ficity, the mechanism underlying the increased rate of shedding caused by agents that activate cells remains poorly understood. Recent work in this area, utilizing a peptide substrate as a probe for cellular TACE activity, indicates that the intrinsic activity of the enzyme is somehow increased.

Introduction

Tumour necrosis factor α (TNFα)-converting enzyme (TACE/ADAM-17, where ADAM stands for a disintegrin and metalloproteinase) plays a role in both inflammation and mammalian development. It therefore remains a subject of intensive investigation, particularly with respect to its substrate specificity and the regulation of its activity. These topics will be the focus of this review.

TACE was identified by its ability to cleave TNF at the physiological processing site [1,2]. TNF is a central mediator of inflammation: it is induced by bacterial wall components and by various immune system regulators, and it causes the production of proteases, reactive oxygen species, vasoactive compounds, adhesion proteins and other cytokines. It is critical for host defensive responses to certain infections, but it can also cause severe damage in autoimmune diseases such as rheumatoid arthritis [3]. Full-length TNF is a type II transmembrane protein, with cytokine activity residing in the C-terminal 157 residues [4]. This portion of the protein is released from cells by a proteolytic cleavage between Ala^{76} and Val^{77}, approx. 20 residues from the predicted transmembrane domain. Both the cell-associated and the released forms of TNF are biologically active, but full inflammatory responses require the soluble form in at least some situations [5].

Studies with cells bearing mutated TACE, lacking its Zn-binding domain (Δ Zn TACE), demonstrated a requirement for this domain for the bulk of TNF release by anti-CD3-stimulated T cells and by lipopolysaccharide (LPS)-stimulated monocytic cells [1]. Surprisingly, however, the phenotype of the knockout mice resembled in several respects the phenotype of transforming growth factor α (TGFα)-deficient mice [6]. TGFα is a ligand of the epidermal growth factor receptor (EGFR) and is required for normal mammalian development. It is made as a type I transmembrane protein, and a soluble form is generated by a proteolytic cleavage nine residues from the predicted transmembrane domain [7]. Approx. 20 N-terminal residues are also sometimes removed by proteolytic processing. The phenotype of Δ Zn TACE mice suggested that TACE is the enzyme that generates soluble TGFα, and that this form is essential for the developmental function of this factor, even though the cell-associated form is active. Subsequent studies confirmed that TACE knockout cells are deficient in the release of soluble TGFα [6,7].

The finding that TACE releases TGFα as well as TNF was surprising, because early studies showed that it did not cleave a TGFα C-terminal processing-site peptide under conditions in which a TNF-based peptide substrate was readily cleaved. However, as discussed in detail below, increasing the amount of enzyme used did result in cleavage of the TGFα peptide, and the P1-P1' sequence is identical with the corresponding sequence in TNF (Ala-Val). What came as an even greater surprise than its role in TGFα processing was the finding that TACE is required for the shedding of a wide variety of membrane protein ecto-domains, from receptors to adhesion proteins, many with cleavage sites very different from Ala-Val. These findings will be discussed in detail in the first section of the review.

The function of ecto-domain shedding is not known in most cases, although various consequences have been proposed. Possible functions include (i) the generation of growth factors that act at a distance from the releasing cell, or that act in an autocrine fashion; (ii) the conversion of transmembrane into soluble receptors, thereby simultaneously desensitizing cells to growth factors and generating binding proteins that compete with the cell-associated receptors; and (iii) the release of adhesion proteins, thereby modifying the ability of cells to migrate. The importance of shedding (although not necessarily TACE-mediated shedding) has been demonstrated in a number of situations in addition to inflammation (TNF) and epithelial development (TGFα), including axonal pathfinding (ephrin-A2) [8], lymphocyte migration (L-selectin) [9], enhanced bacterial virulence (syndecan-1) [10] and evasion of immune system surveillance by tumour cells (class I MHC homologues) [11]. In some of these examples, as well as in several cases where the function is less clear, a striking aspect of the shedding is its inducibility. The inducibility of these events is itself an indication of physiological significance. The regulation of TACE activity will therefore be the focus of the second section of the review.

TACE is a 684-residue member of the ADAM family of zinc-dependent metalloproteinases, which falls within the metzincin superfamily along with the matrix-degrading metalloproteinases [12]. ADAMs are multi-domain, transmembrane proteins (Figure 1). These proteins include, in addition to the protease module found in the ecto-domain, a so-called disintegrin domain, which can bind to integrins, and a cytoplasmic domain. The pro-domain must be removed, most probably by furin or a related enzyme, to generate the active

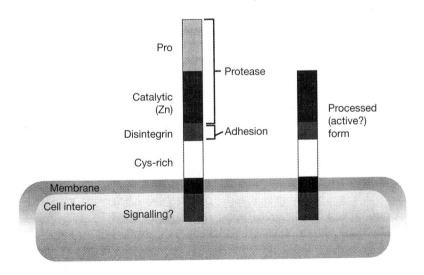

Figure 1 Domain structure of TACE. Removal of the pro-domain is believed to be required for activity, but it may not be sufficient.

form of TACE. The crystal structure of the protease module has been determined [13], revealing a rather deep S1' pocket with an unusual tunnel-like connection to the S3' space. Interestingly, while the structure of the core of the enzyme closely resembles the structure of a member of this family found in snake venom, the TACE catalytic domain has a number of prominently protruding loops that are not found in the snake venom enzyme.

Substrate specificity

Peptide substrates based on TNF and EGFR ligands

Studies with peptide substrates have consistently demonstrated a strong preference of TACE for cleaving at Ala–Val bonds, but also an ability to cleave much less efficiently at a wide variety of sequences. The first indication of specificity was the finding that a 20-residue peptide spanning the cleavage site in TNF (Leu[67]–Asp[86]) was cleaved solely between Ala[76] and Val[77] [14]. This study used partially purified TACE, so no enzyme/substrate ratio was reported. Another early study, in which four matrix metalloproteinases (MMPs) were tested in addition to TACE, demonstrated that this specificity was not due to inaccessibility of other peptide bonds in this sequence [15]. With a 12-residue peptide spanning the TNF processing site (acetyl-SPLAQA↓VRSSSR-NH₂, where ↓ denotes cleavage), TACE again cleaved solely at the Ala–Val bond, but MMPs-1, -2, -3, -7 and -9 all cleaved at multiple sites.

Experiments investigating the effects of discrete variations in the TNF cleavage site sequence further indicated that TACE is highly discriminating. Replacement of the P1 Ala with Ile in the 12-residue peptide described above completely eliminated cleavage by TACE, while cleavage by MMP-1 and MMP-3 was virtually unaffected [15]. A separate study, employing 10-residue peptides with a fluorophore at the N-terminus and a quenching group on the C-terminus, confirmed that TACE is unable to cleave a TNF-based peptide with Ile substituted for Ala at position P1, and the rate of cleavage was drastically reduced with Val (93% reduction) or Gly at P1 (95% reduction) [16]. Cleavage with Phe or Leu at P1 was reduced by approx. 85% (this figure is based on an adjustment of the reported value, to eliminate consideration of secondary cleavages that occurred with these peptides; since no secondary cleavages are found with peptides lacking modified ends [14], the addition of the fluorophore and quencher may affect the way these peptides bind to the enzyme).

TACE is somewhat less discriminating with respect to the P1' residue, based on the study with the fluorophore/quencher substrates. Peptides with Leu or Ile at P1' were cleaved at a rate only about 30% lower than the rate of cleavage with Val at P1'. However, cleavage of a peptide with P1' Ala was reduced by over 85% (again, adjusting the rate downward to eliminate consideration of secondary cleavages). A peptide that showed inversion of the primary scissile bond, i.e. with Val–Ala instead of Ala–Val, was cleaved only 7% as rapidly as the peptide with the native TNF sequence. This decrease could have been due to the P1 Val rather than (or in addition to) the replacement of Val with Ala at P1', but overall it appears that TACE strongly prefers Val or Leu at P1', and a short P1 side chain.

This conclusion is buttressed by studies of TACE's ability to cleave peptides containing the cleavage sites found in other proteins known to be TACE substrates. A peptide spanning the cleavage site in mouse pro-TNF, MAQTLT↓LRSSSR, was cleaved by TACE, although with approx. 5-fold less efficiency than the corresponding human sequence (again, SPLAQA↓VRSSSR) (R.A. Black, unpublished work). The decrease in susceptibility to cleavage could be due either to the different P2–P6 residues or to the presence of Thr-Leu instead Ala-Val, but this result at least confirms that Thr and Leu are acceptable substitutes for Ala and Val. (Whereas this experiment was done with human TACE, we found no better cleavage of the mouse TNF-based peptide by mouse TACE.) Moreover, the only other protein known definitively to be a physiological substrate of TACE, TGFα, is shed by cleavage at an Ala–Val bond, in the sequence ADLLA↓VVAAS. In addition, a peptide with this sequence was cleaved by TACE, albeit only 10% as efficiently as the corresponding TNF-based peptide [6]. The generation of mature TGFα also requires cleavage at an N-terminal site between Ala and Val in the sequence shown in Table 1, and TACE cleaved a peptide representing this site about as well as it cleaved the juxtamembrane cleavage-site peptide.

TGFα is one of a family of proteins that bind to the EGFR, all of which have both soluble and membrane-bound forms. It is likely, although not definitely demonstrated, that TACE also releases one or more of these family members in addition to TGFα, because the phenotype of the TACE knockout mouse is more severe than the phenotype of the TGFα knockout, resembling in several respects the phenotype of EGFR knockout mice [6]. The soluble forms of betacellulin and heparin-binding EGF (HB-EGF) are released from their precursors by cleavage, respectively, of Tyr–Leu and Pro–Val bonds (see Table 1). TACE was found to cleave peptides representing these processing-site sequences at the authentic scissile bond, as well as between Arg and Leu in the HB-EGF peptide. The Val and Leu residues at P1' are consistent with a preference of TACE for aliphatic residues in this position. It was found that 20-fold more TACE was required to observe cleavage of these peptides than was required for cleavage of the TGFα juxtamembrane processing-site peptide. This difference could be due to the lack of Ala or Thr in the P1 position or, of course, to one or more of the many other differences between these peptides. Interestingly, peptides representing the N-terminal processing site of two additional EGFR ligands, amphiregulin and epiregulin, are also cleaved by TACE (and by intact cells) prior to Val, fitting the apparent pattern (Table 1).

The substrate requirements of TACE are clearly more complex than the presence of a Val or Leu residue in the P1' position, however. As already noted, the P1 residue can have considerable influence. In addition, three of the peptides in Table 1 with one or more Val and Leu residues were not cleaved even by a relatively large amount of TACE [betacellulin (BTC) N-terminal, EGF N-terminal and EGF C-terminal peptides]. Conversely, while all the TACE-mediated cleavage of the peptides in Table 1 occurred prior to Val or Leu (with the caveat that the cleavage in the epiregulin C-terminal peptide was not deter-

Table 1 Cleavage by TACE of EGF family processing-site peptides. Peptides representing the cleavage sites of human EGF family members (↓ denotes predicted cleavage site) were incubated with recombinant TACE [1] at 37°C for 4 h and products were analysed by liquid chromatography/MS as described previously [6]. Results of cleavage of TGFα peptides [6] are included for comparison. Observation of cleavage and the relevant site are indicated. The approximate concentration of TACE required for reaction is indicated. Note that cleavage of TGFα peptides required approx. 10 times more TACE than cleavage of a TNF peptide [6]. AR, amphiregulin; EPR, epiregulin; N.D., not determined. Reproduced from [7].

Ligand	N-terminal peptide	Cleaved?	[TACE] (µM)	C-terminal peptide	Cleaved?	[TACE] (µM)
TGFα	PVAAA↓VVSHF	Yes; A/V	1.4	ADLLA↓VVAAS	Yes; A/V	1.4
AR	SVRVEQ↓VVKPPQ	Yes; Q/V	28	ERCGEK↓SMKTHS	No	28
BTC	RSPETN↓LLCGDP	No	28	RVDLFY↓LRGDRG	Yes; Y/L	28
EPR	NPRVAQ↓VSITKC	Yes; Q/V	28	CEHFFL↓TVHQPL	Yes; N.D.	28
HB-EGF	N.D.	–	–	GLSLP↓VENRLYTD	Yes; P/V, R/L	28
EGF	HHYSVR↓NSDSEC	No	28	KVWELR↓HAGHGQ	No	28

mined), the same amount of TACE in the same time period is able to cleave a diverse set of peptides with no similarity at P1 or P1' to TNF or TGFα. This finding was made in the course of studies of other shed proteins, discussed in the next section.

Peptide substrates based on proteins other than TNF and EGFR ligands

In addition to TNF and the EGFR ligands, there are three other proteins which meet the criteria that (i) their shedding is believed to be of physiological significance and (ii) there is strong evidence that TACE is a major shedding enzyme (see Table 2). L-selectin shedding appears to be important in lymphocyte migration [9]. Thymocytes from Δ Zn TACE mice were found to be markedly deficient in PMA-induced L-selectin shedding [6], and an unpublished study found that neutrophils from mice reconstituted with a Δ Zn TACE haematopoietic system fail to shed L-selectin in response to LPS or fMet-Leu-Phe as well as PMA (R.A. Black, K. Charrier and J.J. Peschon, unpublished work). The reported processing site in mouse L-selectin, QETNR↓SFSKI, is not at an Ala/Thr–Leu/Val bond, but it appears that TACE is able to cleave here nonetheless. Indeed, a peptide representing the corresponding site in human L-selectin (QKLDK↓SFSMI) was cleaved by human TACE, albeit with greatly reduced efficiency compared with its preferred substrates discussed above. Cleavage was observed over the course of a 4 h incubation with 7 μM TACE. Taking into account the amount of TACE used and the rate of cleavage, the reaction was 2250-fold less efficient than the cleavage of a TNF processing-site peptide. The cleavage was, however, at the physiological site (between Lys and Ser).

A second protein, other than TNF and the EGFR ligands, whose shedding is of particular interest is the amyloid-β precursor protein (APP). Cleavage of this protein at the α-secretase site releases the bulk of the ecto-domain and, importantly, occurs within the sequence of the amyloid-β peptide that is generated by cleavages at the β and γ sites. Thus shedding of the APP ecto-domain may reduce the production of amyloid-β peptide, a major component of the plaques found in Alzheimer's disease. Primary embryonic fibroblasts from Δ Zn TACE mice were found to be completely deficient in PMA-induced shedding of APP [17], strongly implicating TACE as an APP α-secretase. Consistent with that conclusion, TACE did cleave a peptide representing the α-processing site, acetyl-VHHQKLVFFA-amide, and it did so at

Table 2 Cleavage by TACE of processing-site peptides from other proteins. Procedures were as described for Table 1.

Shed protein	Cleavage site	Cleavage (by 7–28 μM TACE)
L-selectin	QETNR↓SFSKI	Yes
APP	VHHQK↓LVFFA	Yes
p75 TNFR	MGPSPPAEG↓STGDFA	Yes
Type II IL-1R	TLRTTVKEASS↓TFSWG	No
p55 TNFR	PQIEN↓VKGTEDS	Yes
IL-6R	SLPVQ↓DSSSV	No

the physiological site, between Lys and Leu. The P1′ Leu fits the pattern noted above for TNF and EGFR ligands, but the reaction was extremely inefficient, comparable with that found with the L-selectin processing-site peptide.

The third shed protein with both considerable physiological interest and convincing TACE involvement is the p75 TNF receptor (TNFR). Significant amounts of the soluble form of this receptor are found in serum, and it may modulate the effects of TNF by competing with cell-associated receptors. TACE knockout monocytic cells are markedly deficient in PMA-induced shedding of the p75 TNFR, indicating that TACE is probably a physiological sheddase of the receptor [6]. It has been reported that "the major C terminus of the soluble p75 receptor isolated from urine corresponds to Val-[214]" [18] (Val[192] if leader sequence is not counted), and some investigators have inferred from this observation that release of the p75 receptor results from cleavage between Val[214] and His[215] in the sequence SMAPGAVHLPQP. However, since the material sequenced came from urine, it may have resulted from secondary cleavages, and, consistent with that possibility, the site is considerably further from the predicted transmembrane domain of the intact protein than is typical for shed proteins. We therefore revisited the issue of the cleavage site that releases the ecto-domain of the p75 TNFR, using material from a short-term culture of PMA-stimulated C127 cells overexpressing the human receptor. Western analysis of the medium showed that these cells released a 46 kDa form of the receptor, consistent with observations made with various other cells [19]. This protein was purified using an immunoaffinity column, and treatment with cyanogen bromide yielded a peptide, GPSPPAEG (with the serine linked to galactose and N-acetylgalactose), which is from the juxtamembrane region. This analysis suggests that the p75 TNFR is shed by cleavage between Gly[253] and Ser[254], in the sequence MGPSPPAEG-STGDFAL. A peptide with this sequence was cleaved by purified TACE at the apparent physiological site, but with even lower efficiency than observed with the L-selectin and APP processing-site peptides.

Peptides based on the sequences of a number of other shed proteins have also been tested as TACE substrates, even though the biological significance of their shedding is less clear than in the cases discussed thus far. The type II interleukin-1 (IL-1) receptor (IL-1R) is apparently shed by TACE, based on a proteomic analysis of medium conditioned by TACE wild-type compared with Δ Zn TACE monocytic cells stimulated with PMA [20], even though it would serve as a 'decoy' in any case, since it lacks a significant cytoplasmic domain and does not signal upon IL-1 binding. Several years ago we determined the C-terminus of the soluble form by purifying it from the medium of CV-1 cells overexpressing the full-length protein and then analysing the mass of fragments generated by the protease endo-Asp-N. This work suggested that cleavage occurs between Ser and Thr in the sequence TLRTTVKEASS↓TF-SWG. A peptide corresponding to this sequence was not cleaved even by 28 μM TACE, the amount required for cleavage of the more resistant peptides described above. This result is consistent with a study of the shedding of various proteins by Δ Zn TACE cells reconstituted with either TACE or a chimaeric protein composed of the TACE catalytic domain and the ADAM-10 disintegrin/cysteine-rich domain. This study indicated that shedding of the type II IL-1R, but not of the other proteins tested, requires the TACE disintegrin/cysteine-rich domain [21].

The p55 TNFR and the IL-6R provide two final examples of proteins that have served as the basis of substrate peptides in our work. In neither case is there evidence that the endogenous protein is shed by TACE, but Δ Zn TACE fibroblasts co-transfected with these proteins plus TACE release more of the ecto-domains than do cells transfected with the substrate proteins alone [21,22]. Again, however, peptides served as very poor substrates: TACE did cleave a p55 TNFR processing-site peptide and did so at the physiologically correct site, between Asn and Val in the sequence PQIEN↓VKGTEDS, but the reaction was extremely inefficient (comparable with the reaction with the L-selectin- and APP-based peptides). And no cleavage at all was observed with the IL-6R peptide SLPVQ(↓)DSSSV.

A recent study, by another group, reported k_{cat}/K_m values for the reactions between TACE and several of the peptides discussed above [23], and the conclusions were generally consistent with our findings. A peptide representing APP was cleaved correctly by TACE, but with a k_{cat}/K_m approx. 100-fold lower than the value obtained with a TNF-based peptide as substrate. No cleavage at all was observed with IL-6R or p55 TNFR peptides. The complete lack of cleavage with the p55 TNFR peptide, in contrast with the weak cleavage we observed, could have been due to the use of less enzyme, or to the presence of a dinitrophenol group at the N-terminus of the peptide.

The conclusions from the peptide substrate studies are reasonably clear. (i) TACE has a strong preference for the processing site in TNF, and a lesser but still pronounced preference for the processing site in TGFα. The TNF-based peptide was cleaved approx. 10-fold more efficiently than the TGFα peptide, and 100–1000-fold more efficiently than all of the other peptides tested. (ii) The determinants of this specificity are not simple, perhaps including a summation of negative and positive influences of different side chains, but an affinity for Ala/Thr–Val/Leu bonds, and particularly for Val/Leu at the P1′ position, appears likely. The results with substitutions in the TNF sequence are consistent with this interpretation. Moreover, in addition to TGFα, all six of the EGFR ligand peptides that were cleaved by TACE were cleaved prior to Val or Leu, as were the peptides representing APP and the p55 TNFR. However, TACE also cleaved, with similar extremely low efficiency, at sites lacking either Val or Leu (e.g. L-selectin, p75 TNFR), and a Val or Leu was clearly not sufficient for cleavage (note result with BTC and the two EGF peptides in Table 1).

Explanations for TACE's apparent role in the shedding of multiple proteins

These studies raise the question of how the data from TACE-deficient cells indicating a role for TACE in shedding of many proteins can be reconciled with the enzyme's extremely inefficient cleavage of the corresponding processing-site peptides. One possible explanation is simply that interactions with the proteins at sites distal to the cleavage site provide the necessary affinity.

However, we found that TACE also failed to cleave full-length L-selectin, p55 TNFR and p75 TNFR, extracted from cells with detergent (R.A. Black and J. Slack, unpublished work). The detergent could have interfered with the interaction, but the membrane form of TNF extracted under the same conditions was readily cleaved, even by much less TACE than that used with the other proteins [1]. Another caveat in interpreting this experiment, however, is that the transmembrane and cytoplasmic domains, released from the compartmentalization imposed by the plasma membrane, could fold over the cleavage site. Indeed, this complication appears to occur with pro-TGFα extracted from cells. Recent work demonstrated that, while solubilized full-length pro-TGFα is not cleaved by TACE, an ecto-domain construct ending at the transmembrane domain is cleaved, at the physiological processing site [7]. Moreover, it is possible that, in cells, distal interactions are re-inforced by the alignment of TACE and substrate proteins resulting from their membrane anchoring. Another point in favour of the distal-interaction hypothesis is that such interactions have in fact been shown for at least one shed protein (which, however, is apparently not shed by TACE) [24]. In addition, as discussed above, shedding of the type II IL-1R apparently requires the TACE disintegrin/cysteine-rich domain.

Other ways to reconcile the peptide-cleavage data with the cell-based evidence for multiple protein substrates of TACE centre on factors that may alter the conformation of either the enzyme or the substrates in the context of intact cells. Thus it has been proposed that adaptor proteins may increase the affinity of TACE for these substrates [23], and in fact evidence that a protein called ARTS-1 binds to the p55 TNFR and thereby increases its shedding has recently been published [25]. A less interesting possibility is that soluble TACE or detergent-extracted substrate proteins simply lose aspects of their native conformation, even though the enzyme remains active and the receptors still bind their ligands. There is in fact something odd about both recombinant ecto-domain TACE and solubilized native TACE: they are much less active in the presence of physiological salt concentrations than in the absence of salt. As a consequence, all studies with both peptide and solubilized protein substrates have been carried out under low-salt, non-physiological conditions. Moreover, it should be noted that we do not actually know how TACE and its substrates are distributed in the plasma membrane (e.g. they could be concentrated in microdomains), or what the local pH, ionic strength and redox conditions are.

Regulation of TACE activity

The shedding of most protein ecto-domains occurs at a basal rate in resting cells, but can be dramatically up-regulated by activating the cells. Many studies of this phenomenon have been carried out with PMA, a non-physiological activator of protein kinase C and also, directly or indirectly, a number of other kinases. However, several physiological stimuli can also induce shedding, including LPS, growth factors, fMet-Leu-Phe and other G-protein-coupled receptor ligands [26]. It should be emphasized that, in many shedding studies, the sheddase involved is not definitively identified, and ADAM-10, the closest homologue of TACE, is probably the major enzyme in some cases (e.g. [8]). In addition, matrix-degrading metalloproteinases can also act as sheddases [27,28]. This review will focus on regulated shedding that is likely to involve TACE.

Some progress has been made in elucidating the signal transduction components in at least some cases of induced shedding. In a study utilizing TGFα transiently expressed in CHO cells, it was found that inhibitors of the mitogen-activated protein kinase (MAP kinase) cascade blocked fibroblast growth factor-induced shedding of TGFα, and constitutively active forms of the MAP kinase extracellular-signal-regulated kinase 2 (ERK2) or MEK1 (MAP kinase/ERK kinase) increased the shedding [29]. Serum- and PMA-induced shedding also apparently required the MAP kinase cascade in this system. Moreover, the authors found that the fibroblast growth factor-, serum- and PMA-induced shedding of transiently expressed TNF and L-selectin in CHO cells was dependent on ERK MAP kinase signalling. That study also investigated the effects of different kinase inhibitors on the basal shedding of these proteins (TGFα, TNF and L-selectin) in transiently transfected CHO cells, and in all three cases the p38 MAP kinase was found to be essential. Finally, L-selectin shedding from human neutrophils in response to fMet-Leu-Phe was blocked by inhibitors of either ERK or p38 MAP kinase.

An independent study, looking at the shedding of the EGFR ligand HB-EGF stably expressed in CHO cells, also presented convincing evidence that PMA- and serum-induced release require the ERK MAP kinase cascade [30]. Interestingly, the investigators in that study observed a considerable lag between the activation of MAP kinase and the induced shedding, suggesting that additional intermediary steps are required. Consistent with this suggestion, shedding did not occur when the cells were placed in suspension even though MAP kinase activation was unimpaired.

It should be noted that by no means all induced shedding involves the ERK MAP kinase cascade. Stress-induced shedding apparently requires the p38 MAP kinase instead [30,31], and we have found that inhibitors of the ERK MAP kinase pathway had at most a partial inhibitory effect on the shedding of the p75 TNFR by U937 cells.

In any event, the mechanism by which the MAP kinase cascade or other intracellular signalling components affect the shedding machinery remains unknown. Direct modification of TACE by a component of the MAP kinase cascade is of course a possibility, and one study showed that, in HEK-293 cells, ERK2 indeed associated with TACE and phosphorylated a threonine in the cytoplasmic domain upon PMA stimulation of the cells [31]. (The growth factors EGF and nerve growth factor also induced the phosphorylation of TACE in these cells, but an involvement of ERK2 was not demonstrated.) It is not clear, however, whether the phosphorylation of TACE is related to the induced shedding observed in these cells, particularly since only the pro-form of TACE appeared to be phosphorylated. Another caveat is that the only shed protein monitored was the TrkA nerve growth factor receptor, and the assay employed looked only at the generation of the cell-associated remnant resulting from cleavage. It seems plausible that PMA could affect the accumulation of this remnant in multiple ways, not just by increasing shedding. In addition, in studies of TACE-deficient fibroblasts co-transfected with constructs encoding TrkA and either wild-type TACE or TACE in which the threonine in question had been mutated to alanine, PMA-induced accumulation of remnants was only moderately reduced with the Thr→Ala mutant compared with wild-type TACE. It is thus clear that phosphorylation of this threonine residue is not essential for most of the PMA-induced shedding in these cells. Moreover, in a study by other investigators, the shedding of three other proteins by this cell line was found to be just as inducible by PMA when the cells were reconstituted with a truncated form of TACE lacking the entire cytoplasmic domain as when reconstituted with wild-type TACE [21]. Taken altogether, these results indicate that modification of the TACE cytoplasmic domain is not a general mechanism by which the rate of shedding is increased, although it could contribute in some cases.

Other obvious possible explanations for stimulated shedding also appear to be ruled out. We have found no increase in the amount of surface TACE in response to various cell activators [1,32], nor have we found any increase in the extent of TACE processing. It is possible that TACE is activated by displacement of the pro-domain, without its actual removal, and one paper suggested that NO in fact increases shedding by this mechanism [33]. This mechanism could not be a general means by which shedding is increased, however, because at least some proteins are shed from the cell surface, and no pro-TACE has been found there. An increase in co-localization of TACE with its substrates is yet another possible mechanism for induced shedding, but we have not observed such co-localization (in PMA-stimulated THP-1 cells, at least).

One remaining possibility is that the intrinsic activity of TACE is increased, either by removal of an inhibitor or by engagement of an activator. The activator could be a released soluble factor or a membrane-bound factor. We have investigated this possibility using a peptide-based assay for cellular TACE activity. We find that, indeed, TACE-mediated cleavage of a TNF processing-site peptide is dramatically increased by exposing cells to PMA. Moreover, this increase did not require the TACE cytoplasmic domain (J.R. Doedens, R. Mahimkar and R.A. Black, unpublished work).

Conclusion

From peptide substrate studies, it appears that TACE is highly adapted to cleave TNF, and to a lesser extent TGFα, yet somehow it apparently cleaves other substrates in the cellular milieu. Exosite interactions between TACE and its protein substrates could explain this paradox, and future studies should attempt to identify such interactions. The extensive surface loops found on TACE, but not on related enzymes, are obvious starting points for such studies. However, it should also be noted that the peptide substrate studies discussed in this review did not actually test TACE under physiological conditions, perhaps in part because TACE is peculiarly sensitive to even modest concentrations of salt. Under physiological conditions, the disparities in TACE's ability to cleave different substrates might not be so great.

How shedding is activated largely remains a black box. At least some progress has been made in identifying the signalling cascades involved, but very little is known about how these pathways actually alter the shedding machinery.

Several obvious possible mechanisms, such as the classical removal of the pro-domain, do not appear to be involved. One productive area for future research may be an increase in the intrinsic activity of TACE upon cell activation.

References

1. Black, R.A., Rauch, C.T., Kozlosky, C.J., Peschon, J.J., Slack, J.L., Wolfson, M.F., Castner, B.J., Stocking, K.L., Reddy, P., Srinivasan, S. et al. (1997) Nature (London) 385, 729–733
2. Moss, M.L., Jin, S.L., Milla, M.E., Bickett, D.M., Burkhart, W., Carter, H.L., Chen, W.J., Clay, W.C., Didsbury, J.R., Hassler, D. et al. (1997) Nature (London) 385, 733–736
3. Bazzoni, F. and Beutler, B. (1996) N. Engl. J. Med. 334, 1717–1725
4. Kriegler, M., Perez, C., DeFay, K., Albert, I. and Lu, S.D. (1988) Cell 53, 45–53
5. Ruuls, S.R., Hoek, R.M., Ngo, V.N., McNeil, T., Lucian, L.A., Janatpour, M.J., Korner, H., Scheerens, H., Hessel, E.M., Cyster, J.G. et al. (2001) Immunity 15, 533–543
6. Peschon, J.J., Slack, J.L., Reddy, P., Stocking, K.L., Sunnarborg, S.W., Lee, D.C., Russell, W.E., Castner, B.J., Johnson, R.S., Fitzner, J.N. et al. (1998) Science 282, 1281–1284
7. Sunnarborg, S.W., Hinkle, C.L., Stevenson, M., Russell, W.E., Raska, C.S., Peschon, J.J., Castner, B.J., Gerhart, M.J., Paxton, R.J., Black, R.A. and Lee, D.C. (2002) J. Biol. Chem. 277, 12838–12845
8. Hattori, M., Osterfield, M. and Flanagan, J.G. (2000) Science 289, 1360–1365
9. Faveeue, C., Preece, G. and Ager, A. (2001) Blood 98, 688–695
10. Park, P.W., Pier, G.B., Hinkes, M.T. and Bernfield, M. (2001) Nature (London) 411, 98–102
11. Groh, V., Wu, J., Yee, C. and Spies, T. (2002) Nature (London) 419, 734–738
12. Bode, W., Grams, F., Reiemer, P., Gomis-Ruth, F.X., Baumann, U., McKay, D.B. and Stocker, W. (1996) Adv. Exp. Med. Biol. 389, 1–11
13. Maskos, K., Frenandez-Catalan, C., Huber, R., Bourenkov, G.P., Bartunik, H., Ellestad, G.A., Reddy, P., Wolfson, M.F., Rauch, C.T., Castner, B.J. et al. (1998) Proc. Natl. Acad. Sci. U.S.A. 95, 3408–3412
14. Mohler, K.M., Sleath, P.R., Fitzner, J.N., Cerretti, D.P., Alderson, M., Kerwar, S.S., Torrance, D.S., Otten-Evans, C., Greenstreet, T., Weerawarna, K. et al. (1994) Nature (London) 370, 218–220
15. Black, R.A., Durie, F.H., Otten-Evans, C., Miller, R., Slack, J.L., Lynch, D.H., Castner, B., Mohler, K.M., Gerhart, M., Johnson, R.S. et al. (1996) Biochem. Biophys. Res. Commun. 225, 400–405
16. Jin, G., Huang, X., Black, R., Wolfson, M., Rauch, C., McGregor, H., Ellestad, G. and Cowling, R. (2002) Anal. Biochem. 302, 269–275
17. Buxbaum, J.D., Liu, K.-N., Luo, Y., Slack, J.L., Stocking, K.L., Peschon, J.J., Johnson, R.S., Castner, B.J., Cerretti, D.P. and Black, R.A. (1998) J. Biol. Chem. 273, 27765–27767
18. Brakebusch, C., Varfolomeev, E.E., Batkin, M. and Wallach, D.J. (1994) Biol. Chem. 269, 32488–32496
19. Crowe, P.D., Walter, B.N., Mohler, K.M., Otten-Evans, C., Black, R.A. and Ware, C.F. (1995) J. Exp. Med. 181, 1205–1210
20. Guo, L., Eisenman, J.R., Mahimkar, R.M., Peschon, J.J., Paxton, R.J., Black, R.A. and Johnson, R.S. (2002) Mol. Cell. Proteomics 1, 30–36
21. Reddy, P., Slack, J.L., Davis, R., Cerretti, D.P., Kozlosky, C.J., Blanton, R.A., Shows, D., Peschon, J.J. and Black, R.A. (2000) J. Biol. Chem. 275, 14608–14614
22. Alhoff, K., Reddy, P., Voltz, N., Rose-John, S. and Mullberg, J. (2000) Eur. J. Biochem. 267, 2624–2631
23. Mohan, M.J., Seaton, T., Mitchell, J., Howe, A., Blackburn, K., Burkhart, W., Moyer, M., Patel I., Waitt, G.M., Becherer, J.D. et al. (2002) Biochemistry 41, 9462–9469
24. Sadhukhan, R., Sen, G.C., Ramchandran, R. and Sen, I. (1998) Proc. Natl. Acad. Sci. U.S.A. 95, 138–143

25. Cui, X., Hawari, F., Alsaaty, S., Lawrence, M., Combs, C.A., Geng, W., Rouhani, F.N., Miskinis, D. and Levine, S.J. (2002) J. Clin. Invest. **110**, 515–526

26. Schlondorff, J. and Blobel, C.P. (1999) J. Cell Sci. **112**, 3603–3617

27. Haro, H., Crawford, H.C., Finglteton, B., Shinomiya, K., Spengler, D.M. and Matrisian, L.M. (2000) J. Clin. Invest. **105**, 143–150

28. Kajita, M., Itoh, Y., Chiba, T., Mori, H., Okada, A., Kinoh, H. and Seiki, M. (2001) J. Cell Biol. **153**, 893–904

29. Fan, H. and Derynck, R. (1999) EMBO J. **18**, 6962–6972

30. Gechtman, Z., Alonso, J.L., Raab, G., Ingber, D.E. and Klagsbrun, M. (1999) J. Biol. Chem. **274**, 28828–28835

31. Diaz-Rodriguez, E., Montero, J.C., Esparis-Ogando, A., Yuste, L. and Pandiella, A. (2002) Mol. Biol. Cell. **13**, 2031–2044

32. Doedens, J.R. and Black, R.A. (2000) J. Biol. Chem. **275**, 14598–14607

33. Zhang, Z., Kolls, J.K., Oliver, P., Good, D., Schwarzenberger, P.O., Joshi, M.S., Ponthier, J.L. and Lancaster, Jr, J.R. (2000) J. Biol. Chem. **275**, 15839–15844

Biochem. Soc. Symp. **70**, 53–63
(Printed in Great Britain)
© 2003 Biochemical Society

5

Meprin proteolytic complexes at the cell surface and in extracellular spaces

James P. Villa, Greg P. Bertenshaw[1], John E. Bylander and Judith S. Bond[2]

Department of Biochemistry and Molecular Biology, The Pennsylvania State University College of Medicine, Hershey, PA 17033, U.S.A.

Abstract

Meprins are metalloproteinases of the astacin family and metzincin superfamily that are composed of evolutionarily related α and β subunits, which exist as homo- and hetero-oligomeric complexes. These complexes are abundant at the brush border membranes of kidney proximal tubule cells and epithelial cells of the intestine, and are also expressed in certain leucocytes and cancer cells. Meprins cleave bioactive peptides such as gastrin, cholecystokinin and parathyroid hormone, cytokines such as osteopontin and monocyte chemotactic peptide-1, as well as proteins such as gelatin, collagen IV, fibronectin and casein. Database predictions and initial data indicate that meprins are also capable of shedding proteins, including itself, from the cell surface. Membrane-bound meprin subunits are composed of dimeric meprin β subunits or tetrameric hetero-oligomeric $\alpha\beta$ complexes of approx. 200–400 kDa, and can be activated at the cell surface; secreted forms of homo-oligomeric meprin α are zymogens that form high-molecular-mass complexes of 1–6 MDa. These are among the largest extracellular proteases identified thus far. The latent (self-associating) homo-oligomeric complexes can move through extracellular spaces in a non-destructive manner, and deliver a concentrated form of the metalloproteinase to sites that have activating proteases, such as sites of inflammation, infection or cancerous growth. Meprins provide examples of novel ways of concentrating proteolytic activity at the cell surface and in the extracellular milieu, which may be critical to proteolytic function.

[1]Present address: Clearant, Inc., Gaithersburg, MD 20879, U.S.A.
[2]To whom correspondence should be addressed (e-mail jbond@psu.edu).

Introduction

Meprins are thought to play crucial roles in development, as well as in physiological and pathological processes in mature mammals [1–3]. These proteases are highly regulated at the transcriptional and post-translational levels [4,5]. They are tissue-specific and, like most proteases, are synthesized as zymogens [4,6]. The unique structures of meprins have revealed novel mechanisms to concentrate proteolytic potential both at the cell surface and in extracellular spaces. It has been known for a long time that brush border membranes of the kidney proximal tubule and intestine express meprins abundantly (Figure 1) [8]. Meprins are evenly distributed on the luminal surface of the microvilli and in apical invaginations, and comprise approx. 5% of brush border membrane protein [7]. Meprins in the plasma membrane form dimers and tetramers, and associate with other proteases to provide a rich proteolytic environment [9,10]. Recent work has revealed that secreted forms of meprins autocompartmentalize their enzymic activities by self-associating into complexes with masses greater than 1 MDa, and thereby concentrate proteolytic activity [10,11]. Thus both the secreted and membrane-bound forms of meprins exist in concentrated proteolytic complexes, albeit by different mechanisms. The membrane-bound forms are restricted to action at the cell surface. If the secreted forms were active, the high concentration of proteolytic activity could have deleterious effects, as it is known that soluble forms of meprins can be toxic to cells [12]. However, the secreted forms are inactive, and travel through extracelluar spaces in a non-destructive mode. It is only when they encounter trypsin-like proteases (such as plasmin) that they are activated, and

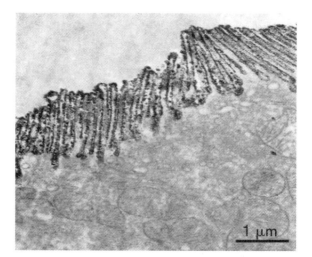

Figure 1 Immunohistochemical detection of meprin α in kidney brush border membranes. Electron micrograph of proximal tubule cells from the juxtamedullary region of kidney cortex, showing immunoperoxidase staining of meprin α in brush border membranes of C57BL/6 mouse kidney. Kidney sections were incubated with anti-meprin α serum and peroxidase-conjugated goat anti-rabbit IgG. Adapted from Craig et al. [7] with permission.

they then can unleash relatively non-specific proteolytic activity to destroy large protein complexes, toxins and other proteinaceous matter [13]. This review will focus on substrates and structure of the meprins.

Substrates, inhibitors and specificity

A wide variety of meprin substrates have been identified [14,15]. Both the α and β subunits hydrolyse peptides and proteins, and both prefer peptides larger than six amino acids. Examples of substrates hydrolysed by the meprin α subunit are angiotensin (I, II and III), azocasein, bombesin, bradykinin, cerulein, cholecystokinin, fibronectin, gelatin, glucagon, gastrin-releasing peptide (GRP), insulin B chain, luliberin, monocyte chemotactic protein-1, melanocyte-stimulating hormone, neuropeptide Y, neurotensin, parathyroid hormone, secretin, substance P, valosin and vasoactive intestinal peptide. Meprin β subunits cleave some of the same substrates as meprin α (such as azocasein, cerulein, cholecystokinin, fibronectin, gelatin, glucagon, GRP, insulin B chain, neuropeptide Y and secretin), but also hydrolyse substrates such as kinetensin, orcokinin, osteopontin and peptide YY. In addition to the above substrates, hetero-oligomeric meprins (containing α and β subunits that are both active) hydrolyse atrial natriuretic peptide, big endothelin 1, collagen IV, cGMP-dependent protein kinase, epidermal growth factor (EGF) receptor, endothelin 1, insulin receptor, lysozyme, protein kinase A and transforming growth factor α. We have found that gastrointestinal peptides (GRP, gastrin) are among the best substrates identified, consistent with our view that meprins have special functions in the intestine.

Detailed comparisons have revealed that mouse, rat and human meprin α enzymes have differences in substrate specificity (Table 1). For example, while GRP, glucagon and parathyroid hormone are excellent substrates for meprin α from all three species, neurotensin and secretin are much better substrates for rat meprin α than for human or mouse meprin α.

Meprins are inhibited by zinc-chelating reagents (EDTA, o-phenanthroline), but there are no known naturally occurring inhibitors in mammals [14]. For example, tissue inhibitor of metalloproteinases-1 (TIMP-1) does not inhibit meprins. In addition, meprins are not inhibited by several other metalloproteinase inhibitors, such as phosphoramidon and captopril. Actinonin and Pro-Leu-Gly-(NHOH), two inhibitors of meprins that contain a hydroxamate moiety, interact with zinc in the active site. Actinonin is a naturally occurring hydroxamate with a bulky pentyl moiety at position P1′, and competitively inhibits mouse meprin A (multimer of α and β subunits) and meprin B (dimer of β subunits) with IC_{50} values of 100 and 400 nM respectively [15]. Crayfish astacin, a homologue of meprin, has been crystallized with Pro-Leu-Gly-(NHOH), and the Pro, Leu and Gly amino acids interact at positions S3, S2 and S1 respectively [16]. Overlay of the astacin crystal structure and the meprin homology model predicts a similar placement of the inhibitor in the meprin active site. This inhibitor binds less efficiently than actinonin to meprin α and β, with IC_{50} values of 30 and 50 μM respectively [15]. Actinonin has been used to elucidate the role of meprins in hypoxic/ischaemic acute renal tubular injury [12].

Table 1 Hydrolysis of bioactive peptides by homologous recombinant meprin α. Hydrolysis was determined by measuring the disappearance of the substrate peak after a 60 min incubation (30 min for GRP) with meprins. Peptides were present at 100 μM (50 μM for GRP) and meprins at 4 nM, based on a subunit molecular mass of 85000 Da. Incubations were carried out at 37°C in 20 mM Tris/HCl and 140 mM NaCl, pH 7.5. PTH, parathyroid hormone; MSH, melanocyte-stimulating hormone.

Peptide	Hydrolysis (%)		
	Human meprin α	Rat meprin α	Mouse meprin α
GRP	>95	>95	>95
Glucagon	93	>95	61
PTH-(13–34) (human)	55	>95	69
Secretin (human)	55	>95	10
α-MSH	38	>95	>95
Bradykinin	28	82	>95
Orcokinin	25	10	<5
Neurotensin	11	95	79
Gastrin I (human)	10	42	<5

Meprin α and β subunits have marked differences in peptide bond specificity (Figure 2) [15]. Peptide library data revealed that meprin β strongly prefers acidic residues (especially Asp) in the P1' position, while meprin α prefers small or aromatic amino acids (e.g. Ser, Phe, Ala, Gly) in the P1' position and Pro in P2'. The pH optima of the meprin α and β subunits also differ: meprin α subunits have optimal activity at neutral or basic pH values, while meprin β subunits have optimal activity at acidic pH values (~5) [17].

Database searches for good meprin substrates based on peptide library data revealed that meprins have the potential ability to shed membrane-bound meprins from the cell surface (Figure 3) [18]. The meprin β subunit was predicted to be released from the membrane, due to cleavage N-terminal to the transmembrane domain near the C-terminus of the meprin β TRAF (tumour necrosis factor receptor-associated factor) domain, by both meprin α and β activities. A secreted His-tagged form of meprin β, containing the putative cleavage site but truncated after the EGF domain, was expressed in HEK 293 cells, purified, and used to investigate the hydrolysis. The secreted His-tagged meprin β was incubated with activated meprins, and Western blots were performed using antibodies to the His tag. The treatments resulted in the complete disappearance of bands, indicating removal of the C-terminal His tag, while Western blots using anti-(meprin α) antibody indicated that only the C-terminus was removed. The presence of the meprin inhibitor actinonin prevented the loss of His-tag band intensity, indicating that the loss of the tag was due to meprin proteolytic activity. N-terminal sequence analysis of treated meprin β demonstrated that the N-terminus was intact and thus that all hydrolysis occurred at the C-terminus. These results show that both meprins α and β are capable of hydrolysing meprin β near the secreted C-terminus and are potentially involved in a novel self-shedding process.

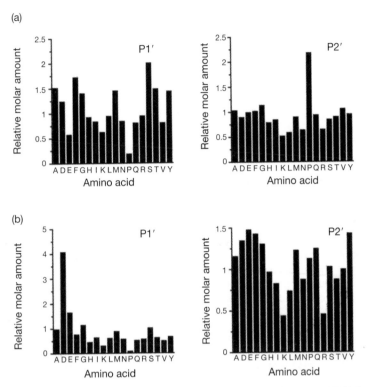

Figure 2 Preferred amino acids in substrate subsites P1′ and P2′ for mouse meprin α (a) and meprin β (b). Meprins were incubated with an acetylated dodecamer peptide library mixture until the library was digested by between 5% and 10%. Incubations were in 25 mM Hepes, 100 mM NaCl and 5 mM $CaCl_2$, pH 7.4, at 37°C. Hydrolysis products were N-terminally sequenced to determine the subsite preferences. The data for each sequencing cycle were normalized to the total amount (mol) of all amino acids present in that cycle, so a value of 1 indicates the average value. The peptide library data represent the averages of two independent experiments. Adapted from Bertenshaw et al. [15] with permission.

Oligomerization of meprins

Meprins are multidomain, oligomeric glycoproteins (Figures 4 and 5) [19,20]. The oligosaccharides are essential for secretion and enzymic activity [20]. The domain structures deduced from the cDNAs of meprin α and β are very similar (Figure 4). Both subunits are synthesized with signal and pro-sequences, which respectively direct the protein to the secretory pathway in the endoplasmic reticulum and keep the protein latent. C-terminal to the catalytic domain are adhesion or interaction domains [MAM (meprin, A5 protein and protein tyrosine phosphatase μ) and TRAF]. These domains are essential to the biosynthesis of activatable, stable proteins and to the interactions between subunits [19]. Both subunits are also synthesized with EGF, transmembrane and short cytoplasmic domains. The major difference between the domain structures is the

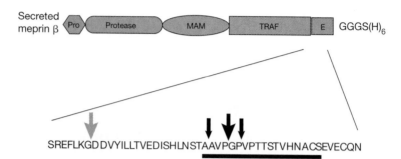

SREFLKGDDVYILLTVEDISHLNSTAAVPGPVPTTSTVHNACSEVECQN

Figure 3 Predicted hydrolysis of meprin β and potential role of meprins in shedding. A secreted form of His-tagged meprin β, which is truncated at the end of the EGF domain, is shown at the top. The amino acid sequence of meprin β corresponding to the C-terminal portion of the TRAF domain and the N-terminal portion of the EGF domain (E) is also displayed. The amino acid sequence in rat meprin β underlined in black corresponds to a region required for shedding in human meprin β; arrows denote the predicted hydrolysis sites for meprin α (black) and meprin β (grey) [18]. The size of the arrowhead corresponds to the strength of the prediction for hydrolysis. Pro, propeptide.

presence of a 56-amino-acid sequence, the I domain, in meprin α. The I domain is essential and sufficient for the C-terminal proteolytic cleavage of meprin α in the lumen of the endoplasmic reticulum [21]. If the I domain is deleted from meprin α, the subunit remains membrane-bound. If the I domain is added to the meprin β subunit between the TRAF and EGF domains, by site-directed mutagenesis, then the β subunit is released from the membrane.

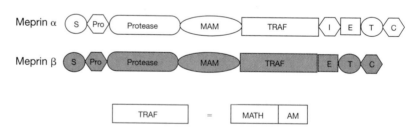

Figure 4 cDNA-deduced domain organization of meprins. Meprins contain a signal peptide (S) and a propeptide (Pro) N-terminal to the protease domain. The protease domain co-ordinates zinc, and shares identity with the astacin family of metalloproteinases and the metzincin superfamily. C-terminal to the protease domain are the MAM domain and the TRAF domain. The TRAF domain was referred to previously as both the MATH and AM domains. Meprin α also contains an inserted (I) domain, an EGF-like domain (E), a transmembrane domain (T) and a cytoplasmic tail (C). Meprin β contains all of these domains with the exception of the I domain. The presence of the I domain allows meprin α to be proteolytically processed to the mature secreted form. The absence of the I domain allows meprin β to remain membrane-bound.

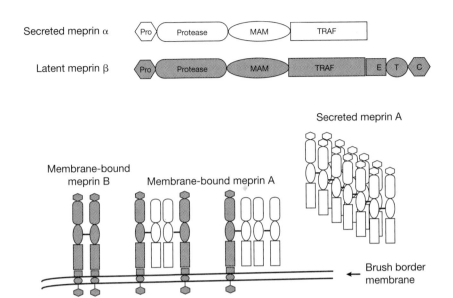

Figure 5 Membrane-bound and secreted forms of meprins. Meprin subunits have different oligomerization potentials. Meprin β (grey) forms homo-oligomers of covalently linked dimers (meprin B). A multimer of meprin α subunits is called Meprin A. Meprin β can also interact covalently with meprin α (white); these dimers form tetramers. During biosynthesis, meprin α is proteolytically processed near the I domain, releasing it from the membrane; this subunit can remain membrane-bound when associated with meprin β. The secreted form of meprin α can form large oligomers, unlike meprin β, through non-covalent interactions.

The MAM domain is essential for oligomerization of meprins, in that the intersubunit S–S bridges of the dimers reside in this domain, and without these disulphide bridges higher-order oligomerization does not occur [22,23]. The MAM domain was identified, based on the identity of meprin with A5 protein and protein tyrosine phosphatase μ, as an adhesive domain [24]. MAM domains generally contain approx. 170 amino acids, including four conserved cysteine residues. Meprin MAM domains contain a fifth cysteine residue that is responsible for intersubunit disulphide-linked homo- and hetero-dimerization [22,23]. This domain was also found to contribute to non-covalent interactions of meprin α subunits in cross-linking studies [11]. Truncation, deletion and site-directed mutagenesis studies revealed that the MAM domain is critical for biosynthesis, structural stability and proteolytic activity [19,20]. Deletion of the MAM domain of meprin α resulted in reverse transport of meprin transcripts into the cytosol and proteolytic degradation of the protein by the proteasome [25]. MAM domains have also been identified in other proteins, such as zonadhesin, neuropilins 1 and 2, enterokinase, and two other astacin family members, a *Hydra* metalloproteinase and a squid metalloproteinase [24,26,27].

Until recently, the meprin TRAF domain was considered to be divided into two domains, MATH ('meprin and TRAF homology') and AM ('After

MATH'). However, a recent study convincingly demonstrated that the meprin MATH and AM domains are part of TRAF domains that have been found in a large number of diverse proteins [28]. These domains are generally associated with intracellular adapter proteins, and were first identified for their binding to tumour necrosis factor family receptors. The TRAF domains are homo- and hetero-philic interaction domains that are found in proteins that are particularly prevalent in processes such as signal transduction and apoptosis [29]. The meprins are the only known TRAF-containing proteins that exist outside cells. Meprin TRAFs are required for the biosynthesis of an activatable protease [19]. These domains may be especially important for the homo- and hetero-typic interactions at the cell surface and extracelluarly.

All forms of meprins contain disulphide-linked dimers (Figure 5). The meprin α and β dimers differ in their propensity to form higher-order multimers. Homo-oligomeric meprin α dimers form large secreted multimers containing a heterogeneous mixture of high-molecular-mass complexes containing 12–100 subunits and with molecular masses of approx. 900 kDa to 6.0 MDa. For rat meprins, latent meprins form the largest multimers; removal of the prosequence results in smaller oligomers of 12–16 subunits [10]. Multimer

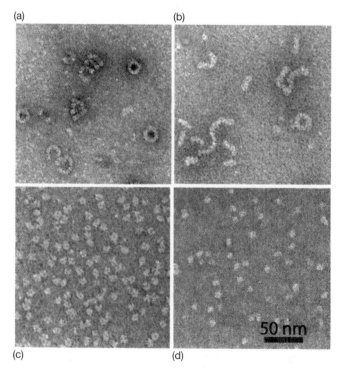

Figure 6 Electron micrographs of rat meprins. Representative fields of negatively stained samples of each recombinant His-tagged meprin isoform are shown: latent (**a**) and active (**b**) homo-oligomeric meprin A (multimer of α subunits); latent (**c**) and active (**d**) homo-oligomeric meprin B (dimer of β subunits). Adapted from Bertenshaw et al. [10] with permission.

formation determined by size-exclusion chromatography light scattering was shown to be a dynamic process dependent upon the activation state, protein concentration and ionic concentration. Electron microscopy revealed that meprin α subunits form rings, crescents and barrel-shaped complexes, reminiscent of the proteasome (Figure 6); the difference in the size of the latent and active meprin α complexes in Figures 6(a) and 6(b) is not obvious from the small dilute samples of the electron micrographs (see [10]). The barrels or tube shapes measured 30–40 nm in length, and contained a large central cavity.

The dynamic oligomerization of meprins potentially serves several roles. The formation of hetero-oligomers allows meprin α to remain membrane-bound through the β subunit. This not only allows meprin α hydrolytic activity to exist at the membrane, but also decreases the amount of secreted meprin α and subsequent activity. Oligomerization of meprin α with meprin β also limits the oligomeric state of meprin. Hetero-oligomeric forms of meprin are predominantly tetramers. The large homo-oligomeric meprin α complexes are seen *in vivo* in both mouse and rat urine [10,30]. Meprins exist in the high micromolar concentration range on the kidney brush border membrane, thus creating a very concentrated and localized proteolytic centre [7].

A diagrammatic summary of the membrane and secreted forms of meprins is shown in Figure 7. Secreted meprin multimers are thought to create concentrated centres of proteolytic activity away from the membrane. Although activated meprin α forms large multimers (12–20 subunits), these oligomers are smaller than the latent forms. This oligomerization difference indicates that meprin activation and multimerization are coupled and regulated through an as

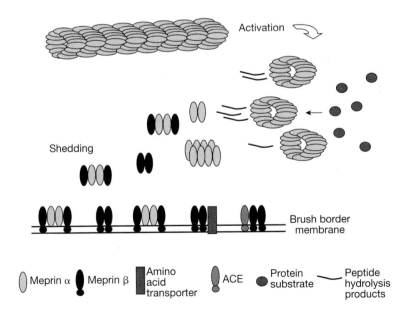

Figure 7 Summary model of meprin homo- and hetero-associations. Schematic representation of meprin homo- and hetero-interactions for secreted and membrane-bound forms. ACE, angiotensin-converting enzyme.

yet undetermined mechanism. Although meprin β cannot form large homo-oligomeric multimers, it is capable of associating with proteins other than meprins, such as angiotensin-converting enzyme, leucine aminopeptidase and amino acid transporters [7,9,31]. Meprin oligomerization thus appears to be a dynamic and complex association of homo- and hetero- interactions.

This work was supported by NIH grants DK19691 and DK54625 to J.S.B.

References

1. Trachtman, H., Valderrama, E., Dietrich, J.M. and Bond, J.S. (1995) Biochem. Biophys. Res. Commun. **208**, 498–505
2. Kumar, J.M. and Bond, J.S. (2001) Biochim. Biophys. Acta **1518**, 106–114
3. Norman, L.P., Matters, G.L., Crisman, J.M. and Bond, J.S. (2003) Curr. Top. Dev. Biol. **54**, 145–166
4. Bond, J.S. and Beynon, R.J. (1995) Protein Sci. **4**, 1247–1261
5. Matters, G.L. and Bond, J.S. (1999) APMIS **107**, 19–27
6. Johnson, G.D. and Bond, J.S. (1997) J. Biol. Chem. **272**, 28126–28132
7. Craig, S.S., Reckelhoff, J.F. and Bond, J.S. (1987) Am. J. Physiol. **253**, C535–C540
8. Sterchi, E.E., Naim, H.Y., Lentze, M.J., Hauri, H.P. and Fransen, J.A. (1988) Arch. Biochem. Biophys. **265**, 105–118
9. Butler, P.E. and Bond, J.S. (1988) J. Biol. Chem. **263**, 13419–13426
10. Bertenshaw, G.P., Norcum, M.T. and Bond, J.S. (2003) J. Biol. Chem. **278**, 2522–2532
11. Ishmael, F.T., Norcum, M.T., Benkovic, S.J. and Bond, J.S. (2001) J. Biol. Chem. **276**, 23207–23111
12. Carmago S., Shah S.V. and Walker, P.D. (2002) Kidney Int. **61**, 959–966
13. Rosmann, S., Hahn, D., Lottaz, D., Kruse, M.N., Stocker, W. and Sterchi, E.E. (2002) J. Biol. Chem. **277**, 40650–40658
14. Bertenshaw, G.P., Bond, J.S. Meprin, A. and Meprin, B. (2003) in Handbook of Proteolytic Enzymes (Barrett, A.J., Woessner, F. and Rawlings, N., eds), Academic Press, in the press
15. Bertenshaw, G.P., Turk, B.E., Hubbard, S.J., Matters, G.L., Bylander, J.E., Crisman, J.M., Cantley, L.C. and Bond, J.S. (2001) J. Biol. Chem. **276**, 13248–13255
16. Grams, F., Dive, V., Yiotakis, A., Yiallouros, I., Vassiliou, S., Zwilling, R., Bode, W. and Stocker, W. (1996) Nat. Struct. Biol. **3**, 671–675
17. Bertenshaw, G.P., Villa, J.P., Hengst, J.A. and Bond, J.S. (2002) Biol. Chem. **383**, 1175–1183
18. Pischitzis, A., Hahn, D., Leuenberger, B. and Sterchi, E.E. (1999) Eur. J. Biochem. **261**, 421–429
19. Tsukuba, T. and Bond, J.S. (1998) J. Biol. Chem. **273**, 35260–35267
20. Kadowaki, T., Tsukuba, T., Bertenshaw, G.P. and Bond, J.S. (2000) J. Biol. Chem. **275**, 25577–25584
21. Marchand, P., Tang, J., Johnson, G.D. and Bond, J.S. (1995) J. Biol. Chem. **270**, 5449–5456
22. Marchand, P., Volkmann, M. and Bond, J.S. (1996) J. Biol. Chem. **271**, 24236–24241
23. Chevallier, S., Ahn, J., Boileau, G. and Crine, P. (1996) Biochem. J. **317**, 731–738
24. Beckmann, G. and Bork, P. (1993) Trends Biochem. Sci. **18**, 40–41
25. Tsukuba, T., Kadowaki, T., Hengst, J.A. and Bond, J.S. (2002) Arch. Biochem. Biophys. **397**, 191–198
26. Yan, L., Fei, K., Zhang, J., Dexter, S. and Sarras, Jr, M.P. (2000) Development **127**, 129–141
27. Yokozawa, Y., Tamai, H., Tatewaki, S., Tajima, T., Tsuchiya, T. and Kanzawa, N. (2002) J. Biochem. (Tokyo) **132**, 751–758

28. Zapata, J.M., Pawlowski, K., Haas, E., Ware, C.F., Godzik, A. and Reed, J.C. (2001) J. Biol. Chem. **276**, 24242–24252

29. Chung, J.Y., Park, Y.C., Ye, H. and Wu, H. (2002) J. Cell Sci. **115**, 679–688

30. Norman, L.P., Jiang, W., Han, X., Squnders, T.L. and Bond, J.S. (2003) Mol. Cell. Biol. **23**, 1221–1230

31. Chestukhin, A., Muradov, K., Litovchick, L. and Shaltiel, S. (1996) J. Biol. Chem. **271**, 30272–30280

Biochem. Soc. Symp. **70**, 65–80
(Printed in Great Britain)
© 2003 Biochemical Society

6

Role of TIMPs (tissue inhibitors of metalloproteinases) in pericellular proteolysis: the specificity is in the detail

Gillian Murphy[*1,2], **Vera Knäuper**[*3], **Meng-Huee Lee**[*1],
Augustin Amour[*4], **Joanna R. Worley**[*], **Mike Hutton**[*], **Susan
Atkinson**[*1], **Magdalene Rapti**[†] **and Richard Williamson**[†]

*School of Biological Sciences, University of East Anglia, Norwich NR4 7TJ, U.K., and
†Department of Biosciences, University of Kent, Canterbury, Kent CT2 7NJ, U.K.

Abstract

Pericellular proteolysis represents one of the key modes by which the cell
can modulate its environment, involving not only turnover of the extracellular
matrix but also the regulation of cell membrane proteins, such as growth factors and their receptors. The metzincins are active players in such proteolytic
events, and their mode of regulation is therefore of particular interest and
importance. The TIMPs (tissue inhibitors of metalloproteinases) are established endogenous inhibitors of the matrix metalloproteinases (MMPs), and
some have intriguing abilities to associate with the pericellular environment. It
has been shown that TIMP-2 can bind to cell surface MT1-MMP (membrane-
type 1 MMP) to act as a 'receptor' for proMMP-2 (progelatinase A), such that
the latter can be activated efficiently in a localized fashion. We have examined
the key structural features of TIMP-2 that determine this unique function,
showing that Tyr^{36} and Glu^{192}-Asp^{193} are vital for specific interactions with
MT1-MMP and proMMP-2 respectively, and hence activation of proMMP-2.
TIMP-3 is sequestered at the cell surface by association with the glycosamino-

[1]Present address: University of Cambridge, Department of Oncology, Cambridge
Institute for Medical Research, Hills Road, Cambridge CB2 2XY, U.K.
[2]To whom correspondence should be addressed: Wellcome Trust/MRC Building,
P.O. Box 139, Hills Road, Cambridge CB2 2XY, U.K. (e-mail gm290@cam.ac.uk).
[3]Present address: Department of Biomedical Tissue Research, Biology, University of
York, York YO10 5YW, U.K.
[4]Present address: GlaxoSmithKline, Medicines Research Centre, Gunnels Wood Road,
Stevenage SG1 2NY, U.K.

glycan chains of proteoglycans, especially heparan sulphate, and we have shown that it may play a role in the regulation of some ADAMs (a disintegrin and metalloproteinases), including tumour necrosis factor α-converting enzyme (TACE; ADAM17). We have established that key residues in TIMP-3 determine its interaction with TACE. Further studies of the features of TIMP-3 that determine specific binding to both ADAM and glycosaminoglycan are required in order to understand these unique properties.

Introduction

Cell–matrix and cell–cell interactions influence a diverse range of cellular functions, including proliferation, differentiation, migration and survival. Proteolytic degradation or activation of cell surface and extracellular matrix proteins can mediate rapid cellular responses to their microenvironment, and hence modulate cell behaviour. Members of the metzincin clan of zinc endopeptidases (metalloproteinases) are key players in such activities, and have been the subject of intense study in relation to the biology of the cell. Of particular interest are the strategies adopted by the cell to focus and control proteolytic activities, which range from directed trafficking and cell surface localization to extracellular matrix binding. Proteolytic activation cascades are also important in the regulation of functional activity at the cell surface, as well as control by natural inhibitors.

Both secreted and membrane-bound forms of metalloproteinases have been implicated in pericellular proteolysis, including the matrix metalloproteinases (MMPs), the adamalysin proteinases with both metalloproteinases and disintegrin-like domains [ADAM (a disintegrin and metalloproteinase) and ADAMTS (ADAM with thrombospondin motifs)] and the astacins [1]. Comprehensive descriptions of all the known members of these families have been published recently [2–4] and will not be repeated here.

The tissue inhibitors of metalloproteinases (TIMPs)-1 to -4 were originally identified as natural regulators of the MMPs in mammalian cells. The four TIMPs have many basic similarities, but they exhibit distinctive structural features and biochemical properties, suggesting that each has specific roles *in vivo* [5,6]. They have an N-terminal domain of three disulphide-bonded loops, and a C-terminal subdomain that also comprises three disulphide-bonded loops that has specific functions in binding to some MMPs. The ability of these proteins to inhibit the MMPs is due largely to the interaction of a wedge-shaped ridge on the N-terminal domain which binds within the active-site cleft of the target MMP, allowing the co-ordination of the catalytic Zn^{2+} of the MMP by the α-amino and carbonyl groups of the N-terminal Cys, while the side chain of the Ser/Thr at position 2 occupies the S1′ pocket. Variations in TIMP binding to MMPs can also occur, e.g. TIMP-1 is a poor inhibitor of a number of the membrane-type MMPs (MT-MMPs) and MMP-19.

This review will summarize our contributions towards the understanding of the functions of TIMP-2 and TIMP-3 in regulating pericellular proteolysis through the analysis of features that form the structural basis of their specificity. The implications of our findings regarding the regulation of cellular function are discussed.

Activation of proMMP-2 and the role of TIMP-2

The activation of MMPs by sequential proteolysis of the propeptide blocking the active-site cleft is regarded as one of the key levels of regulation of these proteinases. Current thinking, based on many studies, is that these events may be orchestrated in a controlled fashion close to the cell surface, with membrane-associated proteinases such as urokinase-type plasminogen activator and MT-MMPs playing key initiating roles (Figure 1) [7]. The activation of the secreted MMP, proMMP-2 (progelatinase A), has proved particularly interesting since, although a soluble MMP, it was found to occur after binding to the surface of the cell. This led to the identification of the membrane-associated MT-MMPs as potential mediators of proMMP-2 activation (reviewed in [8]). It was originally found that proMMP-2 could be activated by soluble forms of MT1-MMP or MT2-MMP in a two-step activation mechanism analogous to the activation of other MMPs by plasmin. The initial propeptide cleavage in proMMP-2 effected by MT-MMP activity is at the Asn^{37}–Leu^{38} peptide bond. The second propeptide cleavage is autoproteolytic, since an inactive proMMP-2 mutant ($proE^{375}A$-MMP-2) is only processed by MT1-MMP to the Leu^{38} intermediate. Processing by MT1-MMP is inhibited by either TIMP-2 or

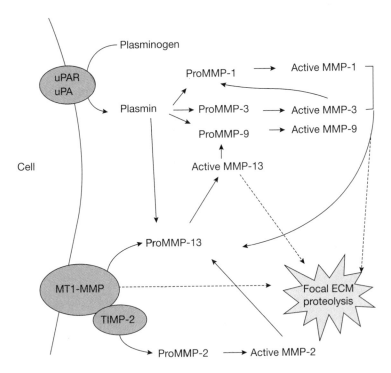

Figure 1 Pericellular activation cascades for MMPs. Pericellular proteolysis is highly regulated by activation cascades initiated by the generation of plasmin by receptor-associated urokinase-type plasminogen activator (uPA) and by the action of MT1-MMP (reviewed in [7]). uPAR, uPA receptor; ECM, extracellular matrix.

TIMP-3, but not by TIMP-1. This agrees with the observation that TIMP-1 is a very poor inhibitor of MT1-MMP, but TIMP-2 and TIMP-3 are extremely efficient [9]. ProMMP-2 activation by cells expressing MT1-MMP also involves a two-step activation mechanism, with an MT1-MMP-mediated 'initiation' cleavage followed by autolytic cleavage of MMP-2 (Figure 2). The process appears to involve binding of proMMP-2 to an MT1-MMP–TIMP-2 complex, which forms a 'receptor' at the surface of the cell through interaction of the haemopexin domain of proMMP-2 with the C-terminal subdomain of TIMP-2. By establishing a trimolecular complex consisting of MT1-MMP, TIMP-2 and proMMP-2, the components of the 'activation cascade' are concentrated on the cell surface. Processing of proMMP-2 to the Leu[38] intermediate may then be effected by an adjacent functional MT1-MMP molecule. This initial cleavage event destabilizes the structure of the propeptide, and autolytic cleavage to generate the fully mature enzyme proceeds in an MMP-2-concentration-dependent manner. In cell culture studies, the enzyme concentrations in solution are very low, and deletion of either the proMMP-2 haemopexin domain or the transmembrane domain of MT1-MMP abrogates

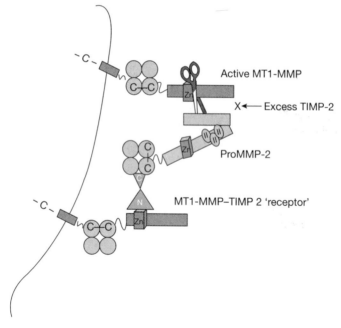

Figure 2 Mechanism of pericellular activation of proMMP-2. ProMMP-2 binds via interaction of its C-terminal haemopexin domain with the C-terminal subdomain of TIMP-2, which is in turn bound by its N-terminal domain to cell surface MT1-MMP, forming a trimolecular complex. An adjacent free molecule of MT1-MMP initiates cleavage of the proMMP-2 propeptide, generating an activation intermediate of proMMP-2. This is susceptible to auto-proteolytic processing to a fully active form. The activation mechanism is subject to tight regulation by the level of TIMP-2: a low concentration of TIMP-2 promotes efficient activation of proMMP-2, but an excess is inhibitory [7].

proMMP-2 processing. This emphasizes that the binding mechanism involving the MT1-MMP–TIMP-2 complex on the cell surface acts as a focusing mechanism for the reactants in this cascade that is crucial for the efficiency of activation (Figure 2).

Addition of small amounts of TIMP-2 to cells expressing MT1-MMP can enhance proMMP-2 activation, because this increases the concentration of the MT1-MMP–TIMP-2 receptor for proMMP-2 on the cell surface. However, at high TIMP-2 concentrations, all of the MT1-MMP molecules are complexed to TIMP-2 and, although binding of proMMP-2 can occur, no active MT1-MMP is available to initiate propeptide processing [8]. This suggests that activation of proMMP-2 is regulated by the amount of TIMP-2 relative to that of MT1-MMP displayed at the cell surface. The importance of TIMP-2 for proMMP-2 activation by MT1-MMP has been highlighted by the finding that TIMP-2$^{-/-}$ mice do not show activation of proMMP-2 [10,11].

Kinetic analysis of the binding of TIMP-2 to MMP-2

Kinetic studies of the TIMP-2-mediated inhibition of various MMPs have demonstrated specific interactions with both MMP-2 and MT1-MMP. We initially analysed the interactions of wild-type TIMP-2, a mutant lacking the C-terminal subdomain (Δ128–194-TIMP-2) and a mutant from which the C-terminal nine-amino-acid extension had been removed (Δ186–194-TIMP-2) with both full-length MMP-2 and a mutant of MMP-2 from which the haemopexin domain had been deleted. Binding of all forms of TIMP-2 to all the forms of MMP-2 was very tight, with binding constants in the low picomolar range, which precluded accurate measurement and the achievement of steady-state kinetic techniques using quenched fluorescent peptide cleavage assays [12]. We employed techniques to overcome this problem, in order to establish a more accurate value for the binding constant, K_i [13]. We found that measurements of the rate of dissociation of MMP-2 and TIMP-2 gave a half-life of approx. 400 days, i.e. binding is effectively irreversible. Using rapid reaction techniques, we were able to measure the rate constant for the inhibition of MMP-2 by TIMP-2, and deduced that the overall K_i value was 0.6 fM. The data suggested that binding of TIMP-2 to MMP-2 occurs in two phases, since the rate-limiting step for inhibition is dependent on TIMP-2 concentration at low inhibitor concentrations, whereas at high concentrations the rate-limiting step is independent of free TIMP-2. Due to the lack of availability of large amounts of the TIMP-2 mutants, most subsequent kinetic analyses were confined to the determination of rate constants for the interactions using steady-state kinetics. We noted that inhibition of MMP-2 was markedly affected by the removal of the C-terminal domain of TIMP-2, with a decrease in k_{on} of 50–100-fold (Table 1a) when the truncated form of enzyme or inhibitor was used. The existence of at least two sites of interaction between TIMP-2 and proMMP-2 through the haemopexin domain has been confirmed in the recently published crystal structure of the proMMP-2–TIMP-2 complex [14]. This showed that the tail sequence of TIMP-2 established five salt bridges with residues within blades III and IV of the haemopexin domain of proMMP-2 that flank hydrophobic interactions established by Phe[188] with a hydrophobic groove within the enzyme.

Table 1 TIMPs: role of domain structures and motifs in (a) rate of binding to and (b) determination of apparent binding constants for MMP-2 and MT1-MMP. Studies were carried out at 25°C, apart from *37°C. nd, not determined.

(a)

TIMP construct	$10^{-6} \times k_{on}$ (M^{-1}·s^{-1})			
	MMP-2	Δ418–631-MMP-2	MT1-MMP	Δ269–559-MT1-MMP
TIMP-2	11.9	0.3*	3.0	2.4
Δ128–194-TIMP-2	0.17	0.3*	3.0	3.6
Δ186–194-TIMP-2	7.2	nd	3.9	nd
TIMP-2-(1–127)–TIMP-4-(129–195)	0.48	nd	nd	nd
Glu^{192}Ile/Asp^{193}Gln-TIMP-2	0.45	nd	nd	nd
TIMP-4	8.7	nd	0.54	0.39
Δ129–195-TIMP-4	0.043	nd	0.17	0.095

(b)

TIMP construct	K_i^{app} (pM)			
	MMP-2	Δ418–631-MMP-2	MT1-MMP	Δ269–559-MT1-MMP
TIMP-2	0.0006	nd	nd	20
Δ128–194-TIMP-2	0.046	nd	3000	1200
TIMP-4	nd	nd	nd	150
Δ129–195-TIMP-4	300	nd	nd	1070

A role for the C-terminal 'tail' of TIMP-2 was also established by kinetic analysis (Table 1). By studying the effect of salt concentration on the rate of association k_{on}, it was shown that ionic interactions are predominant in the association of the TIMP-2 tail with the haemopexin domain of MMP-2 [15]. Indeed, mutation of Glu[192]-Asp[193] to Ile[192]-Asn[193] (the sequence found in TIMP-4; see below and Figure 4) reduced the k_{on} for MMP-2 by 20-fold (Table 1a). In contrast, when the interaction of TIMP-2 with MT1-MMP was analysed similarly, it was found that only the N-terminal domains of both enzyme and inhibitor were critical for initial binding (Table 1a) [16].

Kinetic analysis of binding of Δ128–194-TIMP-2 mutants to MT1-MMP

We carried out kinetic studies on the interaction of Δ128–194-TIMP-2 mutants with MT1-MMP, and found that residues that had been identified at the enzyme–inhibitor interface in the MMP-3–TIMP-1 crystal structure [17] were also important in this case. Mutation of residues Ser[2], Ala[70], Val[71] and Gly[73], which are located on the surface of the TIMP-2 ridge structure, increased the association rate constants and final K_i^{app} values for MT1-MMP as well as MMP-2, MMP-7 and MMP-13 (Table 1a) [16]. However, we demonstrated a specific interaction between the hairpin turn of the A and B β-strands of TIMP-2 (Tyr[36]) and MT1-MMP. This was in agreement with the structural data for the complex, where Tyr[36] of TIMP-2 was seen to fit into a cavity on the surface of MT1-MMP, bordered by the 'MT loop' and the side chains of Asp[212], Ser[189] and Phe[180] [18]. The data from the crystal structure indicated that several other significant interactions occurred between the MT1-MMP surface and residues of the TIMP-2 AB loop. Alignments of the β-strands in the AB hairpins of TIMPs-1–4 were made using the NMR and X-ray diffraction data for TIMP-1 and TIMP-2, and based on the known properties of the OB (oligosaccharide/oligonucleotide binding) protein fold [19]. In free TIMP-2, the strand alignment for the β-hairpin is very well defined in both the NMR and crystal structures, and ends in a type I β-turn with a G1-β bulge. In the crystal structure of the TIMP-2–MT1-MMP complex, the strand alignment is less clear, and there is some evidence of strand realignment in the middle section (Glu[26] to Asp[30]) that may occur as part of the large conformational change seen for this region on binding to MT1-MMP. The structural data for TIMP-1 show no significant differences in strand alignment between the free and MMP-3-bound inhibitor, and the predicted hydrogen-bonding pattern suggests that this hairpin results in a four-residue turn (i.e. four residues in the loop positions). No structural data are currently available for either TIMP-3 or TIMP-4, but their predicted strand alignments and hydrogen-bonding patterns suggest that both of these hairpins may end in type I′ β-turns (i.e. two residues in the loop positions). The predicted β-strand alignment of the ΔAsp[34]–Ile[40] mutant of Δ128–194-TIMP-2 is also shown in Figure 3, and the hydrogen-bonding pattern suggests that this hairpin will end in a longer four-residue loop.

Contribution of the TIMP-2 AB loop to the conformational stability of Δ128–194-TIMP-2

NMR-based structural analysis of the Ile[35]Gly/Tyr[36]Gly-TIMP-2 mutant clearly showed that the glycine substitutions had no effect on the overall folding

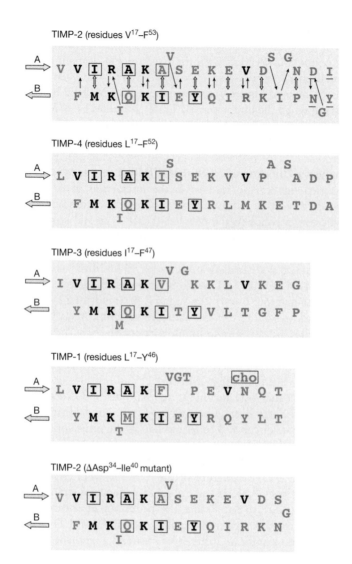

Figure 3 β-Strand alignments for the AB hairpins of TIMPs-1 to -4.
Cross-strand hydrogen-bonding patterns have been predicted from the available NMR and X-ray diffraction data, and based on the known properties of the OB protein fold. Strong Hα–Hα nuclear Overhauser effects seen in the NMR structure analysis of N-TIMP-2 (the N-terminal domain of TIMP-2) are shown by double-headed arrows, and identify residues adjacent to one another in the β-strands. The hydrogen-bonding pattern is shown by single-headed arrows from NH to O. Residues with side chains internalized into the hydrophobic core of N-TIMP-2 are boxed. Residues conserved within the sequences for human TIMPs are shown in black (other residues are shown in grey). The glycosylated Asn residue of TIMP-1 is labelled 'cho'. Residues within the AB hairpin that were mutated are underlined. The predicted strand alignment for the ΔAsp³⁴–Ile⁴⁰ deletion mutant is also shown for comparison [19].

of the protein, except at the tip of the AB hairpin. Although this region is highly solvent-exposed and does not interact with the rest of the protein structure, it was found to make a significant contribution to the conformational stability of the molecule. The change in structure at the tip of the AB loop in the Ile[35]Gly/Tyr[36]Gly mutant resulted in a denaturation midpoint shift of 0.39 M guanidinium chloride, whereas the removal of the entire tip region (Δ34–40) resulted in a larger shift in stability of 0.68 M guanidinium chloride. The lower stability of the deletion mutant may be due in part to the new non-native turn between strands A and B, which is unlikely to be as energetically stable as the well-ordered turn found in the wild-type protein. The findings from the structural studies on Ile[35]Gly/Tyr[36]Gly-TIMP-2 and Δ34–40-TIMP-2 serve to highlight the structural independence of the AB hairpin, and suggest that quite large changes to the sequence can be made in this region without perturbing the structure of the rest of the inhibitor.

The kinetic studies of the binding of Δ128–194-TIMP-2 mutants to MT1-MMP (Table 1) clearly showed that the side chain of Tyr[36] is the most important feature of the AB loop, in terms of both initial association and final binding. Complete deletion of the tip of the AB hairpin (Δ34–40) of Δ128–194-TIMP-2 did not modify these parameters significantly, suggesting that there is no net binding contribution towards association or final binding from other residues in this region of the inhibitor apart from Tyr[36] (results not shown; [19]). This finding further supports our previous suggestion that, although the position of the extended AB hairpin will necessitate its close contact with a proteinase bound at the inhibitory site of TIMP-2, this interaction need not contribute to the overall binding affinity in all cases and could, in some cases, conceivably weaken the overall binding interaction by making unfavourable contacts with the proteinase.

Kinetic analysis of Δ128–194-TIMP-2 mutants with structural features of TIMP-4

The C-terminal deletion mutant Δ129–195-TIMP-4 associates with MT1-MMP at a 20-fold lower rate than Δ128–194-TIMP-2 mutants, which may be due to the lack of a residue comparable with Tyr[36] in the AB β-turn (see Figure 3). However, Δ129–195-TIMP-4 has been shown to have a similar K_i^{app} value for MT1-MMP binding as Δ128–194-TIMP-2 (Table 1b), suggesting that the lack of a binding contribution from Tyr[36] is compensated for by other interactions elsewhere. The charged and polar residues Asp[34] and Asn[38] were considered as potentially important residues for the TIMP-2–MT1-MMP association because they occupy the same positions as Asp[34] and Asp[37] in TIMP-4. Mutation of both residues to Ala did markedly reduce the rate of initial binding to MT1-MMP (the k_{on} value was decreased by 180-fold), but had little effect on the final K_i^{app} value. Interestingly, the modification of Asn[38] to Gln in TIMP-2 (making it comparable with TIMP-4) resulted in a 50-fold decrease in the value of K_i^{app} for binding to MT1-MMP. The side chain of Asn[38] is involved in making a hydrogen bond to its own backbone (amide O to HN) and in making weaker electrostatic interactions across the hairpin with the side chain of Asp[34] and to the proteinase with the side chain of Asn[208] and the backbone HN of

Ile[209]. Substitution of Asn[38] with the negatively charged Asp may cause some structural rearrangement at this site, allowing stronger interactions to form between the tip of the AB hairpin and the catalytic domain of MT1-MMP. It is interesting to speculate that Asp[37] in Δ129–195-TIMP-4 may help to compensate for the lack of a residue equivalent to Tyr[36], allowing the overall binding constant for MT1-MMP to be similar to that measured for Δ128–194-TIMP-2.

The precise biological significance of the unique interactions between the AB loop of TIMP-2 and MT1-MMP can only be speculated upon, but they may play an important role in the stabilization of the TIMP-2–MT1-MMP complex in which the C-terminal region of TIMP-2 is free to bind to the haemopexin-like domain of MMP-2. This would represent the basis of a cell surface MT1-MMP–TIMP-2–proMMP-2 complex, leading to the activation of proMMP-2. It is clear from this work that TIMP-2 may be engineered to abrogate MT1-MMP binding, whereas its binding properties for many other MMPs, including MMP-2, are maintained.

TIMP-2 mutants that affect activation of proMMP-2 by MT1-MMP

We have begun a study on the effects of mutating TIMP-2 in order to modify its interactions with proMMP-2 in the context of the efficiency of cellular MT1-MMP-mediated activation of the latter. It is well documented that TIMP-4 is unable to promote activation of proMMP-2 by MT1-MMP [20,21]. However, the kinetic constants indicate that TIMP-4 is an excellent inhibitor of MMP-2, with a comparable k_{on} value to TIMP-2 or Δ128–194-TIMP-2. Furthermore, TIMP-4 binds to proMMP-2 via the C-terminal domain of the enzyme, the binding site being complementary to the TIMP-2 binding site [20,21a]. However, from the crystal structural data of the TIMP-2–proMMP-2 complex, we can deduce that the C-terminal interactions in a TIMP-4–proMMP-2 complex are likely to be considerably weaker, e.g. only three salt bridges could be formed between the C-terminal tail of TIMP-4 and the haemopexin domain of proMMP-2 [21a].

In the case of MT1-MMP, the k_{on} values are somewhat smaller and the K_i^{app} values slightly larger (Table 1). Using the various domain deletion mutants in the kinetic studies, we found that, interestingly, the C-terminal subdomains of both TIMPs appear to have significant interactions with the catalytic domain of MT1-MMP, but the haemopexin domain of the latter shows no significant interactions with TIMPs (results not shown). We compared the sequences of human TIMP-2 and human and mouse TIMP-4 (Figure 4). Our studies were carried out using mouse TIMP-4, which is 91% identical to the human sequence and therefore unlikely to present significant species differences.

We hypothesized that the interaction between TIMP-2, proMMP-2 and MT1-MMP is tailored to promote the cellular activation of proMMP-2 by MT1-MMP, and that unique features of the TIMP-2 structure relative to TIMP-4 would be a good focus for studies to test this concept. We analysed the ability of modified forms of TIMP-2 to bind both the active catalytic domain of MT1-MMP and the mutant proE[375]A MMP-2. The use of this catalytically inactive form of proMMP-2 prevents full proteolysis of the proMMP-2 propeptide, but allows binding to gelatin–Sepharose. Species associating with the proE[375]A MMP-

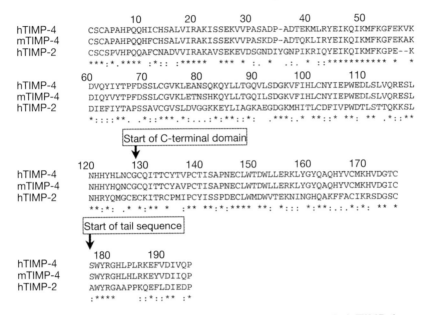

```
              10        20        30        40        50
hTIMP-4   CSCAPAHPQQHICHSALVIRAKISSEKVVPASADP-ADTEKMLRYEIKQIKMFKGFEKVK
mTIMP-4   CSCAPAHPQQHFCHSALVIRAKISSEKVVPASKDP-ADTQKLIRYEIKQIKMFKGFEKAK
hTIMP-2   CSCSPVHPQQAFCNADVVIRAKAVSEKEVDSGNDIYGNPIKRIQYEIKQIKMFKGPE--K
          ***.*.**** :*.:: :***** *** *. * ..:. * :**********  *  *

              60        70        80        90       100       110
hTIMP-4   DVQYIYTPFDSSLCGVKLEANSQKQYLLTGQVLSDGKVFIHLCNYIEPWEDLSLVQRESL
mTIMP-4   DIQYVYTPFDSSLCGVKLETNSHKQYLLTGQILSDGKVFIHLCNYIEPWEDLSLVQRESL
hTIMP-2   DIEFIYTAPSSAVCGVSLDVGGKKEYLIAGKAEGDGKMHITLCDFIVPWDTLSTTQKKSL
          *::::**. .*:.:***.*:...:*:**.:*:  .***.:* **:.* **: ** .*:.**
```

```
                          ┌─────────────────────────┐
                          │ Start of C-terminal domain │
                          └─────────────────────────┘
                                    ↓
              120       130       140       150       160       170
hTIMP-4   NHHYHLNCGCGQITTCYTVPCTISAPNECLWTDWLLERKLYGYQAQHYVCMKHVDGTC
mTIMP-4   NHHYHQNCGCGQITTCYAVPCTISAPNECLWTDWLLERKLYGYQAQHYVCMKHVDGIC
hTIMP-2   NHRYQMGCECKITRCPMIPCYISSPDECLWMDWVTEKNINGHQAKFFACIKRSDGSC
          **.*: .* *;** * ;** **:*:**** **: *::: *:**::..*:*: ** *
```

```
                          ┌──────────────────────┐
                          │ Start of tail sequence │
                          └──────────────────────┘
                                 ↓
              180       190
hTIMP-4   SWYRGHLPLRKEFVDIVQP
mTIMP-4   SWYRGHLHLRKEYVDIIQP
hTIMP-2   AWYRGAAPPKQEFLDIEDP
          :****   ::*::** :*
```

Figure 4 Sequence alignments of human (h) and mouse (m) TIMP-4 with human TIMP-2. Fully conserved residues (*), conservation of strong groups (:) and conservation of weak groups (•) are shown.

2 will hence be retained on gelatin–Sepharose, and were identified by immunoblotting of eluates [21a]. TIMP-2, Δ186–194-TIMP-2, a chimaera of the TIMP-2 N-terminal domain and the TIMP-4 C-terminal domain [TIMP-2-(1–127)–TIMP-4-(129–195)], a chimaera of the TIMP-4 N-terminal domain and the TIMP-2 C-terminal domain [TIMP-4-(1–128)–TIMP-2-(128–194)], TIMP-4 and the Ile^{193}Glu/Gln^{194}Asp-TIMP-4 mutant were assessed for their ability to bind to proMMP-2 in a bimolecular complex and to promote the formation of a trimolecular complex with the active catalytic domain of MT1-MMP. All TIMP forms were able to form a bimolecular complex with proMMP-2 (results not shown; [21a]), and only TIMP-4 and TIMP-2-(1–127)–TIMP-4-(129–195) were unable to promote trimolecular complex formation. The ability of Ile^{193}Glu/Gln^{194}Asp-TIMP-4 to promote trimolecular complex formation led us to conclude that the charged tail sequence of TIMP-2, i.e. Glu^{192}-Asp^{193}, contributes significantly to stability.

Structural determinants of TIMP-2 required for efficient activation of proMMP-2 by MT1-MMP

Finally, we assessed the ability of the different forms of TIMP to promote the cell-based activation of proMMP-2 via the MT1-MMP mechanism. This was effected by the use of membrane fractions from fibroblasts derived from TIMP-2$^{-/-}$ mice stably expressing human MT1-MMP [21a]. These cells do not make detectable levels of TIMP-2 or TIMP-4. Analysis of proMMP-2 processing by zymography showed that the enzyme was processed effectively to the fully active form by TIMP-2 (Figure 5). It was converted only into the intermediate

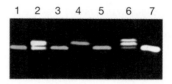

Figure 5 Activation of proMMP-2 by membranes from TIMP-2$^{-/-}$ cells expressing MTI-MMP: effects of TIMP-2 and TIMP-4 constructs.
Recombinant proMMP-2 was incubated at 37°C for 16 h with membranes prepared for TIMP-2$^{-/-}$ mouse fibroblasts transfected with MTI-MMP in either the presence or the absence of wild-type and mutant TIMPs. Lane I, recombinant proMMP-2 buffer control; lane 2, TIMP-2; lane 3, TIMP-4; lane 4, TIMP-4-(1–128)–TIMP-2-(128–194); lane 5, TIMP-2-(1–128)–TIMP-4-(129–195); lane 6, Glu^{192}Ile/Asp^{193}Gln-TIMP-2; lane 7, Δ186–194-TIMP-2. For experimental details, see [21a].

form, i.e. MT1-MMP-mediated cleavage of the propeptide at the Asn^{37}– Leu^{38} peptide bond, in the presence of the chimaera TIMP-4-(1–128)–TIMP-2-(128–194) or the mutant Glu^{192}Ile/Asp^{193}Gln-TIMP-2. TIMP-4, the chimaera TIMP-2-(1–127)–TIMP-4-(129–195) and Δ186–194-TIMP-2 did not promote the MT1-MMP-mediated processing of the proMMP-2 peptide at all (Figure 5). This substantiates the concept that the charged amino acid residues Glu^{192} and Asp^{193} within the TIMP-2 tail sequence are major determinants of the specificity of this TIMP in the establishment of the MT1-MMP–TIMP-2 receptor required for proMMP-2 activation at the cell surface. Removal of this motif does not prevent assembly of the MT1-MMP–TIMP-2–proMMP-2 trimolecular complex or intermediate cleavage of the proMMP-2 propeptide, but the final cleavage of the propeptide that is effected by MMP-2 cannot proceed [16,22]. We hypothesize that MMP-2 is only weakly attached to the cell in the absence of Glu^{192} and Asp^{193}, and therefore autocleavage reactions would be less efficient.

Membrane protein ectodomain proteolysis by metalloproteinases

Metalloproteinases have been linked to the proteolysis of a very large number of structurally and functionally diverse membrane proteins, a process known as ectodomain shedding, which is often essentially a solubilization of the whole ectodomain of the protein. Roles in the shedding of epidermal growth factor receptor ligands, cytokines, their receptors and adhesion molecules have been attributed to both MMPs and the ADAM family of metalloproteinases [23]. The regulation of cell surface proteolytic activities is a key issue, and could occur at a number of levels. Induction of shedding by physiological stimuli leading to intracellular signalling cascades is of importance. The interactions of the enzyme and substrate within the plasma membrane, or at focused locations determined by interactions with other extracellular or intracellular proteins, are thought to be key levels of regulation.

MMPs and ADAMs all require activation by proteolysis of the propeptide that resides in the active-site cleft (cysteine switch mechanism [24]).

Regulation of ADAM and ADAMTS activity by TIMPs

We have investigated the potential role of TIMPs in the regulation of ectodomain shedding events mediated by metalloproteinases. About half of the 33 ADAMs cloned to date are predicted to be active metalloproteinases, based on the presence of the HEXXHXXGXXH zinc-binding motif. Of these, ADAM17, also known as tumour necrosis factor α (TNFα)-converting enzyme (TACE), is the most thoroughly characterized member. In addition to processing precursor membrane-bound TNFα to its soluble form, TACE also cleaves other membrane proteins [25]. Some of the ADAMs predicted to be active metalloproteinases have also subsequently been demonstrated to be able to participate in proteolytic activities similar to those of TACE in cell-based systems [26]. The catalytic activities of purified recombinant ADAMs have been studied using α2-macroglobulin and myelin basic protein, as well as various peptides [27,28]. These assays have allowed their susceptibility towards the TIMPs, potential physiological regulators of ADAM proteolytic activity *in vivo*, to be evaluated, as well as the effects of many low-molecular-mass synthetic inhibitors. Of the TIMPs, only TIMP-3 was found to inhibit TACE, ADAM12 and ADAM19 [29–31], while both TIMP-1 and TIMP-3 could inhibit ADAM10. Furthermore, TIMP-3 also inhibited the aggrecanases ADAMTS-4 and ADAMTS-5 as well as ADAMTS-1, which are members of the related family of disintegrin metalloproteinases with thrombospondin domains [32,33]. In contrast, ADAM8 and ADAM9 were not inhibited by any of the TIMPs [34]. In the few cases where TIMP-2 has been found to be an inhibitor of a proteolytic shedding event, it seems likely that an MMP is involved [35]. In cell-based studies of ectodomain shedding, TIMP-3 has frequently been shown to be an effective inhibitor of the processing of TNFα, L-selectin, the interleukin-6 receptor, CD30 and the p55 TNF receptor 1. In many cases this may be due to the inhibition of TACE, but substantial further study is necessary. Since TIMP-3 is associated primarily with the extracellular matrix, it is effectively localized to the pericellular environment of cells and may represent a significant physiological regulator of membrane metalloproteinases, including those involved in ectodomain shedding. This has not been definitively established, however. Studies of the TIMP-3$^{-/-}$ mouse have indicated the importance of TIMP-3 in the regulation of extracellular matrix turnover in the lung [36] and the involuting mammary gland [37].

Overexpression of TIMP-3 *in vivo* significantly inhibited MMP-driven neointima formation by smooth muscle cells in a pig vein graft model [38]. Cell-based studies showed that this was due to apoptosis of smooth muscle cells [39,40]. Indeed, high levels of TIMP-3 promote apoptosis in many cell types, and this effect has been associated with TNF receptor modulation [41]. TIMP-3 may also modulate the levels of other death receptors [42,43].

Table 2 Association rate constants (a) and apparent binding constants (b) for TIMP-3, TACE and MMP-2. TACE-651 is the whole ectodomain of TACE (Δ652–824-TACE), and TACE-473 is the TACE catalytic domain (Δ474–824-TACE).

(a)

	$10^{-5} \times k_{on}$ (M^{-1}·s^{-1})		
	TACE-651	TACE-473	MMP-2
TIMP-3	0.98	9.94	>100
Δ122–188-TIMP-3	0.45	3.65	0.25

(b)

	K_i^{app} (nM)		
	TACE-651	TACE-473	MMP-2
TIMP-3	0.74	0.20	<0.002
Δ122–188-TIMP-3	1.75	0.22	0.25

Structural features of TIMP-3 that modulate its affinity for TACE

We have started to investigate some of the structural features of TIMP-3 that determine its ability to inhibit cell surface proteinases such as the ADAMs. Initially we showed that full-length TIMP-3 and the construct Δ122–188-TIMP-3, comprising the three N-terminal disulphide-bonded loops, had similar k_{on} and K_i^{app} values for interaction with a soluble form of the TACE catalytic domain (Δ474–824-TACE; TACE-473) or the whole ectodomain of TACE (Δ652–824-TACE; TACE-651) [44]. This is a comparison with its interaction with MMP-2, where the loss of the C-terminal three loops of TIMP-3 markedly abrogates binding (Table 2) [45]. It was noted that TIMP-3 associated rather more efficiently with the TACE catalytic domain than with the full ectodomain, suggesting that the disintegrin and cysteine-rich domains either modify the catalytic domain or interfere with TIMP-3 interactions.

By site-directed mutagenesis, we showed that it was possible to specifically enhance TIMP-3 binding to the TACE active-site cleft by modifying Ser^4 to Met, Tyr, Lys or Arg [46], leading to a >3-fold drop in K_i^{app}. Binding to MMP-2 was concomitantly increased approx. 10-fold; hence an element of selectivity can be introduced into TIMP-3 by the modification of a single residue. This suggests that further mutagenesis studies should allow further specificity to be engineered. However, the problem of interference by the non-catalytic domains of TACE needs to be addressed. Since TIMP-3 is largely a matrix-associated protein, the role of the extracellular matrix in its activity and the effects on TACE function need to be considered. We have shown previously that the rate of association between TIMP-3 and MMP-2 was increased by various glycosaminoglycans, including heparan sulphate [47]. These charged molecules could therefore play a significant role in both focusing and potentiating TIMP-3 activity at the cell surface. Further studies of the precise nature of the extracellular matrix binding sites and the structural features of TIMP-3 involved in binding are needed in order to address the significance of TIMP-3 sequestration on the extracellular matrix.

We thank our many colleagues for their input into the work described, and Paul Soloway for TIMP-2$^{-/-}$ cells. This work was supported by the Wellcome Trust, the Arthritis Research Campaign, the British Heart Foundation, MRC, BBSRC and the Nuffield Foundation.

References

1. Werb, Z. (1997) Cell **91**, 439–442
2. Egeblad, M. and Werb, Z. (2002) Nat. Rev. Cancer **2**, 161–174
3. Seals, D.F. and Courtneidge, S. (2003) Genes Dev. **17**, 7–30
4. Gomis-Rüth, F-.X. (2003) Mol. Biotechnol., in the press
5. Brew, K., Dinakarpandian, D. and Nagase, H. (2000) Biochim. Biophys. Acta **1477**, 267–283
6. Baker, A.H., Edwards, D.R. and Murphy, G. (2002) J. Cell Sci. **115**, 3719–3727
7. Murphy, G., Knäuper, V., Atkinson, S., Gavrilovic, J. and Edwards, D. (2000) Fibrinolysis Proteolysis **14**, 165–174
8. Knäuper, V. and Murphy, G. (1998) in Matrix Metalloproteinases (Parks, W.C. and Mecham, R.P., eds), pp. 199–218, Academic Press, San Diego
9. Will, H., Atkinson, S.J., Butler, G.S., Smith, B. and Murphy, G. (1996) J. Biol. Chem. **271**, 17119–17123
10. Wang, Z., Juttermann, R. and Soloway, P.D. (2000) J. Biol. Chem. **275**, 26411–26415
11. Caterina, J.J., Yamada, S., Caterina, N.C., Longenecker, G., Holmback, K., Shi, J., Yermovsky, A.E., Engler, J.A. and Birkedal-Hansen, H. (2000) J. Biol. Chem. **275**, 26416–26422
12. Willenbrock, F. and Murphy, G. (1994) Am. J. Respir. Crit. Care Med. **150**, S165–S170
13. Hutton, M., Willenbrock, F., Brocklehurst, K. and Murphy, G. (1998) Biochemistry **37**, 10094–10098
14. Morgunova, E., Tuuttila, A., Bergmann, U. and Tryggvason, K. (2002) Proc. Natl. Acad. Sci. U.S.A. **99**, 7414–7419
15. Willenbrock, F., Crabbe, T., Slocombe, P.M., Sutton, C.W., Docherty, A.J.P., Cockett, M.I., O'Shea, M., Brocklehurst, K., Phillips, I.R. and Murphy, G. (1993) Biochemistry **32**, 4330–4337
16. Butler, G.S., Butler, M.J., Atkinson, S.J., Will, H., Tamura, T., Van Westrum, S.S., Crabbe, T., Clements, J., D'Ortho, M.-P. and Murphy, G. (1998) J. Biol. Chem. **273**, 871–880
17. Gomis-Rüth, F.-X., Maskos, K., Betz, M., Bergner, A., Huber, R., Suzuki, K., Yoshida, N., Nagase, H., Brew, K., Bourenkov, G.P. et al. (1997) Nature (London) **389**, 77–79
18. Fernandez-Catalan, C., Bode, W., Huber, R., Turk, D., Calvete, J.J., Lichte, A., Tschesche, H. and Maskos, K. (1998) EMBO J. **17**, 5238–5248
19. Williamson, R.A., Hutton, M., Vogt, G., Rapti, M., Knäuper, V., Carr, M.D. and Murphy, G. (2001) J. Biol. Chem. **276**, 32966–32970
20. Bigg, H.F., Shi, Y.E., Liu, Y.E., Steffensen, B. and Overall, C.M. (1997) J. Biol. Chem. **272**, 15496–15500
21. Hernandez-Barrantes, S., Shimura, Y., Soloway, P.D., Sang, Q.X.A. and Fridman, R. (2001) Biochem. Biophys. Res. Commun. **281**, 126–130
21a. Worley, J.R., Thompkins, P.B., Lee, M.H., Hutton, M., Soloway, P., Edwards, D.R., Gillian Murphy G. and Knäuper, V. (2003) Biochem. J. **372**, 799–809

22. Atkinson, S.J., Crabbe, T., Cowell, S., Ward, R.V., Butler, M.J., Sato, H., Seiki, M., Reynolds, J.J. and Murphy, G. (1995) J. Biol. Chem. **270**, 30479–30485

23. Blobel, C.P. (2000) Curr. Opin. Cell Biol. **12**, 606–612

24. Van Wart, H.E. and Birkedal-Hansen, H. (1990) Proc. Natl. Acad. Sci. U.S.A. **87**, 5578–5582

25. Black, R.A. (2002) Int. J. Biochem. Cell Biol. **34**, 1–5

26. Schlondorff, J. and Blobel, C.P. (1999) J. Cell Sci. **112**, 3603–3617

27. Amour, A., Knight, C.G., Webster, A., Slocombe, P.M., Stephens, P.E., Knäuper, V., Docherty, A.J.P. and Murphy, G. (2000) FEBS Lett. **473**, 275–279

28. Roghani, M., Becherer, J.D., Moss, M.L., Atherton, R.E., Erdjument-Bromage, H., Arribas, J., Blackburn, R.K., Weskamp, G., Tempst, P. and Blobel, C.P. (1999) J. Biol. Chem **274**, 3531–3540

29. Amour, A., Slocombe, P.M., Webster, A., Butler, M., Knight, C.G., Smith, B.J., Stephens, P.E., Shelley, C., Hutton, M., Knäuper, V. et al. (1998) FEBS Lett. **435**, 39–44

30. Loechel, F., Fox, J.W., Murphy, G., Albrechtsen, R. and Wewer, U.M. (2000) Biochem. Biophys. Res. Commun. **278**, 511–515

31. Kang, Q., Cao, Y. and Zolkiewska, A. (2000) Biochem. J. **352**, 883–892

32. Kashiwagi, M., Tortorella, M., Nagase, H. and Brew, K. (2001) J. Biol. Chem. **276**, 12501–12504

33. Rodriguez-Manzaneque, J.C., Westling, J., Thai, S.N., Luque, A., Knäuper,V., Murphy,G., Sandy, J.D. and Iruela-Arispe, M.L. (2002). Biochem. Biophys. Res. Commun. **293**, 501–508

34. Amour, A., Knight, C.G., English, W.R., Webster, A., Slocombe, P.M., Knäuper, V., Docherty, A.J., Becherer, J.D., Blobel, C.P. and Murphy, G. (2002) FEBS Lett. **524**, 154–158

35. Schlondorff, J., Lum, L. and Blobel, C.P. (2001) J. Biol. Chem. **276**, 14665–14674

36. Leco, K.J., Waterhouse, P., Sanchez, O.H., Gowing, K.L., Poole, A.R., Wakeham, A., Mak, T.W. and Khokha, R. (2001) J. Clin. Invest. **108**, 817–829

37. Fata, J.E., Leco, K.J., Voura, E.B., Yu, H.Y., Waterhouse, P., Murphy, G., Moorehead, R.A. and Khokha, R. (2001) J. Clin. Invest. **108**, 831–841

38. George, S.J., Lloyd, C.T., Angelini, G.D., Newby, A.C. and Baker, A.H. (2000) Circulation **101**, 296–304

39. Ahonen, M., Baker, A.H. and Kähäri, V.M. (1998) Cancer Res. **58**, 2310–2315

40. Baker, A.H., Zaltsman, A.B., George, S.J. and Newby, A.C. (1998) J. Clin. Invest. **101**, 1478–1487

41. Smith, M.R., Kung, H.F., Durum, S.K., Colburn, N.H. and Sun, Y. (1997) Cytokine **9**, 770–780

42. Bond, M., Murphy, G., Bennett, M.R., Amour, A., Knäuper, V., Newby, A.C. and Baker, A.H. (2000) J. Biol. Chem. **275**, 41358–41363

43. Bond, M., Murphy, G., Bennett, M.R., Newby, A.C. and Baker, A.H. (2002) J. Biol. Chem. **277**, 13787–13795

44. Milla, M.E., Leesnitzer, M.A., Moss, M.L., Clay, W.C., Carter, H.L., Miller, A.B., Su, J.L., Lambert, M.H., Willard, D.H., Sheeley, D.M. et al. (1999) J. Biol. Chem. **274**, 30563–30570

45. Lee, M.H., Knäuper, V., Becherer, J.D. and Murphy, G. (2001) Biochem. Biophys. Res. Commun. **280**, 945–950

46. Lee, M.H., Verma, V., Maskos, K., Nath, D., Knäuper, V., Dodds, P., Amour, A. and Murphy, G. (2002) Biochem. J. **364**, 227–234

47. Butler, G.S., Apte, S.S., Willenbrock, F. and Murphy, G. (1999) J. Biol. Chem. **274**, 10846–10851

Biochem. Soc. Symp. **70**, 81–94
(Printed in Great Britain)
© 2003 Biochemical Society

7

Collagen–platelet interactions: recognition and signalling

Richard W. Farndale[1], Pia R.-M. Siljander, David J. Onley,

Pavithra Sundaresan, C. Graham Knight and

Michael J. Barnes[2]

Department of Biochemistry, University of Cambridge, Downing Site, Cambridge CB2 1QW, U.K.

Abstract

The collagen–platelet interaction is central to haemostasis and may be a critical determinant of arterial thrombosis, where subendothelium is exposed after rupture of atherosclerotic plaque. Recent research has capitalized on the cloning of an important signalling receptor for collagen, glycoprotein VI, which is expressed only on platelets, and on the use of collagen-mimetic peptides as specific tools for both glycoprotein VI and integrin α2β1. We have identified sequences, GPO and GFOGER (where O denotes hydroxyproline), within collagen that are recognized by the collagen receptors glycoprotein VI and integrin α2β1 respectively, allowing their signalling properties and specific functional roles to be examined. Triple-helical peptides containing these sequences were used to show the signalling potential of integrin α2β1, and to confirm its important contribution to platelet adhesion. Glycoprotein VI appears to operate functionally on the platelet surface as a dimer, which recognizes GPO motifs that are separated by four triplets of collagen sequence. These advances will allow the relationship between the structure of collagen and its haemostatic activity to be established.

Introduction

The collagen gene product requires proteolytic cleavage to remove N- and C-terminal propeptides as the monomeric triple helix is secreted from the cell, allowing its assembly into highly ordered fibrils which constitute the collagen fibre, the primary structural component of the vertebrate organism. This chapter, however, addresses the function of collagen not as a protease substrate

[1]To whom correspondence should be addressed (e-mail rwf10@cam.ac.uk).
[2]Deceased (4 May 2003).

nor as an inert structural protein, but rather as an important modulator of cell function which acts through specific adhesive and signalling receptors expressed on the cell surface. A vivid example of the capacity of collagen to modulate cell function is provided by the human platelet, which must respond over a timescale of a few tens of seconds to prevent loss of blood should there be life-threatening damage to the vasculature. The platelet's haemostatic response is initiated by the dual adhesive and activatory properties of vessel wall collagens, notably the fibrillar collagens I and III, along with other subendothelial collagens such as collagens IV and VI; the circulating platelet scans the vascular endothelium for damage that exposes such collagens.

Significant progress has been made over the past 10 years in understanding the molecular detail of this complex process, including the precise sequences within collagen that are recognized by collagen receptors, the molecular architecture of the receptors themselves and the signals that arise from the interaction between the two (reviewed in [1]). This increasing body of knowledge underpins the development of tools with which to disrupt platelet recruitment and the deposition of platelet thrombi on subendothelial collagens at the site of rupture or fissure of an atherosclerotic plaque. An inappropriate haemostatic response of platelets in this disease setting is a primary cause of sudden heart attack and unstable angina, which gives great impetus to the studies described below.

Collagen structure and primary sequence

The triple-helical structure of collagen has been understood for many years and will not be described in great depth here. Polyproline helix formation in the nascent collagen chains is favoured by the presence of proline residues, and the propensity of collagen to form triple helices is enhanced by post-translational hydroxylation of proline in the X' position of the model GXX' triplet, GPP (Gly-Pro-Pro), taking place in the endoplasmic reticulum. The resulting GPO triplets (where O is hydroxyproline) form approx. 10% of the primary sequence of collagen, although both P and O occur frequently in other X and X' positions respectively and contribute to the secondary structure and stability of the triple helix. Diversity of sequence within this conserved GXX' framework has proved to be both important and specific for the recognition of cell-surface collagen receptors, not least because native triple-helical structure is required for the recognition of collagens by platelets and other cells.

Collagen receptors

Knowledge about collagen receptors is increasing rapidly, especially in relation to the collagen-binding integrins. A subset of those integrins that contain an inserted domain, or I-domain, within their α subunit is now considered to be the only class that interacts directly with the collagens [2,3]. Four $\beta1$ integrins, i.e. $\alpha1\beta1$, $\alpha2\beta1$, $\alpha10\beta1$ and $\alpha11\beta1$, fall into this category, and, although these are expressed in many human tissues, platelets contain only $\alpha2\beta1$. The collagen sequences described below appear to bind to each of these integrins

[4,5], and although there may be selectivity of a given collagen sequence for specific integrin I-domains, as has been established for α1β1 [4,6,7], final understanding of their interaction has yet to be achieved.

The platelet also expresses several other types of collagen receptor, the best characterized being the immunoglobulin-superfamily member glycoprotein VI (GpVI), a key platelet signalling receptor [8]. Work from this laboratory has identified sequences within collagen that bind integrin α2β1 [4,9] or activate GpVI [10,11]; however, apart from these two, the molecular understanding of collagen receptors is at a rudimentary level. GpV has been shown to bind collagen [12], the scavenger receptor CD36 is considered to have an ancillary role in platelet activation by collagen [13], and receptors specific for either collagen I or collagen III have also been identified [14,15]. A specific sequence in collagen III, KOGEOGPK, has been proposed as a recognition motif [14,16]. The linear peptide has been used as a receptor antagonist and affinity ligand, yet, in our hands, this sequence in triple-helical form does not appear to be active. The rest of this chapter will be devoted to collagen-binding integrins and GpVI.

Multi-site model for the platelet–collagen interaction

The ability of collagens to stimulate platelets was discovered over 40 years ago [17], and was shown during the subsequent decades to be an active process that induces cytoskeletal activity, secretion of platelet granule contents and up-regulation of the fibrinogen receptor integrin αIIbβ3, leading to platelet aggregation. A line of evidence originating in platelet deficiencies and autoimmunities in Japanese patients [18,19] revealed the involvement of GpVI in platelet activation by collagen [20,21]. Overall, the interaction must be sufficient to allow collagens of the vessel wall to capture platelets from the circulation under the high shear stress of the arterial circulation, and experiments using platelets from a patient lacking integrin α2β1 [22,23], along with the development of blocking antibodies [24], indicated a role for this integrin in haemostasis and especially in adhesion to collagen under both static [25] and flow [26] conditions. Thus platelet collagen receptors combine to express both adhesive and activatory functions. During the 1980s, chemical modification of collagens [27] and fragmentation using cyanogen bromide [28] revealed that these two properties need not reside at the same locus within collagens. Thus the two-site, two-step model was born, which in its simplest form stated that the platelet–vessel-wall interaction occurs in two stages: an adhesive step, quite quickly shown to require integrin α2β1, and an activatory step, currently identified with GpVI. What is not yet clear is the extent to which these discrete functions may overlap or regulate one another, and whether the other minor collagen receptors alluded to above contribute to thrombus formation under physiological conditions. It should be emphasized that, as a pre-requisite for the capture of platelets under shear stress, indirect interaction between collagen and GpIb is mediated by von Willebrand factor, which allows the platelet to roll across the subendothelium, where it can be arrested by higher-affinity interactions mediated through integrin α2β1. These several interactions are summarized schematically in Figure 1.

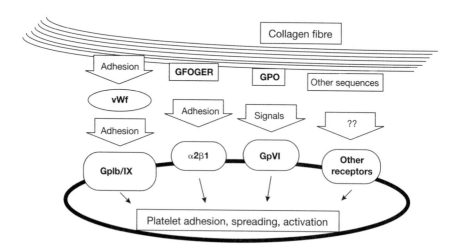

Figure 1 The interaction of collagen with platelets. The known sequences in collagen, i.e. GFOGER and GPO motifs, interact with integrin α2β1 and GpVI respectively, whose major functions are adhesion and signalling. Indirect adhesion is mediated by von Willebrand factor (vWf) and platelet GpIb. Little is known of the sequences in collagen that recognize von Willebrand factor or other, less well characterized, receptors.

GpVI and integrin α2β1 are expressed in similar numbers on the platelet surface, with about 1000–3000 copies of each. Expression of each receptor correlates with polymorphisms: the silent C807T for the integrin α2 gene, and the five linked dimorphisms of GpVI that occur as two common alleles, designated a and b. The significance of the latter will be discussed below. A link between high expression levels of α2β1 [29] or of the b/b genotype of GpVI [30] and thrombosis has been reported for certain populations. On the other hand, deficiencies in either receptor cause haemostatic dysfunction [19,22]. Recently, studies using either GpVI-deficient mouse platelets (achieved either by depletion using an anti-GpVI antibody, JAQ1, or by knockout of the Fc receptor γ-chain with which GpVI is co-expressed) or α2β1-deficient mouse platelets (α2 or β1 knockouts) have re-examined the relative importance of these collagen receptors. Some studies emphasize the role of GpVI in primary adhesion [31–33], whereas others support an adhesive role of α2β1 [34]. Such methods are lacking for the study of human platelets. However, a recent study in our laboratory using a single-chain antibody to block human GpVI, along with specific blockade of α2β1 using the peptide described below, has indicated that a complex interplay between collagen and these receptors, along with GpIb, is required for platelets to be captured by collagen-coated surfaces under flow (P.R.-M. Siljander, I.C.A. Munnix, P.A. Smethurst, H. Deckmyn, T. Lindhout, W.H. Ouwehand, R.W. Farndale and J.W.M. Heemskerk, unpublished work). It seems that in mice there may be a stronger dependence on GpVI for the adhesive as well as the activatory steps than is the case in humans, causing us to question the use of mouse models as the sole basis for investigating the role of collagen receptors in thrombosis.

Recognition of integrin α2β1

The non-uniform distribution of integrin α2β1 recognition sites among triple-helical peptides of collagens I and III produced by cyanogen bromide fragmentation (CB peptides) [35–37] supported a mapping approach using synthetic peptides. These comprised overlapping stretches of the collagen (guest) sequence under test flanked by GPP or GPO (host) polymers [9]. Amino acid substitution allowed the precise detail of each recognition motif to be established. Finally, the sequence GFOGER was identified as the ligand for α2β1 within the CB3 fragment of the collagen I α1 chain [4]. Subsequent work showed weaker recognition of the integrin by the sequences GLOGER and GASGER. The latter in particular appears to be a low-affinity motif, and together these sequences may be sufficient to account for the capacity of recombinant I-domains to bind collagen I, as determined using rotary shadowing as a probe [38].

Co-crystallization of a short GFOGER-containing collagen peptide with the α2 I-domain revealed the molecular detail of the interaction [39], which occurs primarily via the middle strand of the staggered triple helix. Key features are that the glutamate side chain is fully extended to allow co-ordination of its carboxylate group to the bivalent cation, such as Mg^{2+}, in the MIDAS (metal ion-dependent adhesion site), such that the shorter aspartate side chain would be unlikely to serve this function, as confirmed by experimental findings. Other interactions include hydrophobic binding between phenylalanine and a pocket on the I-domain surface between Gln^{215} and Asn^{154}, and a salt bridge between arginine and a negatively charged locus centering on Asp^{219}. Binding is stabilized further by both hydrophobic and electrostatic interaction between the trailing strand and the I-domain, as well as by hydrogen bonds formed between main-chain carbonyls in both peptide strands and the I-domain.

The relatively non-specific nature of the bonds between the two structures outside the MIDAS suggested that GEK sequences might be active. A model peptide did indeed bind the α2 I-domain, but we have not pursued this finding. Other hydrophobic residues might substitute for phenylalanine, as shown already in the sequence GLOGER. Further, alignment of the sequences of the fibrillar collagen α-chains revealed some interesting similarities. Two novel sequences, GAOGER and GLSGER, occupied the same position in collagen III as GLOGER and GASGER in the collagen I α1 chain. The identity between the collagen fragments α1(III)CB4 and α1(I)CB3, both of which recognize integrins, had already been noted [35,40], as had the alignment of GAOGER (collagen III) with GFOGER (collagen I). The cyanogen bromide methodology, however, fails to address another possible integrin-binding sequence, GMOGER, since it would be cleaved at methionine. This was addressed by synthesizing peptides that confirmed the integrin-binding potential of both GMOGER and GLSGER. Again, alignment of sequences shows that GMOGER occurs in the same position relative to GFOGER and GAOGER in collagens I and III respectively, as summarized in Figure 2. Thus it appears that the spatial relationship between integrin sequences within collagen α-chains is conserved, underscoring their importance and suggesting that the organization

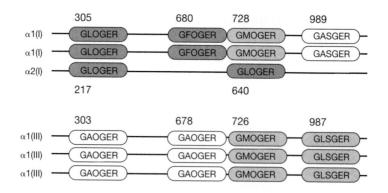

Figure 2 GER sequences from collagens I and III that recognize integrin α2β1. Numbers show the position of the initial Gly residue of each sequence within the corresponding procollagen gene product, including the signal sequence. The capacity of these motifs to bind integrin and platelets was verified as described previously [4], and darker grey shading reflects strength of binding.

of collagen might have an essential impact on integrin function, perhaps in the way that collagen might signal by clustering integrins.

GpVI recognition

Our interest in GpVI recognition arose from the use of synthetic peptides in the search for other receptor-binding motifs. In 1995, control peptides comprising the triple-helical GPO hosts were found to be potent platelet activators without binding to integrin α2β1 [10]. In contrast, GpVI-deficient platelets did not respond to these GPO polymers [41], and a series of studies ensued in which the use of this or other similar peptides, designated collagen-related peptide (CRP), allowed the signalling pathways downstream from GpVI to be elucidated ([42–44]; reviewed in [45,46]). GpVI itself was cloned and sequenced in 1999 [8], paving the way for the use of recombinant protein in a number of contexts, including the development of antibodies against human GpVI.

CRP contains ten GPO triplets and forms a very stable triple helix. The length of the rope-like CRP molecule far exceeds the dimensions of the ectodomain of GpVI, which immediately suggests that the recognition motif for GpVI must correspond to a shorter stretch of collagen sequence. Analogy with integrin α2β1 recognition suggests that two triplets might be ample to span the collagen-binding site of GpVI, an idea supported by modelling based on the homologous natural killer receptor, NKR1. GPO triplets are common within the collagenous triple helix, but seldom occur in tandem. However, peptides comprising repeated GPP triplets are very poor platelet ligands at best [11], showing that hydroxyproline is critical for the CRP–GpVI interaction. At present, there is little direct evidence for the involvement of any collagen triplet other than GPO, an important topic that is discussed further below.

To investigate the minimum GpVI recognition motif in collagen, a set of model peptides (shown in Table 1) was used in platelet adhesion, aggregation and signalling assays. In this set, increasing numbers of GPO triplets were hosted within inert GPP polymers, so that the overall peptide length was conserved at ten triplets. In addition, a Cys-containing triplet was introduced at each end to allow cross-linking (see below). The parent GPP host was silent, supporting no discernible platelet adhesion or activation. Platelet adhesion commenced with the inclusion of a single GPO triplet, rising progressively with up to four GPOs. Only a modest further increase in platelet adhesion occurred with CRP (ten GPO triplets). Model peptides with two separated GPO triplets or GPOGPO motifs showed progressive increases in reactivity compared with peptides containing a single GPO. These unpublished data are summarized in Figure 3.

Given that sub-optimal platelet adhesion can occur to peptides that are sufficiently long to fully occupy the putative collagen-binding surface of the receptor, specifically peptide H, it follows that additional binding must result from the co-operative interaction of more than one copy of GpVI. Thus we deduce that GpVI on the platelet surface operates functionally as a dimer, presumably linked by the disulphide-linked dimeric Fc receptor γ-chain with which it interacts. This concept is supported by the increased adhesion of platelets to peptides with separated GPO triplets. Recently, the use of ectodomains of GpVI expressed as Fc fusion proteins led to a similar proposal [47].

The gain in binding that occurs as the GPO nucleus increases from one to two triplets suggests that the footprint of GpVI upon collagen ideally comprises two adjacent GPO triplets, although the presence of GPP, being of similar structure, adjacent to GPO does not prevent a weaker interaction occurring with a single GPO residue. Consideration of the disposition of hydroxyproline residues between adjacent strands of the triple helix gives further insights into the possible structure of the GpVI-binding surface of collagen. One model that is consistent with the data presented above is that a single GPO triplet within the primary sequence of the peptide will result in GPO triplets in adjacent strands of the triple helix being adjacent to each other. Such a structure will bind GpVI, albeit weakly. Placing a second GPO triplet adjacent to the first in the primary sequence of the peptide will complete a triangle of GPO triplets, which may constitute the optimum GpVI recognition motif, spanning two chains of the triple helix. This concept is shown in Figure 4.

Signalling through GpVI

The same peptide set allowed us to investigate the mechanisms of platelet signalling through GpVI, using aggregation and protein tyrosine phosphorylation as indices of activation. The first and most important conclusion of this work is that triple-helical peptides, even $(GPO)_6$, do not support signalling unless they are cross-linked, with the exception of monomeric CRP, which induces modest tyrosine phosphorylation. The capacity of cross-linked GPO motifs to elicit these responses increased with the length of the GPO motif, in line with the capacity of the peptides to support platelet adhesion (see Figure 3b). These data lead us to the

Table 1 Peptides used to examine GpVI recognition in platelets. Peptides were synthesized that had a conserved length (37 residues) and variable numbers of GPO triplets as indicated. They can be used in monomeric, triple-helical form to coat the surface of 96-well plates for platelet adhesion studies, and when cross-linked they can also be used to elicit signals from platelet suspensions (see Figure 3).

Peptide	Sequence	Name
A	GCP-(GPP)$_{10}$-GCPG	GPP10
B	GCP-(GPP)$_4$ **GPO** (GPP)$_5$-GCPG	
C	GCP-(GPP)$_4$ **GPO GPO** (GPP)$_4$-GCPG	
D	GCP-(GPP)$_3$ **GPO GPO GPO GPO** (GPP)$_3$-GCPG	
E	GCP-(GPP)$_2$ **GPO GPO GPO GPO GPO GPO** (GPP)$_2$-GCPG	
F	GCO **GPO GPO GPO GPO GPO GPO GPO GPO GPO GPO** GCOG	CRP
G	GCP-(GPP)$_2$ **GPO** GPP GPP GPP GPP **GPO** (GPP)$_2$-GCPG	
H	GCP-GPP **GPO GPO** GPP GPP GPP GPP **GPO GPO** GPP-GCPG	

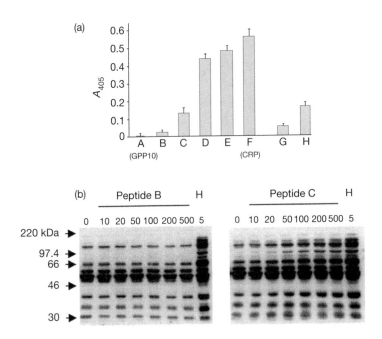

Figure 3 Adhesion and protein tyrosine phosphorylation responses of platelets to peptides with increasing numbers of GPO triplets.
Peptides are defined in Table 1. A–F have 0, 1, 2, 4, 6 and 10 GPO triplets respectively; G and H have 2 and 4 triplets respectively in two separate GPO motifs. (a) Experiments were performed as described in [52] in the presence of a fibrinogen receptor antagonist to prevent activation-dependent adhesion. Adhesion was GpVI-dependent, as demonstrated using blocking antibodies (not shown). (b) Dose of peptide is shown in μg/ml. Peptide B, containing a single GPO, is silent, despite being cross-linked, whereas peptide C, having two GPO triplets, shows modest activity compared with the high levels of tyrosine phosphorylation induced by peptide H (CRP-XL; 5 μg/ml). Experiments were performed by treating platelets in suspension with peptides for 3 min at 37°C, and using monoclonal antibody 4G10 to detect tyrosine phosphorylation in Western blots as described in [53].

conclusion that occupancy of GpVI alone is not a sufficient stimulus to support platelet activation, but that clustering of receptors is necessary, which can be induced by the application of cross-linked peptides. Thus the signals induced by monomeric CRP may in fact reflect the capacity of this quite long structure to accommodate more than one GpVI dimer, allowing it to promote bridging between multiple copies of GpVI and so fulfil the need for receptor clustering. The requirement among the model peptides for this higher-order structure also applies to the parent collagens themselves: monomeric collagens produced by pepsin digestion will support adhesion of platelets and other cells, but not platelet aggregation. Native collagen fibres, however, expressing multiple copies of the GpVI recognition motif upon their surface along with the integrin motifs that firmly anchor the platelet, are potent platelet agonists.

(a)

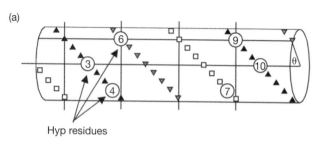

Hyp residues

(b)

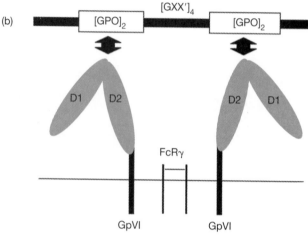

Figure 4 Schematic representation of the interface between triple-helical peptide H and GpVI. (a) Strands are staggered by a single residue, and are shown as a series of squares (leading strand), black triangles (middle strand) and grey inverted triangles (trailing strand). Numbered circles show the positions of hydroxyproline (Hyp; O) residues upon the surface of the cylinder containing the model triple helix. Hydroxyproline residues are numbered from the leading strand; thus 3 and 4 represent the two O residues in the first GPOGPO motif of the middle strand; 6 is the second O residue in the first GPOGPO motif of the trailing strand; 7 is the first O residue in the second GPOGPO motif of the leading strand; and 9 and 10 are the two O residues in the second GPOGPO motif of the middle strand. (b) Proposed way in which the peptide binds to a GpVI dimer, which is stabilized in the plane of the platelet membrane by interaction with the disulphide-bridged dimeric Fc receptor γ-chain (FcRγ). D1 and D2 represent the N- and C-terminal Ig folds respectively of GpVI.

The detail of the signalling pathways lying downstream from GpVI is being worked out at present. CRP has proved a valuable and specific tool in this context, and recent reviews provide ample detail on this topic [1,46,48]. It is sufficient here to say that Src-family kinases that associate with the cytoplasmic tails of GpVI are considered to have a role in the phosphorylation of the Fc receptor γ-chain ITAM (immunoreceptor tyrosine-based activation motif), and that this is a key event in the recruitment of the intracellular tyrosine kinase p72syk.

Downstream lie a series of adaptor proteins [LAT (linker for activation of T cells), SLP76 (SH2-domain-containing leucocyte protein of 76 kDa) and several others] which, with phosphoinositide 3-kinase, lead to the activation of phospholipase Cγ2 and the hydrolysis of inositol lipids, so generating a Ca^{2+} signal.

Recent interest surrounds the five dimorphisms of GpVI, which are grouped as two major alleles, designated a and b. One dimorphism leads to the expression of Pro rather than Ser in the pericellular stem of GpVI, and forms part of the b allele, which has been linked with increased risk of myocardial infarction [30]. Surprisingly, recent data show that b/b platelets are in fact less responsive to collagen and to CRP than the a/a genotype, with the dose–response curve shifted approx. 4-fold to the right, whereas their response to non-collagenous agonists was not impaired [49]. The underlying cause of this effect remains to be determined.

The question of whether integrin α2β1 also signals in the setting of the platelet–collagen interaction is both interesting and topical. A wealth of literature concerning other cells supports the concept of bi-directional signalling through integrins, leading to changes both in the affinity of the receptor and in cell function [50]. Mainly, the outside-in component of integrin signalling is a rather slow process, providing input that is, for example, essential for cell growth or survival. It is thus not obvious that integrin α2β1 has a signalling role in platelets, where rather rapid responses are required.

Recent unpublished data from our group have shown that integrin α2β1 may re-inforce signals that arise through platelet GpVI under arterial shear conditions, observed at the level of Ca^{2+} signalling and the expression of a procoagulant surface (P.R.-M. Siljander, I.C.A. Munnix, P.A. Smethurst, H. Deckmyn, T. Lindhout, W.H. Ouwehand, R.W. Farndale and J.W.M. Heemskerk, unpublished work). These responses were attenuated, but not abolished, by the use of GFOGER-containing peptides or α2β1-specific antibodies to block the integrin in platelets deposited on collagen fibres. The precise relationship between GpVI and integrin α2β1 remains to be defined.

Signals arising directly from integrin occupancy by collagen are much harder to observe, however. The application of the integrin α2β1-specific peptide GFOGER to platelets in suspension causes no obvious general increase in protein tyrosine phosphorylation. However, we have found that the p38 mitogen-activated protein kinase (MAP kinase) becomes phosphorylated (and presumably activated) after treatment of platelets with peptides containing integrin α2β1 recognition motifs. Collagen monomers, fibres and cross-linked GFOGER (the latter unpublished data are shown in Figure 5) were each able to promote p38 phosphorylation, an effect that peaked at 2 min and then declined quickly to basal levels. Moreover, the GpVI-specific CRP caused p38 phosphorylation without inducing subsequent loss of activity, suggesting that the dephosphorylation event is mediated by engagement of α2β1 specifically. The decline in phosphorylation levels correlates with the association of protein phosphatase 2A with the p38 MAP kinase [51], and can be blocked by integrin-specific antibodies (Figure 5) as well as by cell-permeant peptides whose sequence corresponds to the cytoplasmic tail of the α2 integrin subunit. Despite these findings that convergent signals (the activation of p38) may arise through either of the two collagen receptors, GpVI

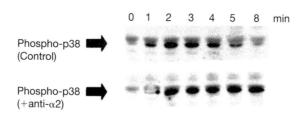

Figure 5 Activation of p38 MAP kinase by integrin ligation. Platelet suspensions were treated with cross-linked triple-helical GFOGER for the indicated times, with and without preincubation with the anti-α2 monoclonal antibody, 6F1. Phosphorylation levels, measured as the phospho-form of the enzyme in Western blots [51], peaked at 2 min then declined in controls, whereas integrin blockade prevented the dephosphorylation of p38.

and integrin α2β1, discrimination of their roles can be observed at the level of signal termination by protein phosphatase 2A.

Conclusion

The data discussed above emphasize the importance of the platelet–collagen interaction in thrombosis and haemostasis, and the peptide tools described here may prove useful in identifying specific signalling pathways and thus specific targets for anti-thrombotic therapy.

This work was supported by the Medical Research Council and the British Heart Foundation. We are indebted to Dr W. Ouwehand and Dr P. Smethurst (Department of Haematology, University of Cambridge) for advice, and to L.F. Morton and A.R. Peachey for technical support.

References

1. Siljander, P.R.-M. and Farndale, R.W. (2002) in Platelets in Thrombotic and Non-thrombotic Disorders: Pathophysiology, Pharmacology and Therapeutics (Gresele, P., Page, C.P., Fuster, V. and Vermylen, J., eds), pp. 158–178, Cambridge University Press, Cambridge
2. Heino, J. (2000) Matrix Biol. **19**, 319–323
3. Gullberg, D.E. and Lundgren-Akerlund, E. (2002) Prog. Histochem. Cytochem. **37**, 3–54
4. Knight, C.G., Morton, L.F., Peachey, A.R., Tuckwell, D.S., Farndale, R.W. and Barnes, M.J. (2000) J. Biol. Chem. **275**, 35–40
5. Zhang, W.-M., Käpylä, J., Puranen, J.S., Knight, C.G., Tiger, C.F., Pentikäinen, O.T., Johnson, M.S., Farndale, R.W., Heino, J. and Gullberg, D. (2003) J. Biol. Chem. **278**, 7270–7277
6. Kern, A., Eble, J., Golbik, R. and Kuhn, K. (1993) Eur. J. Biochem. **215**, 151–159
7. Golbik, R., Eble, J.A., Ries, A. and Kuhn, K. (2000) J. Mol. Biol. **297**, 501–509
8. Clemetson, J.M., Polgar, J., Magnenat, E., Wells, T.N.C. and Clemetson, K.J. (1999) J. Biol. Chem. **274**, 29019–29024
9. Knight, C.G., Morton, L.F., Onley, D.J., Peachey, A.R., Messent, A.J., Smethurst, P.A., Tuckwell, D.S., Farndale, R.W. and Barnes, M.J. (1998) J. Biol. Chem. **273**, 33287–33294

10. Morton, L.F., Hargreaves, P.G., Farndale, R.W., Young, R.D. and Barnes, M.J. (1995) Biochem. J. **306**, 337–344

11. Knight, C.G., Morton, L.F., Onley, D.J., Peachey, A.R., Ichinohe, T., Okuma, M., Farndale, R.W. and Barnes, M.J. (1999) Cardiovasc. Res. **41**, 450–457

12. Moog, S., Mangin, P., Lenain, N., Strassel, C., Ravanat, C., Schuhler, S., Freund, M., Santer, M., Kahn, M., Nieswandt, B. et al. (2001) Blood **98**, 1038–1046

13. Nakamura, T., Jamieson, G.A., Okuma, M., Kambayashi, J.-I. and Tandon, N.N. (1998) J. Biol. Chem. **273**, 4338–4344

14. Monnet, E. and Fauvel-Lafeve, F. (2000) J. Biol. Chem. **275**, 10912–10917

15. Chiang, T.M., Rinaldy, A. and Kang, A.H. (1997) J. Clin. Invest. **100**, 514–521

16. Bevers, E.M., Karniguian, A., Legrand, Y.J. and Zwaal, R.F.A. (1985) Thromb. Res. **37**, 365–370

17. Bounameaux, Y. (1959) C.R. Séances Soc. Biol. Fil. **153**, 865–869

18. Sugiyama, T., Okuma, M., Ushikubi, F. and Uchino, H. (1987) Blood **69**, 1712–1720

19. Moroi, M., Jung, S.M., Okuma, M. and Shinmyozu, K. (1989) J. Clin. Invest. **84**, 1440–1446

20. Ishibashi, T., Ichinohe, T., Sugiyama, T., Takayama, H., Titani, K. and Okuma, M. (1995) Int. J. Hematol. **62**, 107–115

21. Ichinohe, T., Takayama, H., Ezumi, Y., Arai, M., Yamamoto, N., Takahashi, H. and Okuma, M. (1997) J. Biol. Chem. **272**, 63–68

22. Nieuwenhuis, H.K., Akkerman, J.W.N., Houdijk, W.P.M. and Sixma, J.J. (1985) Nature (London) **318**, 470–472

23. Nieuwenhuis, H.K., Sakariassen, K., Houdijk, W.P.M., Nievelstein, P.F.E.M. and Sixma, J.J. (1986) Blood **68**, 692–695

24. Coller, B.S., Beer, J.H., Scudder, L.E. and Steinberg, M.H. (1989) Blood **74**, 182–192

25. Santoro, S.A. (1986) Cell **46**, 913–920

26. Saelman, E.U.M., Nieuwenhuis, H.K., Hese, K.M., de Groot, P.G., Heijnen, H.F.G., Sage, E.H., Williams, S., McKeown, L., Gralnick, H.R. and Sixma, J.J. (1994) Blood **83**, 1244–1250

27. Santoro, S.A., Walsh, J.J., Staatz, W.D. and Baranski, K.J. (1991) Cell. Regul. **2**, 905–913

28. Zijenah, L.S., Morton, L.F. and Barnes, M.J. (1990) Biochem. J. **268**, 481–486

29. Santoso, S., Kunicki, T.J., Kroll, H., Haberbosch, W. and Gardemann, A. (1999) Blood **93**, 2449–2453

30. Croft, S.A., Samani, N.J., Teare, M.D., Hampton, K.K., Steeds, R.P., Channer, K.S. and Daly, M.E. (2001) Circulation **104**, 1459–1463

31. Nieswandt, B., Brakebusch, C., Bergmeier, W., Schulte, V., Bouvard, D., Mokhtari-Nejad, R., Lindhout, T., Heemskerk, J.W.M., Zirngibl, H. and Fassler, R. (2001) EMBO J. **20**, 2120–2130

32. Nieswandt, B., Schulte, V., Bergmeier, W., Mokhtari-Nejad, R., Rackebrandt, K., Cazenave, J.-P., Ohlmann, P., Gachet, C. and Zirngibl, H. (2001) J. Exp. Med. **193**, 459–469

33. Holtkotter, O., Nieswandt, B., Smyth, N., Muller, W., Hafner, M., Schulte, V., Krieg, T. and Eckes, B. (2002) J. Biol. Chem. **277**, 10789–10794

34. Chen, J., Diacovo, T.G., Grenache, D.G., Santoro, S.A. and Zutter, M.M. (2002) Am. J. Pathol. **161**, 337–344

35. Zijenah, L.S. and Barnes, M.J. (1990) Thromb. Res. **59**, 553–566

36. Morton, L.F., Zijenah, L.S., McCulloch, I.Y., Knight, C.G., Humphries, M.J. and Barnes, M.J. (1991) Biochem. Soc. Trans. **19**, 439S

37. Saelman, E.U.M., Morton, L.F., Barnes, M.J., Gralnick, H.R., Hese, K.M., Nieuwenhuis, H.K., de Groot, P.G. and Sixma, J.J. (1993) Blood **82**, 3029–3033

38. Xu, Y., Gurusiddappa, S., Rich, R.L., Owens, R.T., Keene, D.R., Mayne, R., Hook, A. and Hook, M. (2000) J. Biol. Chem. **275**, 38981–38989

39. Emsley, J., Knight, C.G., Farndale, R.W., Barnes, M.J. and Liddington, R.C. (2000) Cell **101**, 47–56

40. Morton, L.F., Peachey, A.R. and Barnes, M.J. (1989) Biochem. J. **258**, 157–163

41. Kehrel, B., Wierwille, S., Clemetson, K.J., Anders, O., Steiner, M., Knight, C.G., Farndale, R.W., Okuma, M. and Barnes, M.J. (1998) Blood **91**, 491–499

42. Gibbins, J., Asselin, J., Farndale, R.W., Barnes, M.J., Law, C.L. and Watson, S.P. (1996) J. Biol. Chem. **271**, 18095–18099

43. Gibbins, J.M., Okuma, M., Farndale, R.W., Barnes, M.J. and Watson, S.P. (1997) FEBS Lett. **413**, 255–259

44. Asselin, J., Gibbins, J.M., Achison, M., Lee, Y.H., Morton, L.F., Farndale, R.W., Barnes, M.J. and Watson, S.P. (1997) Blood **89**, 1235–1242

45. Watson, S.P. and Gibbins, J.M. (1998) Immunol. Today **19**, 260–264

46. Watson, S.P. (1999) Thromb. Haemostasis **82**, 365–376

47. Miura, Y., Takahashi, T., Jung, S.M. and Moroi, M. (2002) J. Biol. Chem. **277**, 46197–46204

48. Watson, S., Berlanga, O., Best, D. and Frampton, J. (2000) Platelets **11**, 252–258

49. Joutsi-Korhonen, L., Smethurst, P.A., Rankin, A., Gray, E., Ijsseldijk, M., Onley, C.M., Watkins, N.A., Williamson, L.M., Goodall, A.H., de Groot, P.G. et al. (2003) Blood, **101**, 4372–4379

50. Hynes, R.O. (2002) Cell **110**, 673–687

51. Sundaresan, P. and Farndale, R.W. (2002) FEBS Lett. **528**, 139–144

52. Onley, D.J., Knight, C.G., Tuckwell, D.S., Barnes, M.J. and Farndale, R.W. (2000) J. Biol. Chem. **275**, 24560–24566

53. Achison, M., Elton, C.M., Hargreaves, P.G., Morton, L.F., Knight, C.G., Barnes, M.J. and Farndale, R.W. (2001) J. Biol. Chem. **276**, 3167–3174

Biochem. Soc. Symp. **70**, 95–106
(Printed in Great Britain)
© 2003 Biochemical Society

8

Control of the expression of inflammatory response genes

Jeremy Saklatvala[1], Jonathan Dean and Andrew Clark

Kennedy Institute of Rheumatology Division, Imperial College London,
1 Aspenlea Rd, London W6 8LH, U.K.

Abstract

The expression of genes involved in the inflammatory response is controlled both transcriptionally and post-transcriptionally. Primary inflammatory stimuli, such as microbial products and the cytokines interleukin-1 (IL-1) and tumour necrosis factor α (TNFα), act through receptors of either the Toll and IL-1 receptor (TIR) family or the TNF receptor family. These cause changes in gene expression by activating four major intracellular signalling pathways that are cascades of protein kinases: namely the three mitogen-activated protein kinase (MAPK) pathways, and the pathway leading to activation of the transcription factor nuclear factor κB (NFκB). The pathways directly activate and induce the expression of a limited set of transcription factors which promote the transcription of inflammatory response genes. Many of the mRNAs are unstable, and are stabilized by the p38 MAPK pathway. Instability is mediated by clusters of the AUUUA motif in the 3′ untranslated regions of the mRNAs. Control of mRNA stability provides a means of increasing the amplitude of a response and allows rapid adjustment of mRNA levels. Not all mRNAs stabilized by p38 contain AUUUA clusters; for example, matrix metalloproteinase-1 and -3 mRNAs lack these clusters, but are stabilized. Inflammatory gene expression is inhibited by glucocorticoids. These suppress MAPK signalling by inducing a MAPK phosphatase. This may be a significant mechanism additional to that by which the glucocorticoid receptor interferes with transcription factors.

Introduction: the inflammatory response

Proteinases mediate the tissue destruction that is an important feature of chronic inflammatory diseases such as rheumatoid arthritis. The extensive damage to articular cartilage, bone, tendons and ligaments that occurs in rheumatoid disease has been attributed to multiple attack by a variety of pro-

[1]To whom correspondence should be addressed (e-mail j.saklatvala@imperial.ac.uk).

teinases that are the subject of other contributions to this symposium; these include the matrix metalloproteinases (MMPs), the aggrecanases and the lysosomal proteinases, particularly those of osteoclasts, and the neutral serine proteinases of polymorphonuclear leucocytes. The synthesis of a number of these enzymes, for example MMPs-1, -3, -8 and -13 and the aggrecanases ADAMTS-1 and ADAMTS-4 [where ADAMTS denotes ADAM (a disintegrin and metalloproteinase) with thrombospondin motifs], is strongly induced by inflammatory stimuli. The purpose of this review is to summarize current knowledge of how the expression of the many proteins of the inflammatory response, including the proteinases, is controlled.

Inflammation is the response of tissues to injury. It is a defence mechanism against microbial invasion, its immediate function being to attract leucocytes from the circulation to sites of tissue damage. Incoming leucocytes deal with the threat and the inflammation resolves. If there is tissue damage, dead cells and their associated matrix are removed, and repair takes place by proliferation of fibroblasts and capillaries to form granulation tissue. This eventually forms a fibrous scar. In chronic inflammatory diseases of unknown cause, such as rheumatoid arthritis, the inflammation fails to resolve appropriately and what should be a repair process of limited duration becomes invasive and destructive.

Over the last 20 years there has been a huge expansion in knowledge about the complex cellular processes of inflammation and their molecular mechanisms. Important advances have been made in understanding the signals that attract leucocytes and the molecular interactions that guide them. Insights have been gained into the microbicidal and phagocytic mechanisms of leucocytes. And last but not least, complex controlling networks of cytokines and growth factors have been discovered. These chemical messengers control not only the early events of inflammation, but also the later reparative phase of cellular proliferation, as well as the final stage of production of collagen to form scar tissue.

Inflammatory receptor signalling

The primary initiating stimuli of inflammation signal through either of two major classes of cell surface receptor: the Toll and interleukin-1 (IL-1) receptor (TIR) family and the tumour necrosis factor (TNF) receptor family. These are unrelated, but both types of receptor activate a common set of four well characterized intracellular signalling pathways. These are the protein kinase system that leads to the activation of the transcriptional regulator nuclear factor κB (NFκB), and the three mitogen-activated protein kinase (MAPK) cascades. When the receptors bind their ligand, there is oligomerization that creates the signal. For example, the IL-1 receptor forms heterodimers with an accessory protein (Figure 1). TNF is a trimer and induces trimerization of its receptor. The intracellular parts of these dimers and trimers interact with adaptor molecules to form signalling complexes that activate the downstream pathways [1].

The Toll-like receptors (TLRs) of mammalian cells were discovered relatively recently [2], and have provided us with a unified view of the innate immune response and inflammation. The cytoplasmic part of the type I IL-1 receptor (the signalling receptor) is a homologue of *Drosophila* Toll, a trans-

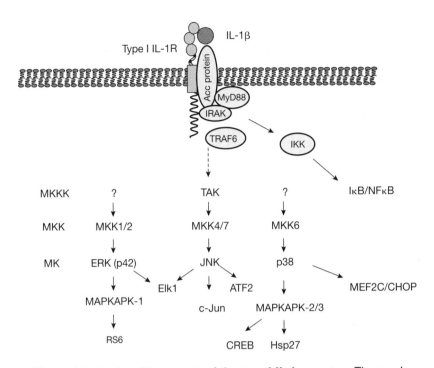

Figure 1 Early signalling events of the type I IL-1 receptor. The type I IL-1 receptor (IL-1R) in the cell membrane binds IL-1 and interacts with the IL-1R accessory (Acc) protein. The adaptor MyD88 interacts with the complex and allows IRAK to dock transiently. IRAK becomes hyperphosphorylated and dissociates, interacting with TRAF6. This complex activates the downstream protein kinases. The MAPK (MK) cascades are shown: ERK, JNK and p38 MAPK. These are activated by MKKs, which in turn are activated by MKK kinases (MKKKs). Transforming growth factor β-activated kinase (TAK) is a candidate MKKK in the IL-1 signalling pathway. The downstream kinases MAP-KAPK-1 and -2 phosphorylate ribosomal protein S6 (RS6) and Hsp27 respectively. IκB is the inhibitor of NFκB and IKK is IκB kinase. Elk1, ATF2, c-Jun, MEF2C, CHOP and CREB (cAMP response element-binding protein) are transcription factors that become phosphorylated.

membrane protein involved in dorso-ventral patterning in embryogenesis, and in anti-fungal responses. Toll is at the head of a signalling cascade analogous to the mammalian NFκB pathway. NFκB regulates many inflammatory and immune response genes. Mammalian homologues of Toll, called TLRs, detect and signal the presence of microbial products and are at the centre of the innate immune response. TLR4 is crucial for recognition of bacterial lipopolysaccharide (LPS), TLR2 for detecting peptidoglycans and bacterial lipopeptides, TLR3 senses double-stranded RNA, TLR5 detects bacterial flagellin and TLR9 regonizes bacterial DNA containing unmethylated CpG dinucleotides. There are ten TLRs, but not all of their recognition targets are known as yet.

The TLR-mediated innate immune response is phylogenetically ancient, and permits rapid responses to invading pathogens. Induction of IL-1 and

TNFα is an early event in this process. Because these pro-inflammatory cytokines share intracellular signalling pathways with the TLRs, they serve to amplify the innate immune response. The acquired response evolved later to enable specific rapid recognition of micro-organisms and to focus powerful microbicidal responses. The two processes are closely linked, with the innate response contributing to the generation of acquired immunity, and the acquired immune system triggering inflammatory responses as an effector mechanism.

Figure 1 shows a simple version of the current model for signalling by the IL-1 receptor. The signalling complex involves formation of a heterodimer between the receptor and the IL-1 receptor accessory protein. The intracellular parts of the heterodimer interact with the adaptor protein MyD88. IL-1 receptor-associated kinase 1 (IRAK1) and IRAK4 dock on to the complex. IRAK1 becomes hyperphosphorylated during a transient association with the receptor and then interacts with TNF receptor-associated factor 6 (TRAF6) [3]. TRAF6 activates the downstream cascades via MAPK kinase kinases.

The TLRs create similar signalling complexes, but there are additional and/or alternative adaptor proteins involved, giving rise to variations in downstream signalling. Examples include the recently described MyD88-like adaptor (MAL) or TIRAP [4,5]. TRAF6 is essential for TIR signalling, and is also implicated in signalling by IL-17 and CD40, which are not members of the TIR family [1].

Downstream mechanisms

At the bottom of the cascades are the proteins that regulate expression of the inflammatory response genes: until recently these were mainly considered to act on transcription, but some act post-transcriptionally, particularly in the control of mRNA stability. Many of the proteins of the inflammatory response have potent effects and need to be tightly controlled. It is therefore not surprising that there are complex regulatory mechanisms that permit rapid and substantial changes in expression.

Transcriptional control

Figure 2 shows a diagram of the proximal promoter of the cyclo-oxygenase-2 (COX-2) gene, a typical inflammatory response gene. COX-2 is thought to be responsible for much of the increased prostaglandin production at inflammatory sites, and is the target of many anti-inflammatory drugs. Prostaglandins have important effects on local blood flow and perception of pain.

The COX-2 promoter has binding sites for transcription factors that are typical for inflammatory response genes: NFκB, AP1 (activator protein 1), C/EBPβ (CAAT/enhancer-binding protein; also called NFIL-6), Ets and CREB (cAMP response element-binding protein). NFκB lies at the bottom of a specific activation cascade (Figure 1), and the majority of inflammatory response genes are NFκB-dependent. AP1 sites bind various heterodimers and homodimers of the c-Jun and c-Fos families, and ATF2 (activating transcrip-

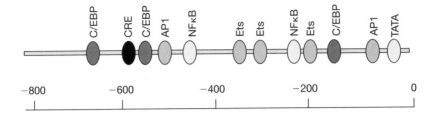

Figure 2 The human COX-2 promoter. The beads represent transcription factor binding sites in the proximal promoter. 'API' indicates sites for activator API complexes (e.g. c-Fos/c-Jun/ATF2 heterodimers). C/EBP is CAAT/enhancer-binding protein (C/EBPβ is also called nuclear factor-IL-6). CRE is the cAMP response element that binds CRE-binding protein (CREB). Ets is a transcription factor family whose members include Elk1 and SAP-1 (serum response factor accessory protein-1), which are phosphorylated by MAPKs. See Figure 1 for activating phosphorylation of the transcription factors.

tion factor-2). The MAPK family member c-Jun N-terminal kinase (JNK) phosphorylates and transcriptionally activates c-Jun and ATF2. Ets family members such as [Elk1 and SAP-1 (serum response factor accessory protein-1), as well as C/EBPβ, are activated by the original MAPK, extracellular-signal-regulated kinase (ERK) [1].

There is an extensive literature on the regulation of transcription factors by these signalling pathways, and some generalizations can be made (see [1] for a review). First, binding sites for the same limited set of transcription factors are usually found in the proximal promoters of inflammatory genes. Secondly, there are often multiple sites, and their arrangements vary widely. Thirdly, there may be sites for binding of transcription factors activated by other cytokine pathways (e.g. by interferons or by IL-6 and its related cytokines). Fourthly, the transcription factors act in concert to promote the binding of cofactors such as histone acetylases and RNA polymerase. Fifthly, the NFκB, ERK and JNK pathways are all probably crucial for the transcription of inflammatory response genes. The transcriptional role of the p38 MAPK cascade is less well understood, but it is involved in the regulation of NFκB [6,7]. Finally, the transcription factors may be activated directly by phosphorylation by the signalling pathways, or their expression may be induced. Examples of the former are NFκB, which translocates from cytoplasm to nucleus on cell activation, and AP1-binding proteins such as c-Jun, which are phosphorylated and so transcriptionally activated upon cell stimulation. Examples of the latter are induction of expression of c-Jun and c-Fos, which then secondarily activate, or sustain the activation of, inflammatory genes.

Post-transcriptional regulation

The stability of the mRNA transcripts of inflammatory response proteins is regulated. Many contain AU-rich elements (AREs) in their 3′ untranslated region (UTR). These AREs characteristically comprise clusters of the motif AUUUA and confer instability [8]. Activity in the p38 MAPK pathway counters this destabilizing effect and so stabilizes the mRNA in question. This was first shown for COX-2 mRNA [9,10], IL-3 mRNA [11], IL-6 mRNA [12], IL-8 mRNA [13], and later TNFα mRNA [14] and a number of others. The mechanism has been investigated by using a tetracycline-regulated post-transcriptional reporter plasmid [14–16]. The COX-2 destabilizing element was localized to a cluster of six AUUUA repeats just 3' of the open reading frame. This destabilizing element could be counteracted by activating the p38 MAPK pathway, and this was dependent upon the downstream kinase MAPK-activated protein kinase-2 (MAPKAPK-2). The relevant substrate(s) of MAPKAPK-2 remain(s) to be identified.

The stability of an mRNA is closely related to its translation. In some situations the p38 MAPK pathway appears to be needed for translation of an mRNA, because blocking the pathway inhibits production of protein without causing disappearance of the mRNA. For example, blocking p38 MAPK in murine macrophages stimulated with LPS almost completely inhibits production of TNFα protein, but only partially inhibits induction of its mRNA [17]. LPS induces TNFα mRNA in the MAPKAPK-2 knockout mouse, but little or no protein is produced [18]. It is likely that blocking the p38 pathway uncouples translation, and, depending on the mRNA species and the capacity of the degradative machinery, the uncoupled mRNA may be rapidly degraded or may persist.

The regulation of mRNA stability has three functions. First, it provides additional amplification of a response; secondly, it enables rapid adjustment of mRNA levels; and thirdly, it allows rapid termination of the production of a protein whose expression is no longer required.

It is too early to say how general the stabilization of inflammatory mRNAs by the p38 cascade is. It is also not known to what extent other signalling pathways are involved. The JNK pathway stabilizes IL-2 mRNA [19], and phorbol ester that activates protein kinase C and the ERK pathway stabilizes granulocyte/macrophage colony-stimulating factor mRNA [8]. The phosphoinositide 3-kinase pathway affects IL-3 mRNA stability [20].

The presence of an ARE in the 3' UTR of an mRNA does not necessarily confer regulation by p38 MAPK. The 3' UTR of the gene encoding c-Fos contains a typical cluster of AU repeats which, when placed in a reporter, confer instability and can be stabilized in a p38-dependent manner [16]. However, c-Fos mRNA is not stabilized by p38. This is probably because an instability determinant in the coding sequence overrides any effect of the ARE [21]. Another example of an unstable oncogene mRNA not stabilized by p38 acting on the 3' UTR ARE is c-Myc [15].

The mechanism by which mRNA stability is controlled via AREs is unknown. Several heterogeneous nuclear ribonucleoproteins bind tightly to AREs, including the prototype AU-binding protein AUF1 [22,23], its close relative CARG-box-binding factor-A [24], HuR [25] and tristetraprolin (TTP) [26]. Of these, only TTP definitely plays a role in cytokine expression. It is a member of a family of early-response genes, and TTP knockout mice have a diffuse inflammatory syndrome, including an arthritis [27]. The absence of TTP results in overproduction of TNFα (and probably other cytokines) due to an increased half-life of its mRNA. TTP binds to the ARE and destabilizes the mRNA [26].

TTP is not expressed in HeLa cells, which have been used for the post-transcriptional reporter studies, so the basic p38-dependent mechanism is independent of it. Interestingly, the expression of TTP (in macrophages) is p38-dependent [28]. Thus TTP is induced by LPS (and probably generally by inflammatory stimuli) and is a negative regulator of cytokine expression. It appears to represent another level of regulation on top of the basic destabilizing mechanism controlled by p38. It is paradoxical that p38 MAPK should have a stabilizing role but also mediate the expression of a destabilizing protein such as TTP.

It is possible that the p38 MAPK pathway directly regulates the proteins that bind to the ARE. Figure 3 shows a current model of mRNA structure. The poly(A)-binding protein binds the 3' poly(A) tail and eukaryotic initiation factor 4G, a component of the initiation complex that binds to the 5' cap structure. This circularizes the mRNA and facilitates translation. The circular structure is stable as long as the poly(A) tail is present. When it is lost, the mRNA is degraded [29], probably mainly 3'→5' by the exosome, a large complex comprising at least 10 exonucleases [30]. In this model, ARE-binding proteins could direct deadenylation and be under the control of the p38 pathway. Alternatively, the p38 pathway may act on the decay machinery, instructing it to disregard mRNAs 'tagged' by ARE-binding proteins. Decay of mRNA also occurs due to the action of 5'→3' exonucleases following decapping [31,32]. The significance of this pathway in mammalian cells and its importance in the control of mRNA decay by AREs and p38 MAPK remains to be established [33].

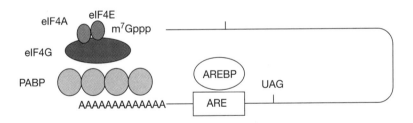

Figure 3 Post-transcriptional regulation of gene expression. When being translated, mRNA forms a circular structure because the poly(A)-binding protein (PABP) binds both the 3' poly(A) tail and eukaryotic initiation factor 4G (eIF4G). eIF4G, eIF4A and eIF4E are a trimeric complex (eIF4F) which binds to the 7-methylguanosine 5' cap structure (m⁷Gppp). Start (AUG) and stop (UAG) codons are shown, and the 3' UTR contains an ARE which typically might contain five or six copies of the motif AUUUA. ARE-binding proteins (AREBPs) target the mRNA for degradation, and the p38 MAPK counteracts this effect.

Control of the MMPs

The expression of MMP-1 and MMP-3 is induced in cells of connective tissue by inflammatory stimuli such as IL-1 and TNF, and by growth factors. The MMP-1 promoter has an AP1 site that has been much studied [34,35]. There is also evidence that NFκB is involved in MMP-1 expression, because there is an NFκB-binding site far upstream in the promoter region. In addition, IL-1 may stabilize the mRNA via three AUUUA motifs [35]. It is important to stress that, in practice, an inflammatory stimulus such as IL-1 does not act alone, but in conjunction with other cytokines, such as, for example, the IL-6 family (which they induce). The IL-6 family member oncostatin M synergizes with IL-1 in the production and activation of MMP-1, and this phenomenon is discussed in Chapter 11 by Cawston et al. in this volume.

Expression of both MMP-1 and MMP-3 in gingival and dermal fibroblasts is strongly dependent on p38 MAPK activity [37]. It was shown recently that the p38 MAPK pathway stabilizes the MMP mRNAs [38]. Activation of the pathway by transfecting an active MAPK kinase 3 (MKK3) mutant (MKK3 is a p38 activator) stabilized the MMP-1 and MMP-3 mRNAs. This is interesting and surprising, because neither MMP has a typical destabilizing, p38-responsive ARE in its 3' UTR. The 3' UTR of MMP-3 has only a single AUUUA motif, and that of MMP-1 has three which are dispersed [35]. From mapping experiments on the COX-2 ARE, we would not predict that these MMP 3' UTR sequences would be regulated by p38 MAPK. Thus the pathway may also stabilize mRNAs that do not have typical AREs. This implies that the p38 MAPK cascade controls the stability of mRNA by mechanisms additional to the recognition of AREs and their binding proteins, or that the latter also recognize sequences other than AUUUA clusters.

Other proteinases regulated by inflammatory stimuli are ADAM family members. ADAMTS-1 is up-regulated by LPS [39], and ADAMTS-4 (also called aggrecanase 1) is up-regulated by IL-1 [40,41]. However, so far neither transcriptional or post-transcriptional mechanisms of their regulation have been reported. The control of MMP activation mechanisms by inflammatory stimuli has yet to be explored. Urokinase-type plasminogen activator and plasminogen activator inhibitor 2 are both regulated by inflammatory stimuli [1]. Regulation of membrane-type MMPs in inflammation has not been reported.

Glucocorticoid suppression of inflammation

The expression of many genes of the inflammatory response is suppressed by glucocorticoids, and this may be due in part to actions on the intracellular signalling pathways used by inflammatory stimuli. Synthetic glucocorticoids are used widely for anti-inflammatory therapy in diseases such as rheumatoid arthritis, Crohn's disease and asthma, and as immunosuppressants in transplantation. They are also used in certain malignancies. Although very potent in suppressing inflammation, they have a variety of highly undesirable side effects that limit their use. These include osteoporosis, diabetes, cataracts, hypertension, the characteristic features of Cushing's syndrome and suppression of the hypothalamic/pituitary/adrenal axis.

Physiologically, glucocorticoids are produced in increased amounts in response to physical stress such as infection. The inflammatory cytokines and microbial products act on the pituitary gland, via the hypothalamus, to stimulate the production of corticotropin ('ACTH'), which stimulates the adrenal cortex. The glucocorticoids produced act as a negative feedback on the production of inflammatory cytokines and mediators.

Glucocorticoids act by binding to the glucocorticoid receptor, a transcription factor that is a member of the nuclear receptor family [42]. The liganded glucocorticoid receptor binds as a dimer to target DNA sequences called glucocorticoid response elements (GREs) in the promoter regions of glucocorticoid-sensitive genes. While GRE-dependent genes are thought to underlie some of the side effects of glucocorticoids (e.g. diabetes), none have been identified that account for the profound anti-inflammatory actions of glucocorticoids. Examples of inflammatory response proteins that are suppressed by glucocorticoids are IL-1, TNF, chemokines, IL-6, COX-2, MMP-1 and MMP-3. All are inhibited at the level of mRNA expression. One mechanism behind this global suppression is transrepression, or transcriptional interference [42,43]. The liganded glucocorticoid receptor, in addition to binding to the GREs, also binds to AP1 and NFκB transcription factors, and in so doing interferes with their transcriptional activity.

Glucocorticoids also have post-transcriptional effects on the expression of genes of the inflammatory response: they destabilize the mRNAs of COX-2 and several cytokines. This is likely to be because they prevent the p38 MAPK-mediated stabilization of mRNAs. We found that treating cells with dexamethasone, a synthetic glucocorticoid, reversed the mRNA-stabilizing effect of p38 MAPK [44]. Furthermore, dexamethasone strongly inhibited the activation of p38 MAPK and JNK by inflammatory stimuli. It had no effect on the activation of MKK6 (the kinase upstream of p38) or on the downstream kinase MAPKAPK-2. For these reasons, and because JNK and p38 have different activators (see Figure 1), it seemed likely that dexamethasone was inducing a molecule that countered the effect of the MKKs by maintaining the MAPKs in their inactive, unphosphorylated state. A phosphatase was a good candidate. MAPKs are inactivated by removal of either their threonine or tyrosine phosphates, so a large number of phosphatases potentially could be involved. Among these are the MAPK phosphatases (MKPs)/dual-specificity phosphatases (DUSPs), of which more than 10 have been identified. Gene chip experiments revealed that the major phosphatase induced by dexamethasone was MKP1/DUSP1, and this was confirmed by Northern and Western blotting [45]. MKP1 is an early response gene that is transiently activated by very many stimuli, and it has a preference for p38 MAPK and JNK over ERK [46]. Most interestingly, dexamethasone caused a sustained increase in the expression of MKP1 – thus, within 1 h of dexamethasone treatment, and even after 24 h exposure to the steroid, the protein was strongly induced [45]. IL-1 itself induces MKP1 transiently (Figure 4a). IL-1 causes two phases of p38 MAPK activation (Figure 4b). An initial strong transient phase peaks at 30 min whose

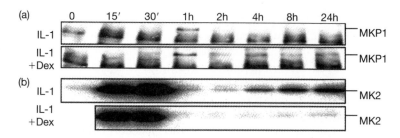

Figure 4 Dexamethasone causes prolonged induction of MKP1 and inhibits late-phase activation of p38 MAPK by IL-1. HeLa cells were treated with IL-1 (10 ng/ml) or IL-1 plus dexamethasone (Dex; 1 μM) for the course of the experiment. **(a)** MKP1 was detected by Western blotting. **(b)** The activity of p38 MAPK in the cells was measured at the indicated times by immuno-precipitating it from cell lysates and incubating the precipitates with recombinant MAPKAPK-2 substrate and ATP (10 μM) containing [γ-^{32}P]ATP. Reaction products were separated by SDS/PAGE, and the gel was autoradiographed to show phosphorylated MAPKAPK-2 (MK2). From [46].

subsidence corresponds with the induction of MKP1. By 2 h the transiently induced MKP1 has gone, and IL-1-dependent p38 activation returns at a lower level and is sustained. Adding dexamethasone together with IL-1 blunts the initial burst of signalling slightly as MKP1 is being induced, but the sustained MKP1 expression caused by the dexamethasone prevents the second phase of activation (Figure 4). It is important to point out that, in order to inhibit the first peak of signalling, dexamethasone would need to be added 1–2 h prior to IL-1 in order to induce expression of the phosphatase [44]. Recently, overexpression of MKP1 in murine cells has been shown to inhibit cytokine expression [47]. It remains to be proved that MKP1 induction is the mechanism by which glucocorticoids inhibit MAPK signalling, and is of physiological significance for glucocorticoid action.

We are grateful to the Arthritis Research Campaign and the Medical Research Council for their support.

References

1.   Kracht, M. and Saklatvala, J. (2002) Cytokine **20**, 91–106
2.   Medzhitov, R. and Janeway, Jr, C.A. (2002) Science **296**, 298–300
3.   Cao, Z., Xiong, J., Takeuchi, M., Kurama, T. and Goeddel, D.V. (1996) Nature (London) **383**, 443–446
4.   Fitzgerald, K.A., Palsson-McDermott, E.M., Bowie, A.G., Jefferies, C.A., Mansell, A.S., Brady, G., Brint, E., Dunne, A., Gray, P., Harte, M.T. et al. (2001) Nature (London) **413**, 78–83
5.   Horng, T., Barton, G.M. and Medzhitov, R. (2001) Nat. Immunol. **2**, 835–841
6.   Saccani, S., Pantano, S. and Natoli, G. (2002) Nat. Immunol. **3**, 69–75
7.   Vermeulen, L., De Wilde, G., Damme, P.V., Vanden Berghe, W. and Haegeman, G. (2003) EMBO J. **22**, 1313–1324

8. Shaw, G. and Kamen, R. (1986) Cell **46**, 659–667
9. Ridley, S.H., Dean, J.L., Sarsfield, S.J., Brook, M., Clark, A.R. and Saklatvala, J. (1998) FEBS Lett. **439**, 75–80
10. Dean, J.L., Brook, M., Clark, A.R. and Saklatvala, J. (1999) J. Biol. Chem. **274**, 264–269
11. Ming, X.F., Kaiser, M. and Moroni, C. (1998) EMBO J. **17**, 6039–6048
12. Miyazawa, K., Mori, A., Miyata, H., Akahane, M., Ajisawa, Y. and Okudaira, H. (1998) J. Biol. Chem. **273**, 24832–24838
13. Holtmann, H., Winzen, R., Holland, P., Eickemeier, S., Hoffmann, E., Wallach, D., Malinin, N.L., Cooper, J.A., Resch, K. and Kracht, M. (1999) Mol. Cell. Biol. **19**, 6742–6753
14. Brook, M., Sully, G., Clark, A.R. and Saklatvala, J. (2000) FEBS Lett. **483**, 57–61
15. Lasa, M., Mahtani, K.R., Finch, A., Brewer, G., Saklatvala, J. and Clark, A.R. (2000) Mol. Cell. Biol. **20**, 4265–4274
16. Winzen, R., Kracht, M., Ritter, B., Wilhelm, A., Chen, C.Y.A., Shyu, A.B., Muller, M., Gaestel, M., Resch, K. and Holtmann, H. (1999) EMBO J. **18**, 4969–4980
17. Prichett, W., Hand, A., Sheilds, J. and Dunnington, D. (1995) J. Inflamm. **45**, 97–105
18. Kotlyarov, A., Neininger, A., Schubert, C., Eckert, R., Birchmeier, C., Volk, H.D. and Gaestel, M. (1999) Nat. Cell Biol. **1**, 94–97
19. Chen, C.Y., Del Gatto-Konczak, F., Wu, Z. and Karin, M. (1998) Science **280**, 1945–1949
20. Ming, X.F., Stoecklin, G., Lu, M., Looser, R. and Moroni, C. (2001) Mol. Cell. Biol. **21**, 5778–5789
21. Schiavi, S.C., Wellington, C.L., Shyu, A.B., Chen, C.Y., Greenberg, M.E. and Belasco, J.G. (1994) J. Biol. Chem. **269**, 3441–3448
22. Brewer, G. (1991) Mol. Cell. Biol. **11**, 2460–2466
23. Xu, N., Chen, C.Y. and Shyu, A.B. (2001) Mol. Cell. Biol. **21**, 6960–6971
24. Dean, J.L.E., Sully, G., Wait, R., Rawlinson, L., Clark, A.R. and Saklatvala, J. (2002) Biochem. J. **366**, 709–719
25. Fan, X.C. and Steitz, J.A. (1998) EMBO J. **17**, 3448–3460
26. Lai, W.S., Carballo, E., Strum, J.R., Kennington, E.A., Phillips, R.S. and Blackshear, P.J. (1999) Mol. Cell. Biol. **19**, 4311–4323
27. Carballo, E., Lai, W.S. and Blackshear, P.J. (1998) Science **281**, 1001–1005
28. Mahtani, K.R., Brook, M., Dean, J.L., Sully, G., Saklatvala, J. and Clark, A.R. (2001) Mol. Cell. Biol. **21**, 6461–6469
29. Xu, N., Chen, C.Y. and Shyu, A.B. (1997) Mol. Cell. Biol. **17**, 4611–4621
30. Chen, C.Y., Gherzi, R., Ong, S.E., Chan, E.L., Raijmakers, R., Pruijn, G.J., Stoecklin, G., Moroni, C., Mann, M. and Karin, M. (2001) Cell **107**, 451–464
31. Wang, Z., Jiao, X., Carr-Schmid, A. and Kiledjian, M. (2002) Proc. Natl. Acad. Sci. U.S.A. **99**, 12663–12668
32. Van Dijk, E., Cougot, N., Meyer, S., Babajko, S., Wahle, E. and Seraphin, B. (2002) EMBO J. **21**, 6915–6924
33. Gao, M., Wilusz, C.J., Peltz, S.W. and Wilusz, J. (2001) EMBO J. **20**, 1134–1143
34. Brenner, D.A., O'Hara, M., Angel, P., Chojkier, M. and Karin, M. (1989) Nature (London) **337**, 661–663
35. Vincenti, M.P., Coon, C.I., Lee, O. and Brinckerhoff, C.E. (1994) Nucleic Acids Res. **22**, 4818–4827
36. Reference deleted
37. Ridley, S.H., Sarsfield, S.J., Lee, J.C., Bigg, H.F., Cawston, T.E., Taylor, D.J., DeWitt, D.L. and Saklatvala, J. (1997) J. Immunol. **158**, 3165–3173
38. Reunanen, N., Li, S.P., Ahonen, M., Foschi, M., Han, J. and Kahari, V.M. (2002) J. Biol. Chem. **277**, 32360–32368
39. Kuno, K., Kanada, N., Nakashima, E., Fujiki, F., Ichimura, F. and Matsushima, K. (1997) J. Biol. Chem. **272**, 556–562

40.   Tortorella, M.D., Burn, T.C., Pratta, M.A., Abbaszade, I., Hollis, J.M., Liu, R., Rosenfeld, S.A., Copeland, R.A., Decicco, C.P., Wynn, R. et al. (1999) Science **284**, 1664–1666

41.   Curtis, C.L., Hughes, C.E., Flannery, C.R., Little, C.B., Harwood, J.L. and Caterson, B. (2000) J. Biol. Chem. **275**, 721–724

42.   Newton, R. (2000) Thorax **55**, 603–613

43.   Karin, M. (1998) Cell **93**, 487–490

44.   Lasa, M., Brook, M., Saklatvala, J. and Clark, A. (2001) Mol. Cell. Biol. **21**, 771–780

45.   Lasa, M., Abraham, S.M., Saklatvala, J. and Clark, A.R. (2002) Mol. Cell. Biol. **22**, 7802–7811

46.   Franklin, C.C. and Kraft, A.S. (1997) J. Biol. Chem. **272**, 16917–16923

47.   Chen, P., Li, J., Barnes, J., Kokkonen, G.C., Lee, J.C. and Liu, Y. (2002) J. Immunol. **169**, 6408–6416

Biochem. Soc. Symp. **70**, 107–114
(Printed in Great Britain)
© 2003 Biochemical Society

9

Use of anti-neoepitope antibodies for the analysis of degradative events in cartilage and the molecular basis for neoepitope specificity

John S. Mort*†[1], Carl R. Flannery‡[2], Joe Makkerh*[3], Joanne C. Krupa* and Eunice R. Lee*†

*Shriners Hospital for Children, Montreal, Quebec, Canada, †Department of Surgery, McGill University, Montreal, Quebec, Canada, and ‡Connective Tissue Biology Laboratory, Cardiff School of Biosciences, University of Cardiff, Cardiff, Wales, U.K.

Abstract

Degradation of the cartilage proteoglycan, aggrecan, is an essential aspect of normal growth and development, and of joint pathology. The roles of different proteolytic enzymes in this process can be determined from the sites of cleavage in the aggrecan core protein, which generates novel termini (neoepitopes). Antibodies specific for the different neoepitopes generated by such cleavage events provide powerful tools with which to analyse these processes. The same approach can be used to differentiate the processed, active forms of proteases from their inactive pro-forms. Since the proteolytic processing of these enzymes requires the removal of the inhibitory pro-region, it also results in the generation of N-terminal neoepitopes. Using the newborn rat long bone as a model system, it was shown that the active form of ADAMTS-4 [ADAM (a disintegrin and metalloproteinase) with thrombospondin motifs-4], but not ADAMTS-5, co-localizes with the aggrecan cleavage neoepitopes known to be produced by this metalloproteinase. Thus,

[1]To whom correspondence should be addressed: Joint Diseases Laboratory, Shriners Hospital for Children, 1529 Cedar Avenue, Montreal, Quebec, Canada H3G 1A6 (e-mail jmort@shriners.mcgill.ca).
[2]Present address: Wyeth Research, 200 Cambridge Park Drive, Cambridge, MA, U.S.A.
[3]Present address: Montreal Neurological Institute, McGill University, Montreal, Quebec, Canada.

in long bone growth, aggrecan turnover seems to be dependent on ADAMTS-4 activity. To demonstrate the molecular basis of the specificity of anti-neoepitope antibodies, the Fv region of a monoclonal antibody specific for a neoepitope generated by the ADAMTS-4-mediated cleavage of aggrecan has been modelled and the binding of the peptide epitope simulated. In the docked structure, the N-terminus of the peptide antigen is clearly buried in the binding-site cavity. The absence of an open cleft makes it impossible for the intact substrate to pass through the binding site, providing a rationale for the specificity of this class of antibodies.

Introduction

Articular cartilage turnover is a critical feature of normal growth and development, and of degenerative diseases such as arthritis. Cartilage is a bio-composite material, and in simple terms can be considered to consist of a type II collagen network into which an underhydrated mass of the large aggregating proteoglycan, aggrecan, is embedded, giving the tissue its shock-absorbing properties. This proteoglycan is present in multimolecular complexes of 50 to 100 aggrecan molecules non-covalently bound through a specialized domain (G1) to a central hyaluronate filament, these interactions being stabilized by a small glycoprotein termed link protein (Figure 1a) [1]. While the turnover of cartilage collagen appears to be dependent on the action of the true collagenases (and perhaps cathepsin K), aggrecan is a substrate for most proteases [2]. However, analysis of natural aggrecan cleavage products indicates that two protease families mediate most of its degradation *in vivo*. Of particular importance is cleavage in the region close to the point of attachment of the aggrecan molecule to the hyaluronate filament. The region termed the interglobular domain (Figure 1b) is particularly susceptible to proteolysis, and cleavage of the aggrecan core protein at sites along this domain allows the release of the glycosaminoglycan-containing portion of the molecule from the cartilage matrix, while leaving the globular region still associated with hyaluronate in the tissue. Matrix metalloproteinases (MMPs) such as stromelysin (MMP-3) have been shown to cleave aggrecan in this region at a specific site (Figure 1b) [3], and MMP-3-dependent cleavage products have been found in cartilage [4]. In addition, a second cleavage within the interglobular domain is found in both normal and diseased cartilage [5,6]. This is produced by three members of a novel family of metalloproteinases that have the unusual specificity of cleavage following glutamic acid residues. The activities, termed aggrecanases, are due to the action of ADAMTS-4 [ADAM (a disintegrin and metalloproteinase) with thrombospondin motifs-4] [7], ADAMTS-5 [8] and possibly ADAMTS-1 [9] (Figure 1c). These multidomain proteases are closely related to the ADAM family [10], but in addition contain a variable number of thrombospondin type 1 motifs that have been shown to assist in substrate binding (Figure 1c) [11]. As with the ADAM family of proteases, the pro-forms of the ADAMTS are activated, prior to secretion, by furin (or furin-like enzymes) [12].

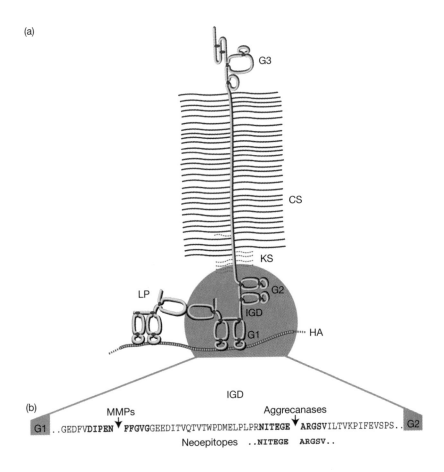

(a)

(b)

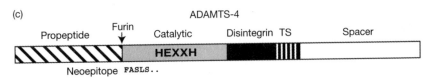

Figure 1 Aggrecan, aggrecanases and associated neoepitopes. (a) Diagrammatic representation of a monomer of the proteoglycan aggregate, consisting of an aggrecan molecule with chondroitin sulphate (CS) and keratan sulphate (KS) glycosaminoglycan side chains and the three globular domains, G1, G2 and G3. As indicated, the G1 domains of aggrecan monomers interact with hyaluronate (HA) and link protein (LP) to form stable complexes, resulting in the production of proteoglycan aggregates consisting of between 50 and 100 aggrecan monomers. **(b)** Details of the aggrecan interglobular domain (IGD), showing the cleavage sites for MMPs and aggrecanases (in bold), and the sequences against which anti-neoepitope antibodies were prepared. **(c)** Schematic representation of the domain structure of ADAMTS-4 (aggrecanase 1). TS, thrombospondin type 1 motif. The position of the furin processing site and the resulting neoepitope of the active enzyme are indicated.

Our understanding of the cleavage of aggrecan in health and disease has been facilitated by the development of anti-peptide antibodies that are specific for the new termini generated following protease-mediated cleavage events [13], but are unreactive with the intact substrate. These are commonly referred to as anti-neoepitope antibodies. Such reagents allow detection of the aggrecan neoepitopes ...NITEGE and ARGSV... generated by ADAMTS-4 and ADAMTS-5 *in vivo*. The monoclonal antibody BC3 [14] produced against the ARGSV... epitope played a critical role in the characterization of these enzymes. In addition to mediating protein destruction, proteolysis is also an important aspect of biosynthesis, for example in the activation of the pro-form of ADAMTS-4 by the action of furin. Such processing steps also generate neoepitopes. As shown below, antibodies generated against peptides representing the new N-terminus of the ADAMTS family can be used to correlate the presence of active forms of these enzymes with the products of their activity.

Co-localization of aggrecan cleavage products and active enzyme

During long bone development, a cartilage model is replaced by bone, involving extensive turnover of the cartilage matrix. In order to allow for growth of the animal, the ends of the bone (the epiphyses) are remodelled separately. The cartilage primordium is invaded by blood vessels, and marrow spaces are established in the middle of the bone, resulting in the establishment of two zones of cartilage, termed growth plates. Chondroproliferation in these regions allows for elongation of the bone as the animal matures. In addition, a secondary centre of ossification is established in the epiphyses, allowing for growth in the end regions of the bone. At the cartilage/bone-marrow interface, rapid turnover of aggrecan occurs (Figure 2a), making this tissue a very good system in which to study the role of specific proteases in the degradation of this proteoglycan. Using the proximal tibia of 21-day-old rat pups as a model system, a decrease in metachromasia (purple colour) in Toluidine Blue-stained sections, indicative of proteoglycan loss, is seen at the junction between the cartilage of the growth plate and the bone marrow space of a newly developed ossification centre (Figures 2b and 2c, arrows). As shown previously [15], the aggrecanase-generated neoepitope ...NITEGE, representing the G1-containing product of aggrecanase cleavage, is immunolocalized at this site, as indicated by the brown reaction product shown in Figures 2(e) and 2(h).

Both ADAMTS-4 and ADAMTS-5 have been implicated in the turnover of aggrecan in cartilage. In order to evaluate the involvement of these enzymes in aggrecan turnover in the secondary centre of ossification, antibodies were prepared against the N-terminal neoepitopes generated following furin-mediated cleavage of the ADAMTS-4 and -5 proenzymes, and their specificity demonstrated using the recombinant proteases. While no reactivity was observed for ADAMTS-5, staining for active ADAMTS-4 was found to co-localize with that of the aggrecanase-generated G1 product (Figures 2f and 2i), clearly indicating a cause-and-effect mechanism for the cleavage of aggrecan in this region.

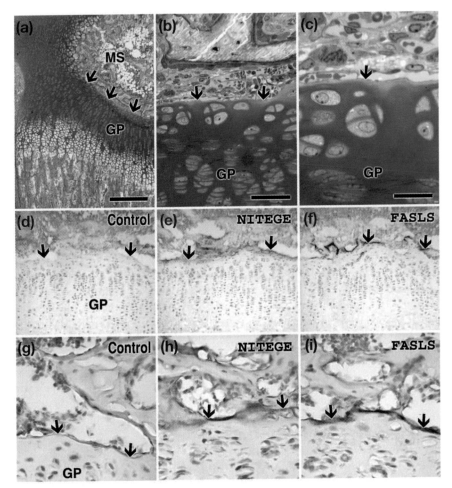

Figure 2 Aggrecan degradation in growth plate cartilage. Aggrecan degradation is revealed in tissue sections prepared from tibial epiphyses from a 21-day-old rat and stained with Toluidine Blue (a–c), with anti-NITEGE antibodies (e, h), or, as a control, with non-immune IgG (d and g). In contrast, in the remaining panels, active aggrecanase 1 enzyme is revealed with anti-FASLS antibodies (f and i). The arrows denote the same surface of the cartilaginous epiphysis at low (bar = 200 μm; a, d–f), intermediate (bar = 50 μm; b, g–i) or high (bar = 20 μm; c) magnification. The anti-NITEGE and anti-FASLS antibodies (compare panels h and i) produce an intense brown reactivity precisely where the cartilage is subjected to resorption, as indicated by the arrows in all panels. MS, marrow space of the secondary ossification centre; GP, growth plate.

Modelling of the Fv region of BC3 and peptide docking

Anti-neoepitope antibodies have been prepared using both mouse monoclonal and rabbit polyclonal approaches. A major advantage of the monoclonal method is that molecular details of the specific immunoglobulin can be determined. The antigen-binding unit (the Fv region) can be cloned, expressed as a recombinant protein and subjected to protein engineering methods to modify specificity. Such approaches depend on knowledge of the three-dimensional structure of the Fv region, preferably as a complex with the corresponding antigen. In the absence of an X-ray crystal structure, molecular modelling can provide this information, since the three-dimensional framework of the immunoglobulin fold is extremely well conserved and a large database of known structures is available. We have used this approach to investigate the molecular basis for the specificity of anti-neoepitope antibodies.

Although good monoclonal antibodies have not been successfully prepared against the ...NITEGE epitope, the monoclonal antibody BC3 was prepared against the N-terminal epitope (ARGSV...) generated at this cleavage site (Figure 1b). It has been an extremely valuable reagent in the study of cartilage degradation. It was thus selected for detailed investigation of the molecular basis underlying its specificity.

Using appropriate degenerate primers, the variable regions of the light and heavy chains of the mouse monoclonal antibody BC3 were amplified by PCR using as a template cDNA reverse-transcribed from RNA prepared from the hybridoma cells. The derived amino acid sequences were used as the basis for homology modelling of the Fv three-dimensional structure using the program AbGen [16], and the resulting structure was energy minimized using the program Discover (InsightII; MSI). When viewed as an accessible surface, the Fv region of the BC3 antibody model shows a closed-in deep pocket rather than a channel, suggesting that a terminal epitope, but not the uncleaved substrate, could fit into the binding site. In addition, the negative electrostatic character of the binding site would be expected to accommodate the N-terminal amino group of the peptide and the positive side chain of the arginine, the second residue of the neoepitope (Figure 3a).

The peptide epitope Ala-Arg-Gly-Ser-Val-amide was docked into the antigen-binding site of the BC3 Fv region using Autodock3 [17]. This program allows for the flexibility of the ligand side chains, and yields different putative conformations of the bound peptide and corresponding estimated dissociation constants. Several clusters of putative binding conformations were obtained; however, those with the highest affinity (K_i values in the low nanomolar range) showed similar conformations, as illustrated in Figure 3(b). As predicted, the N-terminus of the peptide is buried deep in the binding site and the arginine side chain is well accepted. The remainder of the peptide then winds through and out of the binding site, as would be the case for the true aggrecan cleavage product.

These results show that insights into the binding of anti-neoepitope antibodies to their antigen can be obtained by molecular modelling approaches. Through protein engineering techniques, the affinity of these antibodies can be increased as is required for their use in immunoassays. In addition, since it has been difficult to produce antibodies specific for some neoepitopes, if may be possible to adapt the established frameworks to generate the required specificities.

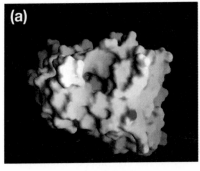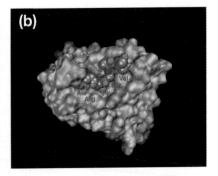

Figure 3 Model of the three-dimensional structure of the anti-neoepitope antibody BC3, and docking of the peptide epitope. (a) Accessible surfaces of the Fv region of monoclonal antibody BC3, showing areas of negative (red) and positive (blue) charge. The variable region from the light chain is on the left and that from the heavy chain on the right. The molecule is oriented so that the antigen-binding site is located in centre of the image. The Figure was prepared using the program Grasp [18]. (b) Accessible surface of the BC3 Fv region in a similar orientation as in panel (a), showing the highest-affinity conformation obtained by docking the peptide epitope Ala-Arg-Gly-Ser-Val-amide (space-filling representation without hydrogen atoms), using the program Autodock3. The Figure was prepared using the Python-based Molecular Viewing Environment (PMV) and AutodockTools (http://www.scripps.edu/pub/olson-web/doc/autodock/tools.html).

Conclusions

Anti-neoepitope antibodies are not only valuable reagents for the analysis and quantification of cartilage matrix degradation, but, as shown here, can also be used to correlate the presence of active matrix-degrading enzymes with their specific degradation products. Molecular modelling results provide a rationale for the often exquisite selectivity of these antibodies to discriminate between the same sequence as a cleavage product over that in the intact protein. Insights gained from these studies can be used to improve the repertoire of these reagents.

This work was supported by the Shriners of North America. We thank Guylaine Bedard and Beata Kluczyk for preparing Figures 1 and 2, and Dr A. Recklies for critical reading of the manuscript.

References
1.   Hascall, V.C. (1988) ISI Atlas Sci. Biochem. **1**, 189–198
2.   Hascall, V.C., Sandy, J.D. and Handley, C.J. (1999) in Biology of the Synovial Joint (Archer, C.W., Caterson, B., Benjamin, M. and Ralphs, J.R., eds), pp. 101–120, Harwood Academic Publishers, Amsterdam
3.   Flannery, C.R., Lark, M.W. and Sandy, J.D. (1992) J. Biol. Chem. **267**, 1008–1014
4.   Singer, I.I., Kawka, D.W., Bayne, E.K., Donatelli, S.A., Weidner, J.R., Williams, H.R., Ayala, J.M., Mumford, R.A., Lark, M.W., Glant, T.T. et al. (1995) J. Clin. Invest. **95**, 2178–2186

5. Sandy, J.D., Neame, P.J., Boynton, R.E. and Flannery, C.R. (1991) J. Biol. Chem. **266**, 8683–8685

6. Sztrolovics, R., Alini, M., Roughley, P.J. and Mort, J.S. (1997) Biochem. J. **326**, 235–241

7. Tortorella, M.D., Burn, T.C., Pratta, M.A., Abbaszade, I., Hollis, J.M., Liu, R., Rosenfeld, S.A., Copeland, R.A., Decicco, C.P., Wynn, R. et al. (1999) Science **284**, 1664–1666

8. Abbaszade, I., Liu, R.Q., Yang, F., Rosenfeld, S.A., Ross, O.H., Link, J.R., Ellis, D.M., Tortorella, M.D., Pratta, M.A., Hollis, J.M. et al. (1999) J. Biol. Chem. **274**, 23443–23450

9. Kuno, K., Okada, Y., Kawashima, H., Nakamura, H., Miyasaka, M., Ohno, H. and Matsushima, K. (2000) FEBS Lett. **478**, 241–245

10. Wolfsberg, T.G., Straight, P.D., Gerena, R.L., Huovila, A.P., Primakoff, P., Myles, D.G. and White, J.M. (1995) Dev. Biol. **169**, 378–383

11. Tortorella, M.D., Pratta, M., Liu, R.Q., Abbaszade, I., Ross, H., Burn, T. and Arner, E. (2000) J. Biol. Chem. **275**, 25791–25797

12. Rodríguez-Manzaneque, J.C., Milchanowski, A.B., Dufour, E.K., Leduc, R. and Iruela-Arispe, M.L. (2000) J. Biol. Chem. **275**, 33471–33479

13. Mort, J.S. and Buttle, D.J. (1999) J. Clin. Pathol. Mol. Pathol. **52**, 11–18

14. Hughes, C.E., Caterson, B., Fosang, A.J., Roughley, P.J. and Mort, J.S. (1995) Biochem. J. **305**, 799–804

15. Lee, E.R., Lamplugh, L., Davoli, M.A., Beauchemin, A., Chan, K., Mort, J.S. and Leblond, C.P. (2001) Dev. Dyn. **222**, 52–70

16. Mandral, C., Kingery, B.D., Anchin, J.M., Subramaniam, S. and Linthicum, D.S. (1996) Nat. Biotechnol. **14**, 323–328

17. Morris, G.M., Goodsell, D.S., Halliday, R.S., Huey, R., Hart, W.E., Belew, R.K. and Olson, A.J. (1998) J. Comp. Chem. **19**, 1639–1662

18. Nicholls, A., Sharp, K.A. and Honig, B. (1991) Proteins **11**, 281–296

Biochem. Soc. Symp. **70**, 115–123
(Printed in Great Britain)
© 2003 Biochemical Society

10

Proteolysis of the collagen fibril in osteoarthritis

**A. Robin Poole*[1], Fred Nelson*[2], Leif Dahlberg*[3],
Elena Tchetina*, Masahiko Kobayashi*[4], Tadashi Yasuda*[5],
Sheila Laverty*†, Ginette Squires*[6], Toshihisa Kojima*[7],
William Wu*[8] and R. Clark Billinghurst*[9]**

*Joint Diseases Laboratory, Shriners Hospitals for Children and Departments of Surgery and Medicine, McGill University, 1529 Cedar Avenue, Montreal, Quebec H3G 1A6, Canada, and †Faculty of Veterinary Sciences, University of Montreal, Ste-Hyacinthe, Quebec, Canada

Abstract

The development of cartilage pathology in osteoarthritis involves excessive damage to the collagen fibrillar network, which appears to be mediated primarily by the chondrocyte-generated cytokines interleukin-1 and tumour necrosis factor α and the collagenases matrix metalloproteinase-1 (MMP-1) and MMP-13. The damage to matrix caused by these and other MMPs can result in the production of sufficient degradation products that can themselves elicit further degradation, leading to chondrocyte differentiation and eventually matrix mineralization and cell death. Knowledge of these MMPs, cellular receptors and cytokine pathways, and the ability to selectively antagonize them by

[1]To whom correspondence should be addressed (e-mail rpoole@shriners.mcgill.ca).
[2]Present address: Henry Ford Hospital, Detroit, MI, U.S.A.
[3]Present address: Institute of Orthopaedics, University Hospital, Malmo, Sweden.
[4]Present address: Department of Orthopaedic Surgery, Kyoto City Rehabilitation Hospital, Kyoto, Japan.
[5]Present address: Department of Orthopaedic Surgery, Kyoto University Medical Center, Kyoto, Japan.
[6]Present address: GlaxoSmithKline Research, Stevenage, Herts., U.K.
[7]Present address: Department of Orthopaedic Surgery, Nagoya University Medical Center, Nagoya, Japan.
[8]Present address: University of Tennessee Health Sciences, VA Medical Center, Memphis, TN, U.S.A.
[9]Present address: Department of Clinical Sciences, Colorado State University, Fort Collins, CO, U.S.A.

A.R. Poole et al.

selective blockade of function, may provide valuable therapeutic opportunities in the treatment of osteoarthritis and other joint diseases involving cartilage resorption, such as rheumatoid arthritis. The ability to detect the products of these degradative events released into body fluids of patients may enable us to monitor disease activity, predict disease progression and determine more rapidly the efficacy of new therapeutic agents.

Introduction

Along with synovial fluid, the articular cartilages of diarthrodial joints ensure an almost frictionless articulation during normal joint movement. Cartilage contains an extensive extracellular matrix that is elaborated by the chondrocyte (Figure 1). This matrix endows cartilage with its capacity to resist deformation and maintain strength. The large proteoglycan aggrecan binds hyaluronic acid to form macromolecular aggregates that can bind a considerable quantity of water. This is a consequence of the hydration of many

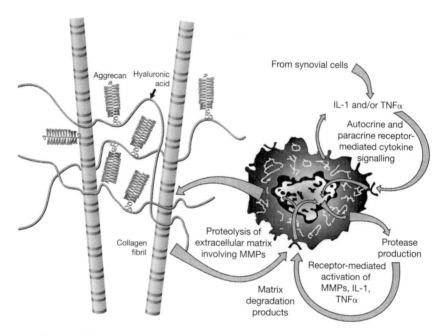

Figure 1 Diagrammatic representation of a chondrocyte in its extracellular matrix. The extracellular matrix contains, among many other molecules, the collagen fibril and the large proteoglycan aggrecan bound to hyaluronic acid. In OA there is excessive production by chondrocytes of MMPs and the cytokines IL-1 and TNFα. Recent research has revealed that matrix degradation products of type II collagen and fibronectin can induce further gene expression and synthesis of MMPs. Induction of these cytokines also occurs, and these have also been shown to be involved in matrix degradation in OA cartilage. Such cytokines, originating from synovial cells, can also induce these changes in chondrocytes.

glycosaminoglycans, which in adult cartilages are composed mainly of chondroitin sulphate, but also of keratan sulphate. Hydration leads to the creation of a swelling pressure that results from the swelling constraints imposed by an 'endoskeleton' of collagen fibrils that ramify throughout the extracellular matrix [1] (Figure 1). The fibrils are composed largely of type II collagen: they also contain type XI collagen within and on the fibril, and type IX collagen on the surface of the fibril. Also present on the fibril surface are small proteoglycans such as decorin, fibromodulin, and biglycan.

The fibril provides tensile strength for the matrix; this is most pronounced at the articular surface, where the shear forces of articulation are at their greatest. The tensile properties drop off rapidly with increasing cartilage depth [2]. At the surface, collagen fibrils, richly decorated with decorin and where higher concentrations of biglycan are present, are much thinner (~25 nm) than elsewhere in the matrix (~75–120 nm). Moreover, fibrils at the articular surface are arranged in bundles parallel to the surface.

Type II collagen, which is the prime focus of this review, is composed of three identical α chains, arranged in a triple helix. The short non-helical telopeptide domains present at the N- and C-termini are primary sites of cross-linking between molecules involving lysine residues both in these locations and in the triple helix. This provides stability for the formation of microfibrils, whereby individual collagen units $[\alpha_1(II)_3]$ self-associate in a quarter-stagger. Further microfibril association leads to fibril formation, such as can be seen by electron microscopy. Fibrils are very thin in pericellular sites, in contrast with the much larger diameters seen in the surrounding territorial matrix and the more remote interterritorial matrix present in the mid and deep zones.

Aging

As collagen ages, it is modified further by non-enzymic mechanisms that lead to the formation of advanced glycation end-products. This produces a stiffer fibril, more resistant to deformation [3]. In humans after the age of 30 years, there is evidence for a loss of tensile properties in large joints such as the knee and hip [4]. In contrast, articular cartilages in the ankle, which show much less evidence of degeneration with age, do not ordinarily exhibit a loss in strength [5].

The age-related loss of tensile properties is associated with an increase in the denaturation of type II collagen, as revealed by immunohistochemistry [6]. This becomes common after about 35 years of age, and is first observed at and near the articular surface in territorial and interterritorial sites around chondrocytes [6,7]. This damage extends progressively into the cartilage with increasing age.

Development of osteoarthritis (OA)

With the advent of OA, denaturation of type II collagen in the fibril is increased [8]. This leads to a loss of stiffness of the fibril and increased swelling and deformation of the cartilage, with a lack of effective containment of the proteoglycan aggrecan by the collagen fibrillar network [9]. Water content increases as a result of the increased swelling of the aggrecan molecules [10]. As collagen molecules at and near the articular surface are degraded, so tensile properties are markedly reduced [2]. This is accompanied by a loss of the proteoglycans decorin and biglycan, which are closely associated with these collagen fibrils [11]. With the damage to the collagen fibril, the surface of the cartilage starts to fibrillate and vertical splits develop involving the superficial and mid zones. The increased denaturation of collagen results in a loss of type II collagen [8], as observed previously for collagen content [12]. There is a clear inverse relationship between type II collagen denaturation and content (Figure 2). Clearly, the less damage there is to collagen, the more collagen is retained. This is an important pointer for therapeutic intervention in the control of collagen degradation in the management of OA.

Collagenases and other matrix metalloproteinases (MMPs)

The increased damage to type II collagen is a consequence, at least in part, of increased expression and activity of collagenases that cleave intially at the classic glycine-775–leucine-776 bond, approximately $\frac{3}{4}$ distant from the N-terminus of the triple-helical molecules. Using an immunoassay that is specific for the collagenase-generated cleavage neoepitope at the C-terminus of what is called the TC^A or $\frac{3}{4}$ fragment, one can demonstrate the increased cleavage in osteoarthritic cartilages [13]. This is usually seen at the same sites where increased denaturation of collagen occurs, starting at and close to the articular surface, and extending progressively into the territorial and interterritorial sites and deeper cartilage [7].

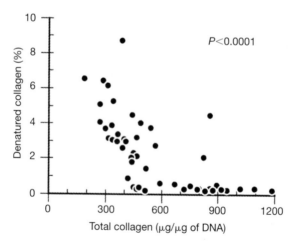

Figure 2 Analyses of knee articular cartilages from 50 patients with OA to show the relationship between denaturation of type II collagen and the content of this collagen in articular cartilage. Clearly, the less the damage to collagen (denaturation), the higher the content of collagen in the tissue. The assays used to determine this relationship are described in [8] (F. Nelson, L. Dahlberg and A.R. Poole, unpublished work).

Studies of explant cultures of osteoarthritic and normal cartilages have revealed that there is a rise in collagenase activity in OA [13,14]. Profiling of collagenase activity using an inhibitor of MMP-13 (collagenase-3) (to differentiate from MMP-1, also known as collagenase-1) has pointed to the involvement of MMP-13 in this increased proteolysis. Not all activity is inhibited by this inhibitor: in fact, in some cases no inhibition is observed. So, although MMP-13 may play a dominant role in the cleavage of collagen, it is not the only collagenase involved. The inhibition profile suggests the involvement of MMP-1 rather than other collagenases such as MMP-8, which is also detectable in cartilage [15].

Associated with the increased proteolysis by collagenases, there is a marked increase in the synthesis of type II collagen [16] and its type IIA variant, which is not expressed in healthy adult cartilage [17]. Using a broad-spectrum inhibitor, which can inhibit MMP-1 as well as MMP-13, we showed that these newly synthesized molecules are rapidly degraded. This was revealed by the increased incorporation of newly synthesized collagen in the presence of this inhibitor [14]. This degradation was unaffected by an MMP-13-selective inhibitor. Increased proteolysis of newly synthesized collagen may thus involve MMP-1 rather than MMP-13. Clearly, for effective control of proteolysis, we need to suppress cleavage not only of resident (existing) molecules but also of molecules involved in attempts at cartilage repair.

In OA cartilage there is increased expression of cathepsin K [14] as well as other proteases, including the gelatinases MMP-2 and MMP-9 [1,18]. Cathepsin K attracts attention since it can also act as a collagenase, cleaving the triple helix of type II collagen, although in a manner different from that of other collagenases [19]. While there is interest in the potential involvement of cathepsin K, we have no direct evidence to implicate this protease in the pathobiology of OA.

Regulation of collagenase activity

Up-regulation of collagenase activity similar to that observed in OA [13] can be induced experimentally in culture by the cytokines interleukin-1 (IL-1) and tumour necrosis factor α (TNFα) [1]. The cytokine oncostatin M can potently synergize with IL-1 to enhance collagen cleavage [20]. However, there is no clear evidence to suggest that oncostatin M is generated by chondrocytes or is involved in the degradative process. In contrast, IL-1 and TNFα are expressed and synthesized by chondrocytes and are up-regulated in OA [18] (Figure 1).

We recently investigated whether IL-1 and/or TNFα are involved in the increased collagen cleavage by collagenases and in aggrecan degradation in OA cartilage. Using IL-1 receptor antagonist protein (IL-1ra) and a soluble p55 TNFα receptor in explant culture, we discovered that collagen cleavage can be arrested by blockade of either cytokine in a majority of cases (10 out of 15): aggrecan degradation could also be arrested in around half (eight of 15) of the cases (M. Kobayashi, G. Squires, M. Tanzer, D.J. Zukor, J. Antoniou, U. Feige and A.R. Poole, unpublished work). This is associated with suppression of MMP-1 expression by IL-1ra and suppression of MMP-13 and MMP-3 (stromelysin-1) expression by IL-1ra combined with the p55 TNFα receptor.

The p55 receptor alone did not suppress these MMPs, and IL-1ra alone did not inhibit MMP-3 or MMP-13 expression. Clearly these MMPs (MMP-1, MMP-3 and MMP-13) are differentially regulated in OA cartilage, as suggested by other earlier work. These studies do, however, point to the involvement of these cytokines in the regulation of matrix degradation in OA, as summarized in Figure 1.

Early events – the development of OA

The development of cartilage degeneration is poorly understood at the tissue and molecular levels. Recent work on focal lesions of articular cartilage in aging individuals has revealed that these lesions embody all the features of excessive collagen and proteoglycan damage and increased matrix turnover that we see in more advanced pathology [21]. There is also evidence of chondrocyte hypertrophy as well as collagenase and IL-1/TNFα up-regulation (E. Tchetina, M. Kobayashi, T. Yashuda and A.R. Poole, unpublished work). Thus it is likely that, in idiopathic OA, lesions develop focally and enlarge progressively over the years to involve surrounding tissue. Whether similar changes occur in post-traumatic OA remains to be established.

Matrix degradation products and chondrocyte differentiation

The relationship between the chondrocyte and its environment is a fine balance if the chondrocyte phenotype and normal gene expression are to be maintained. It is worth stressing the importance of maintaining a matrix with intact type II collagen, which ensures normal chondrocyte gene expression. When type II collagen is cleaved excessively in normal development (as in the growth plate), this is accompanied by chondrocyte hypertrophy and matrix mineralization, with the up-regulation of MMP-13 [22,23]. However, this is a carefully controlled physiological event. If the activity of MMP-13 is arrested by use of a specific non-toxic inhibitor, chondrocyte hypertrophy is suppressed in addition to the suppression of MMP-13 expression [23]. There would therefore appear to be a relationship between MMP-13 activity, collagen proteolysis and chondrocyte hypertrophy.

Recently we have investigated whether degradation products of type II collagen may themselves act in a positive-feedback loop to stimulate collagenase gene expression. We discovered that cleaved, denatured type II collagen is indeed capable of inducing collagenase activity; particularly effective is a 24-residue-long peptide contained within the triple helix of this molecule, such as can be generated by proteolysis (T. Yasuda, W. Wu and A.R Poole, unpublished work). This peptide induces not only MMP-1 and MMP-13, but also IL-1 and TNFα (M. Kobayashi, G. Squires, M. Tanzer, D.J. Zukor, J. Antoniou, U. Feige and A.R. Poole, unpublished work) (Figure 1). Moreover, blockade of these cytokines also arrests the activity of the peptide. Of special interest is the capacity to induce chondrocyte hypertrophy, characterized by induction of expression of COL10A (type X collagen) and Cbfa1, the MMP-13 transcription factor that is up-regulated in hypertrophic chondrocytes (E. Tchetina, M. Kobayashi, T. Yashuda and A.R. Poole, unpublished work).

Chondrocyte hypertrophy is commonly observed in OA cartilage, being seen initially at and close to the articular surface [18]. It occurs in sites of more severe type II collagen damage, not where increased cleavage is first observed at the pathology/normal interface seen in articular cartilage (T. Meijers, R.C. Billinghurst and A.R. Poole, unpublished work). Chondrocyte differentiation seems, therefore, to be a secondary response to the advanced damage to the extracellular matrix, in particular the fibrillar collagen network. Just as hypertrophy leads to chondrocyte apoptosis and cartilage matrix mineralization in development and fracture repair [1], so we see evidence for chondrocyte apoptosis [24] and calcification of cartilage matrix [18] in OA. The mechanism of action of the collagen peptide involves a receptor-mediated signalling pathway that should be inhibitable and thereby provide a means of examining the importance of this pathway in cartilage degradation in OA.

Interestingly, another matrix molecule called fibronectin also has the capacity to induce matrix degradation, but again only when it is degraded. Following the demonstration that fragments of fibronectin can induce up-regulation of MMPs in synovial cells [25], similar activity was observed in cartilage cultures [26]. Subsequently, other domains have been shown to have activity, including the alternatively transcribed domains [27] and the heparin-binding C-terminal fragments [28]. Interestingly, different fibronectin fragments induce the proteolysis of both type II collagen and aggrecan, whereas the collagen peptide has no effect on aggrecan degradation, only stimulating collagen cleavage (M. Kobayashi, G. Squires, M. Tanzer, D.J. Zukor, J. Antoniou, U. Feige and A.R. Poole, unpublished work).

The effects of these degradation products are mainly dose-dependent, with fibronectin fragments being more potent than the collagen peptide. Clearly it is likely that, once a certain level of matrix damage is reached, this is when these autodegradative cycles are activated. Such positive-feedback pathways leading to up-regulation of MMP gene expression (Figure 1) may be very important in the pathology of chronic OA, where the cartilage chondrocyte appears to be fully capable of driving the pathological changes.

Much work has revealed that altered mechanical loading of cartilage can also alter matrix turnover [1]; cartilage matrix resorption can be suppressed or up-regulated by static or cyclic loading respectively. Thus damage to the extracelullar matrix of cartilage would result in changes in loading on chondrocytes. This would no doubt contribute significantly to altered gene expression favouring pathological change. Mechanical loading may also have more direct effects on collagen fibril integrity, as suggested by recent studies [27].

Use of biomarkers of collagen cleavage to assess disease activity *in vivo*

The ability to generate antibodies to the collagenase cleavage site in type II collagen has allowed the creation of immunoassays that can be used to detect and quantify the presence of collagen cleavage products in body fluids. These may reflect disease activity and be of use in the assessment of disease progression and its treatment *in vivo*. To specifically study cleavage of type II collagen, we developed a monoclonal antibody that recognizes a fragment of type II collagen generated by collagenase. Using this antibody (now called C2C, but known previously as COL2-3/4C$_{\text{Long mono}}$) [30,31] in conjunction with the antibody C1, 2C (known previously as COL2-3/4C$_{\text{short}}$), which recognizes a collagenase cleavage fragment generated from type I as well as type II collagens [13], we have found that the ratio of C1, 2C to C2C in sera is predictive of disease progression in patients with knee OA [32]. This may reflect increased cleavage by a protease closer to the collagenase cleavage site (the C1, 2C epitope is the shorter of the two epitopes). This indication of 'altered proteolysis' may be of value in assessing the therapeutic modulation of disease activity in clinical trials, where biomarker assessment over weeks or months may eventually replace clinical trials for OA that require a 2–3-year period for radiological evaluation of disease progression. Moreover, these assays could be used in preclinical studies of experimental OA and rheumatoid arthritis in animals, where their use may also hasten preclinical assessment of new therapies designed to be chondroprotective.

These studies were funded by Shriners Hospitals for Children, The National Institute of Aging, National Institutes of Health, Canadian Institutes of Health Research and the Canadian Arthritis Network, as well as Amgen, Wyeth and Roche Biosciences.

References

1. Poole, A.R. (2001) in Arthritis and Allied Conditions. A Textbook of Rheumatology, 14th edn (Koopman, W., ed.), pp. 226–284, Lippincott, Williams and Wilkins, Philadelphia
2. Kempson, G.E., Muir, H., Pollard, C. and Tuke, M. (1973) Biochim. Biophys. Acta 297, 465–472
3. Verzijl, N., DeGroot, J., Zaken, C.B., Braun-Benjamin, O., Maroudas, A., Bank, R.A., Mizrahi, J., Schwalkwijk, C.G., Thorpe, S.R., Baynes, J.W. et al. (2002) Arthritis Rheum. 46, 114–123
4. Kempson, G.E. (1982) Ann. Rheum. Dis. 41, 508–511
5. Kempson, G.E. (1991) Biochim. Biophys. Acta 1075, 223–230
6. Hollander, A.P., Pidoux, I., Reiner, A., Rorabeck, C., Bourne, R. and Poole, A.R. (1995) J. Clin. Invest. 96, 2859–2869
7. Wu, W., Billinghurst, R.C., Pidoux, I., Antoniou, J., Zukor, D., Tanzer, M. and Poole, A.R. (2002) Arthritis Rheum. 46, 2087–2094
8. Hollander, A.P., Heathfield, T.F., Webber, C., Iwata, Y., Rorabeck, C., Bourne, R. and Poole, A.R. (1994) J. Clin. Invest. 93, 1722–1732
9. Bank, R.A., Soudry, M., Maroudas, A., Mizrahi, J. and TeKoppele, J.M. (2000) Arthritis Rheum. 43, 2202–2210
10. Basser, P.J., Schneiderman, R., Bank, R.A., Wachtel, E. and Maroudas, A. (1998) Arch. Biochem. Biophys. 351, 207–219
11. Poole, A.R., Rosenberg, L.C., Reiner, A., Ionescu, M., Bogoch, E. and Roughley, P.J. (1996) J. Orthop. Res. 14, 681–689
12. Venn, M. and Maroudas, A. (1977) Ann. Rheum. Dis. 36, 121–129
13. Billinghurst, R.C., Dahlberg, L., Ionescu, M., Reiner, A., Bourne, R., Rorabeck, C., Mitchell, P., Hambor, J., Diekmann, O., Tschesche, H. et al. (1997) J. Clin. Invest. 99, 1534–1545
14. Dahlberg, L., Billinghurst, C., Manner, P., Ionescu, M., Reiner, A., Tanzer, M., Zukor, D., Chen, J., Van Wart, H. and Poole, A.R. (2000) Arthritis Rheum. 43, 673–682

15. Aurich, M., Poole, A.R., Reiner, A., Mollenhauer, C., Margulis, A., Kuettner, K.E. and
     Cole, A.A. (2002) Arthritis Rheum., **46**, 2903–2910

16. Nelson, F., Dahlberg, L., Laverty, S., Reiner, A., Pidoux, I., Ionescu, M., Fraser, G., Brooks,
     E., Tanzer, M., Rosenberg, L.C. et al. (1998) J. Clin. Invest. **102**, 2115–2125

17. Aigner, T., Zhu, Y., Chansky, H.H., Matsen, III, F.A., Maloney, W.J. and Sandell, L.J.
     (1999) Arthritis Rheum. **42**, 1443–1450

18. Poole, A.R. and Howell, D.S. (2001) in Osteoarthritis: Diagnosis, and Management, 3rd edn
     (Moskowitz, R., Howell, D.S., Goldberg, V.M. and Mankin H.J., eds), pp. 29–47, WB
     Saunders Co., Philadelphia

19. Kafienah, W., Brömme, D., Buttle, D.J., Croucher, L.J. and Hollander, A.P. (1998)
     Biochem. J. **331**, 727–732

20. Cawston, T.E., Curry, V.A., Summers, C.A., Clark, I.M., Riley, G.P., Life, P.F., Spaull,
     J.R., Goldring, M.G., Koshy, P.J.T., Rowan, A.D. and Shingleton, W.D. (1998) Arthritis
     Rheum. **41**, 1760–1771

21. Squires, G.R., Okouneff, S., Ionescu, M. and Poole, A.R. (2003) Arthritis Rheum.
     48, 1261–1270

22. Mwale, F., Tchetina, E. and Poole, A.R. (2002) J. Bone Miner. Res. **17**, 275–283

23. Wu, W., Tchetina E., Mwale, F., Hasty, K., Pidoux, I., Reiner, A., Chen, J., van Wart, H.E.
     and Poole, A.R. (2002) J. Bone Miner. Res. **17**, 639–651

24. Hashimoto, S., Ochs, R.L., Komiya, S. and Lotz, M. (1998) Arthritis Rheum. **41**, 1632–1638

25. Werb, Z., Tremble, P.M., Behrendtsen, O., Crowley, E. and Damsky, C.H. (1989) J. Cell
     Biol. **109**, 877–889

26. Homandberg, G.A., Meyers, R. and Xie, D.L. (1992) J. Biol. Chem. **267**, 3597–3604

27. Saito, S., Yamaji, N., Yasunaga, K., Saito, T., Matsumoto, S.-I., Katoh, M., Kobayashi, S.
     and Masuho, Y. (1999) J. Biol. Chem. **274**, 30756–30763

28. Yasuda, T. and Poole, A.R. (2002) Arthritis Rheum. **46**, 138–148

29. Thibault, M., Poole, A.R. and Buschmann, M.D. (2002) J. Orthop. Res. **20**, 1265–1273

30. Kojima, T., Mwale, F., Yasuda, T., Girard, C., Poole, A.R. and Laverty, S. (2001) Arthritis
     Rheum. **44**, 120–127

31. Song, X.-Y., Zeng, L., Jin, W., Thompson, J., Mizel, D.E., Lei, K.-J., Billinghurst, R.C.,
     Poole, A.R. and Wahl, S.M. (1999) J. Exp. Med. **190**, 535–542

32. Cerejo, R., Poole, A.R., Ionescu, M., Lobanok, T., Song, J., Cahue, S., Dunlop, D. and
     Sharma, L. (2002) Arthritis Rheum. **46**, S144

Biochem. Soc. Symp. **70**, 125–133
(Printed in Great Britain)
© 2003 Biochemical Society

11

Cytokine synergy, collagenases and cartilage collagen breakdown

Tim E. Cawston[1], Jenny M. Milner, Jon B. Catterall and Andrew D. Rowan

Rheumatology: School of Clinical and Medical Sciences, The Medical School, University of Newcastle Upon Tyne, Framlington Place, Newcastle Upon Tyne NE2 4HH, U.K.

Abstract

We have investigated proteinases that degrade cartilage collagen. We show that pro-inflammatory cytokines act synergistically with oncastatin M to promote cartilage collagen resorption by the up-regulation and activation of matrix metalloproteinases (MMPs). The precise mechanisms are not known, but involve the up-regulation of c-*fos*, which binds to MMP promoters at a proximal activator protein-1 (AP-1) site. This markedly up-regulates transcription and leads to higher levels of active MMP proteins.

Cartilage composition and breakdown

Cartilage is composed of a unique mixture of collagens, proteoglycans and glycoproteins that are precisely assembled and maintained by chondrocytes [1,2]. The collagen fibres give strength and rigidity to the tissue, while the proteoglycans, trapped within this network, swell as they hydrate. This allows the tissue to provide easy movement under load, to distribute load and to retain shape after compression [3].

In joint cartilage, tissue is broken down after cytokines and other stimuli act on chondrocytes and synovial cells to up-regulate and activate proteinases [4]. In healthy tissue these enzymes maintain the balance between synthesis and degradation of the extracellular matrix of cartilage within the joint. Their inappropriate up-regulation and subsequent activation in disease leads to joint destruction and the loss of joint function. Our aim has been to discover the key

[1]To whom correspondence should be addressed (e-mail T.E.Cawston@ncl.ac.uk).

cytokines, cells and enzymes in this process, in order to develop new therapies that can prevent joint destruction.

Collagen, a key component of cartilage

Proteoglycan and glycoproteins can be degraded rapidly, but the chondrocytes within the cartilage can resynthesize these molecules within a short space of time [5]. In contrast, collagen is often released more slowly; however, once the fibrillar structure is lost then it is difficult for the tissue to recover, and irreversible tissue destruction is often the result [6]. We have therefore focused our attention on the endopeptidases that cleave fibrillar collagen.

Endopeptidases are classified by the chemical group at the active site responsible for peptide bond cleavage [7]. In general, cysteine and aspartate proteinases act intracellularly at low pH, whereas serine and metalloproteinases act extracellularly at neutral pH. Both intracellular and extracellular routes are involved in turnover of the extracellular matrix [8].

The matrix metalloproteinases (MMPs) together cleave all extracellular matrix components, and have important roles to play in the body under normal physiological conditions, e.g. in growth and development [9]. However, in pathological states such as cancer, periodontal disease, rheumatoid arthritis (RA) and osteoarthritis (OA), MMPs are overproduced and/or inadequately inhibited. Although collagen fibres are relatively resistant to proteolytic cleavage, the MMPs are proposed to be the principle enzyme group responsible for collagen turnover [10].

Proteinases that break down collagen

The first vertebrate collagenase was described in 1962 by Gross and Lapierre [11], and was shown to specifically cleave the triple-helical collagen molecule at a single point across all three chains, three-quarters of the way along the molecule from the N-terminus. This collagenase was the first described member of the MMP family, and other members, namely MMP-2, MMP-8, MMP-13 and MMP-14, also cleave collagen in this specific fashion [12–16]. All MMPs are synthesized in a proenzyme form that requires activation, and other MMPs (e.g. MMP-3) can activate the collagenases. Some MMPs (e.g. the membrane-type MMPs; stromelysin 3) have a furin recognition site that allows intracellular activation, such that these enzymes can be secreted in an active form [17]. They can then initiate the activation of other MMPs at the cell surface. MMPs can also be activated by serine proteinases, such as plasmin generated from plasminogen by plasminogen activators. Inhibition of furin or plasminogen activators can block cartilage collagen destruction [18]. All MMPs are inhibited by a family of proteins called the tissue inhibitors of metalloproteinases (TIMPs-1–4), which bind active MMPs stoichiometrically to block substrate digestion [4].

Early work on MMP-1

My (T.E.C.) early work involved the purification of pig collagenase [19], and subsequent studies established the full-length structure of the recombinant pig enzyme [20]. This showed that the N-terminal catalytic domain is ellipsoidal, with an active-site cleft which contains the catalytic zinc ion at the bottom. This is attached, via a flexible hinge region, to a C-terminal domain that has a unique four-bladed β-propellor structure. In spite of these advances, the structure gives little clue as to the mechanism of action of this collagenase. It is known that the C-terminal domain is essential for binding to collagen [21], and different mechanisms have been put forward to explain the cleavage of collagen. The structure of the active-site cleft shows clearly that it is too narrow for the full triple helix to enter. It is proposed that perturbation of the structure occurs at the cleavage site, caused by binding of the C-terminal domain to the triple helix; this exposes one chain to the active-site cleft [22].

Evidence for involvement of collagenases in joint diseases

There is strong evidence that these enzymes are involved in the destruction of cartilage collagen within the diseased joint. Many studies show that collagenases are found in synovial fluid [23], synovial tissue [24] and cartilage tissue [25]. Specific epitopes, only produced after cleavage of collagen by the collagenases, are found within cartilage tissue in disease, and co-localization of MMP and cleavage product has been shown [26]. In animal models, joint destruction is associated with increased levels of the collagenases, and *in vitro* addition of TIMP-1 to resorbing cartilage blocks the release of collagen. Recent work has investigated the interaction between synovial and cartilage tissue within the joint. In RA in particular, the proliferation of synovial cells and the invasive nature of their penetration into cartilage determines the extent of cartilage and bone destruction. Studies using invasive synovial cells from RA patients implanted into the SCID (severe combined immunodeficiency) mouse model have identified new associations between MMP production and this invasive behaviour [27]. In addition, considerable interest in large multinucleated cells within this synovium have indicated that cathepsin K may also be involved in joint destruction, particularly with regard to bone [28]. RA is thus synovial-driven destruction, whereas in OA the cartilage destruction is chondrocyte-driven. Different therapies are likely to target different parts of the enzyme cascades involved in tissue destruction in these two diseases of the joint.

Cytokines act together in the joint

A variety of cytokines and growth factors are found within the inflamed joint [29]. These include the pro-inflammatory cytokines interleukin-1 (IL-1) and tumour necrosis factor α (TNFα), both of which are known to initiate the release of proteoglycan, and sometimes collagen, from cartilage [30]. In addition, it is known that a variety of growth factors, such as transforming growth factor β, can protect cartilage from the effects of these pro-inflammatory cytokines [31]. Within the inflamed joint there are a number of cytokines present, and part of our work has involved investigation of the effects of such mixtures of cytokines on cartilage collagen proteolysis. We have found that both IL-1 and TNFα can synergize with a member of the IL-6 family of cytokines, oncostatin M (OSM), to promote cartilage collagen destruction [32]. Collagen release is associated with a large increase (>20-fold) in collagenolytic activity, a significant proportion of which is in the active form. When chondrocytes or synovial cells are stimulated by mixtures of these cytokines, MMP-1, MMP-3, MMP-8 and MMP-13 are up-regulated [33] (Figure 1). For example,

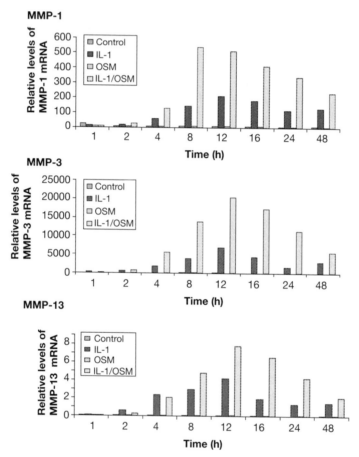

Figure 1 Effects of IL-1 and OSM on the gene expression of MMP-1, MMP-3 and MMP-13 in human chondrocytes. Human chondrocytes (T/C28a4) were treated for 2–48 h with medium alone (control), IL-1 (1 ng/ml), OSM (10 ng/ml) or IL-1+OSM. Following treatment, total RNA was isolated from cells and reverse transcribed, and the resulting cDNA was used in separate real-time quantitative PCR assays with specific primers and probes for MMP-1 (**A**), MMP-3 (**B**) and MMP-13 (**C**). Results are normalized to 18 S rRNA and presented graphically as relative RNA levels.

the level of MMP-1 mRNA produced when IL-1 and OSM are combined is more than 200 times the control level (Figure 1).

OSM is produced within the joint by synovial macrophages, and overexpression of OSM within murine joints causes profound cellular infiltration into the synovium, with subsequent damage to the cartilage and bone within the joint [34]. OSM acts via a specific β receptor [35] which we have found to be expressed by chondrocytes [36], and treatment of mice with collagen-induced arthritis with an antibody to OSM leads to a marked reduction in disease activity [37]. OSM can also synergize in a similar way with the T cell cytokine IL-17 [38]. Initially it was thought that OSM was unique within the IL-6 family in being able to synergize in this way with IL-1 and TNFα. However, subsequent studies have shown that both IL-6 and IL-11 can also synergize if they are present along with their respective soluble receptors.

Recent work using adenoviral vectors to deliver both OSM and IL-1 to mouse joints has shown a very marked response by day 7 after injection (A.D. Rowan, W. Hui, C.D. Richards and T.E. Cawston, unpublished work). When both cytokines are present together, then a marked infiltration of cells into the joint is seen, followed by the destruction of cartilage and bone. Proliferating synovial cells penetrate through into the marrow cavity, and marked swelling of treated joints occurs. Immunolocalization studies show that MMP-13 and MMP-3 levels within cartilage are high compared with those in control joints or joints treated with a single cytokine (A.D. Rowan, W. Hui, C.D. Richards and T.E. Cawston, unpublished work).

Mechanism of synergy between IL-1 and OSM

The mechanism of action to explain this synergy is not known. We have shown that the levels of cell surface receptors are not altered after treatment with cytokines, and consider that an overlap of intracellular signalling pathways is responsible. A previous study [40] proposed that there is an OSM response element in the MMP-1 promoter that was responsible for the increased transcription of MMP-1 when astrocytes were treated with combinations of IL-1 and OSM. However, this OSM response element is not operative in chondrocytes or synovial cells. One of the early genes up-regulated by this combination is c-*fos*, which also responds synergistically to this cytokine combination, and our working hypothesis is that alteration of the Fos/Jun partners at the proximal activator protein-1 (AP-1) site on the MMP-1 promoter dramatically alters transcription of the MMP-1 gene [41] (Figure 2). We continue to investigate the overlap of signalling pathways, and have used both microarray and proteomic techniques to identify both intracellular and secreted proteins that are up-regulated specifically by this combination of cytokines.

Activation of MMPs – a neglected control point

The transcriptional control of collagenases is complex, and considerable effort has gone into determining mechanisms and pathways. In many situations

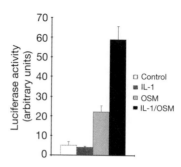

Figure 2 Activation of the c-*fos* promoter in chondrocytes after stimulation by IL-1 and/or OSM. Confluent chondrocytes (T/C28a4) were transiently transfected with a −711/+45 c-*fos* promoter/luciferase construct. Transfected cells were split and grown in acid-treated medium for 24 h before being stimulated with IL-1 (1 ng/ml) and/or OSM (10 ng/ml) for another 24 h. Cells were lysed and the luciferase activity was assayed. Plasmid incorporation was determined using Hirt's assay and used to correct for transfection efficiency.

it is possible to up-regulate collagenolytic MMPs but not see collagen destruction. Human cartilage, if exposed to IL-1 and OSM, up-regulates the levels of MMP-1 to a large extent, but this is not always accompanied by the release of collagen. However, if an activator such as active MMP-3 is added to the cartilage in culture, then early release of collagen is seen. This is illustrated for bovine cartilage in Figure 3, where release of collagen is seen by day 7 when MMP-3 is added. This demonstrates that the up-regulation of MMP-1 in this system occurs early (by day 2), but no release of collagen is seen until after day 7 because there is no activator of MMP-1 present at the early time points. Similar results are obtained when APMA (aminophenyl mercuric acetate) is added to the culture medium at early time points [42]. These results demonstrate the importance of activation as a control step, and illustrate the potential of blocking collagen turnover by the inhibition of activation cascades.

Synthetic inhibitors of MMPs

Highly specific MMP inhibitors have been made. Initial attempts to produce synthetic MMP inhibitors were based on the collagen cleavage site, with the scissile bond being replaced by a chelating group, such as hydroxamate to co-ordinate the catalytic zinc ion. Carboxylic acid, thiol and phosphorous ligands have been used to bind to zinc in other compounds. These inhibitors initially had problems of oral availability, but now compounds have been designed that can be absorbed by the gut without modification, and this has dramatically improved their oral availability. Non-peptidic inhibitors have also been designed [43]. The solving of crystal structures for MMPs [44] has allowed the fine tuning of inhibitors with increased specificity, and crystal structures have been obtained for the catalytic domains of many MMPs (MMP-1, -2, -3, -7, -8, -9, -11, -13 and -14). At least some of the variation in

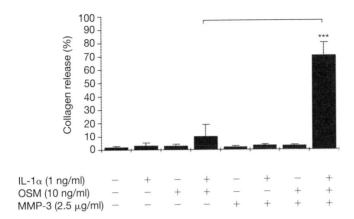

Figure 3 Addition of active MMP-3 to IL-1/OSM-treated bovine nasal cartilage. Bovine nasal cartilage discs were cultured in medium with or without active MMP-3, IL-1 and/or OSM. Medium was removed on day 7 and replenished with identical reagents. On day 14, medium was removed and the remaining cartilage was digested with papain. As a measure of collagen, the levels of hydroxyproline in day 7 and day 14 media and in cartilage digests were assayed. Collagen release by day 7 was calculated as a percentage of total collagen. Values are means ± S.D.; ***$P < 0.001$

substrate specificity among the MMPs can be explained by differences in the six specificity subsites in the active-site cleft and surrounding sequences. The S1′ pocket is particularly important, e.g. being deeper in MMP-3 and larger in MMP-8 than in MMP-1. This offers target sites for engineering specificity into synthetic MMP inhibitors.

Most early studies involved the treatment of cancer. For example Marimastat (BB-2516), an orally available hydroxamate inhibitor of MMPs with limited ability to inhibit sheddase activity, is in clinical development [45]. The inhibitor BAY 12-9566 targets MMP-3 (with low activity against MMP-1) for the treatment of OA [46]. It also inhibits MMP-2, -8, -9 and -13, and was effective in models of OA [42]. However, BAY 12-9566 was withdrawn from a phase III trial involving 1800 patients with arthritis following negative effects in cancer trials of the same drug [47].

As irreversible cartilage damage appears to occur only after the collagen network has been destroyed, it is important to ensure that the collagenases are a potential therapeutic target. Trocade (Ro 32-3555) has a low nanomolar K_i against MMP-1, -8 and -13, with approx. 10–100-fold lower potency against MMP-2, -3 and -9. It blocks IL-1α-induced collagen release from cartilage explants and, *in vivo*, prevented cartilage degradation in a rat granuloma model, a *Propionibacterium acnes*-induced rat arthritis model and an OA model using the SRT/ORT mouse (a well recognized breed susceptible to OA) [48]. However, large-scale trials of Trocade in patients with RA were terminated, presumably because of a lack of efficacy, and its failure does leave the future of therapies targeted at the collagenases in question.

It will be interesting to see whether the blocking of one enzyme in the MMP family with some inhibition of the others is sufficient to halt the progressive and chronic destruction of connective tissue seen in the arthritides. It may be necessary to combine proteinase inhibitors, either in sequence or with other agents that target other specific steps in the pathogenesis, before the chronic cycle of joint destruction found in these diseases can be broken [49].

The help and encouragement of Dr Alan Barrett during my (T.E.C.) postdoctoral training at the Strangeways Research Laboratory in Cambridge is very gratefully acknowledged. The work described has resulted from significant contributions by members of past and present research teams, which has been fun and is gratefully acknowledged.

References

1. Prockop, D.J. (1998) Matrix Biol. **16**, 519–528
2. Hardingham, T.E. and Fosang, A.J. (1992) FASEB J. **6**, 861–870
3. Eikenberry, E.F. and Bruckner, P. (1999) in Dynamics of Cartilage and Bone Metabolism (Seibel, M.J., Robins, S.P. and Bilezikian, J.P., eds), pp. 289–300, Academic Press, San Diego
4. Cawston, T.E. (1996) Pharmacol. Ther. **70**, 163–182
5. Page, T.D.P., King, B., Stephen, T. and Dingle, J.T. (1991) Ann. Rheum. Dis. **50**, 75–80
6. Jubb, R.W. and Fell, H.B. (1980) J. Pathol. **130**, 159–162
7. Barrett, A.J., Rawlings, N.D. and Woessner, J.F. (1998) in Handbook of Proteolytic Enzymes, pp. xxv–xxix, Academic Press, London
8. Everts, V., Van Der Zee, E., Creemers, L. and Beertsen, W. (1996) Histochem. J. **28**, 229–245
9. Nagase, H. and Woessner, Jr, J.F. (1999) J. Biol. Chem. **274**, 21491–21494
10. Shingleton, W.D., Hodges, D.J., Brick, P. and Cawston, T.E. (1996) Biochem. Cell Biol. **74**, 759–775
11. Gross, J. and Lapierre, C.M. (1962) Proc. Natl. Acad. Sci. U.S.A. **54**, 1197–1204
12. Ohuchi, E., Imai, K., Fujii, Y., Sato, H., Seiki, M. and Okada, Y. (1997) J. Biol. Chem. **272**, 2446–2451
13. Cole, A.A., Chubinskaya, S., Schumacher, B., Huch, K., Szabo, G., Yao, J., Mikecz, K., Hasty, K.A. and Kuettner, K.E. (1996) J. Biol. Chem. **271**, 11023–11026
14. Lindy, O., Konttinen, Y.T., Sorsa, T., Ding, Y.L., Santavirta, S., Ceponis, A. and López-Otín, C. (1997) Arthritis Rheum. **40**, 1391–1399
15. Ames, R.T. and Quigley, J.P. (1995) J. Biol. Chem. **270**, 5872–5876
16. Cawston, T.E. (1998) in Handbook of Proteolytic Enzymes (Barrett, A.J., Rawlings, N.D. and Woessner, J.F. eds), pp. 1155–1162, Academic Press, London
17. Pei, D. and Weiss, S.J. (1995) Nature (London) **375**, 244–247
18. Milner, J.M., Rowan, A.D., Elliott, S.-F. and Cawston, T.E. (2003) Arthritis Rheum. **48**, 1057–1066
19. Cawston, T.E. and Tyler, J.A. (1979) Biochem. J. **183**, 647–656
20. Li, J., Brick, P., O'Hare, M.C., Skarzynski, T., Lloyd, L.F., Curry, V.A., Clark, I.M., Bigg, H.F., Hazleman, B.L., Cawston, T.E. et al. (1995) Structure **3**, 541–549
21. Clark, I.M. and Cawston, T.E. (1989) Biochem. J. **263**, 201–206
22. Bode, W. (1995) Structure **3**, 527–530
23. Cawston, T.E., Mercer, E., De Silva, M. and Hazleman, B.L. (1983) Arthritis Rheum. **27**, 641–646

24. Konttinen, Y.T., Ainola, M., Valleala, H., Ma, J., Ida, H., Mandelin, J., Kinne, R.W., Santavirta, S., Sorsa, T., López-Otín, C. and Takagi, M. (1999) Ann. Rheum. Dis. **58**, 691–697

25. Tetlow, L.C. and Woolley, D.E. (1998) Br. J. Rheumatol. **37**, 64–70

26. Billinghurst, R.C., Dahlberg, L., Ionescu, M., Reiner, A., Bourne, R., Rorabeck, C., Mitchell, P., Hambor, J., Diekmann, O., Tschesche, H., Chen, J. et al. (1997) J. Clin. Invest. **99**, 1534–1545

27. Muller-Ladner, U., Kriegsmann, J., Franklin, B.N., Matsumoto, S., Geiler, T., Gay, R.E. and Gay, S. (1996) Am. J. Pathol. **149**, 1607–1615

28. Hummel, K.M., Petrow, P.K., Franz, J.K., Mullner-Ladner, U., Aicher, W.K., Gay, R.E., Bromme, D. and Gay, S. (1998) J. Rheumatol. **25**, 1887–1894

29. Van den Berg, W.B. (1999) Z. Rheumatol. **58**, 136–141

30. Ellis, A.J., Curry, V.A., Powell, E.K. and Cawston, T.E. (1994) Biochem. Biophys. Res. Commun. **201**, 94–101

31. Hui, W., Rowan, A.D. and Cawston, T.E. (2000) Cytokine **12**, 765–769

32. Cawston, T.E., Curry, V.A., Summers, C.A., Clark, I.M., Riley, G.P., Life, P.F., Spaull, J.R., Goldring, M.B., Koshy, P.J., Rowan, A.D. and Shingleton, W.D. (1998) Arthritis Rheum. **41**, 1760–1771

33. Koshy, P.J.T., Lundy, C.J., Rowan, A.D., Porter, S., Edwards, D.R., Hogan, A., Clark, I.M. and Cawston, T.E. (2002) Arthritis Rheum. **46**, 961–967

34. Langdon, C., Kerr, C., Hassen, M., Hara, T., Arsenault, A.L. and Richards, C.D. (2000) Am.J.Pathol. **157**, 1187–1196

35. Mosley, B., DeImus, C., Friend, D., Boiani, N., Thoma, B., Park, L.S. and Cosman, D. (1996) J. Biol. Chem. **271**, 32635–32643

36. Rowan, A.D., Koshy, P.J.T., Shingleton, W.D., Degnan, B.A., Heath, J., Vernallis, A.B., Spaull, J.R., Life, P.F., Hudson, K. and Cawston, T.E. (2001) Arthritis Rheum. **44**, 1620–1632

37. Plater-Zyberk, C., Buckton, J., Thompson, S., Spaull, J., Zanders, E., Papworth, J. and Life, P.F. (2001) Arthritis Rheum. **44**, 2697–2702

38. Koshy, P.J.T., Henderson, N., Logan, C., Life, P.F., Cawston, T.E. and Rowan, A.D. (2002) Ann. Rheum. Dis. **61**, 704–713

39. Reference deleted

40. Korzus, E., Nagase, H., Rydell, R. and Travis, J. (1997) J. Biol. Chem. **272**, 1188–1196

41. Catterall, J.B., Carrere, S., Koshy, P.J.T., Degnan, B.A., Shingleton, W.D., Brinckerhoff, C.E., Rutter, J., Cawston, T.E. and Rowan, A.D. (2001) Arthritis Rheum. **44**, 2296–2310

42. Milner, J.M., Elliott, S.-F. and Cawston, T.E. (2001) Arthritis Rheum. **44**, 2084–2096

43. Whittaker, M., Floyd, C.D., Brown, P. and Gearing, A.J.H. (1999) Chem. Rev. **99**, 2735–2776

44. Bode, W., Fernandez-Catalan, C., Tschesche, H., Grams, F., Nagase, H. and Maskos, K. (1999) Cell. Mol. Life Sci. **55**, 639–652

45. Brown, P.D. (1999) APMIS **107**, 174–180

46. Leff, R.L. (1999) Ann. N.Y. Acad. Sci. **878**, 201–207

47. Bayer drug casts shadow over MMP inhibitors in cancer (1999) Scrip **2476**, 7

48. Lewis, E.J., Bishop, J., Bottomley, K.M.K., Bradshaw, D., Brewster, M., Broadhurst, M.J., Brown, P.A., Budd, J.M., Elliott, L., Greenham, A.K. et al. (1997) Br. J. Pharmacol. **121**, 540–546

49. Mengshol, J.A., Mix, K.S. and Brinckerhoff, C.E. (2002) Arthritis Rheum. **46**, 13–20

Biochem. Soc. Symp. **70**, 135–146
(Printed in Great Britain)
© 2003 the Authors

12

Proteolytic 'defences' and the accumulation of oxidized polypeptides in cataractogenesis and atherogenesis

Roger T. Dean[*][1]**, Rachael Dunlop**[†]**, Peter Hume**[†]
and Ken J. Rodgers[†]

*University of Canberra, Canberra, ACT 2601, Australia, and †Heart Research
Institute, Sydney, NSW 2050, Australia

Abstract

Over the last few years, it has been clearly established that normal plasma
contains low levels of oxidized polypeptides, and that these accumulate in tissues during several age-related pathologies. In contrast, normal mammalian
aging, contrary to conventional dogma, is not clearly associated with enhanced
levels of oxidized proteins, except in extracellular connective tissues, whose
proteins can, for example, be oxidized by the neutrophil oxidative burst. Since
mildly oxidized proteins are susceptible to accelerated degradation in most
experimental systems, the question arises as to how the accumulation of oxidized proteins can take place. Such accumulation requires an excess of
production (or deposition) over removal, which might reflect alterations in
capacity or rate of production or removal. This chapter discusses our presently
limited knowledge of rates and control of proteolysis of oxidized proteins in
two pathologies, cataractogenesis and atherogenesis. It commences with a brief
summary of current understanding of the mechanisms of protein oxidation,
and of the observed accumulation of oxidized proteins in several pathologies.

Introduction

In the discussion that follows, we will very briefly summarize current
knowledge of protein oxidation, and of the accumulation of oxidized proteins in
physiology and pathology. Our main focus is on the proteolytic degradation of

[1]To whom correspondence should be addressed (e-mail roger.dean@canberra.edu.au).

such molecules, which can be construed as a defence reaction, particularly in the light of our observations of long-lived reactive species included among the spectrum of these oxidation products. Such reactive species can damage most, if not all, other biomolecules, and hence their removal from biological systems might well be protective. The selective references quoted are chosen mainly from the very recent literature, together with our own contributions, but we refer readers to some more detailed reviews that are pointers to the wider literature [1–4].

Mechanisms of protein oxidation

A variety of pathways of protein oxidation, one-electron (radical-mediated) or two-electron (for example by some of the reactions of hypochlorite), have now been characterized. In addition, there are related adduction reactions, such as between aldehydes and other carbonyl functions and amino groups of proteins; these may be ancillary to other oxidation pathways, in that the carbonyls may themselves be the result of prior radical-mediated events (as in the case of some aldehydes generated by sugar autoxidation [5]). While radical-mediated pathways are important in both lipid and protein oxidation, certain oxidants, such as hypochlorite and peroxynitrite, show selectivity for protein in some circumstances, such as in low-density lipoprotein (LDL) or nucleoproteins (e.g. [6,7]). The myeloperoxidase/hypochlorite pathway is important in neutrophil-mediated damage to bacteria [8]. An unusual cysteine–lysine sulphinamide cross-link has recently been identified as a product of the hypochlorite-mediated damage to protein [9].

The chemistries of several of these pathways have been gradually elucidated; for example, the Maillard reaction, between sugars and proteins, has been under intensive study for many decades, and radical/autoxidative components have been defined in the reactions both of free sugars and of previously formed sugar–protein adducts. In diabetes induced in rats by streptozotocin, levels of kidney protein carbonyls are elevated, suggesting that oxidative events may be important [10], and this also seems to be true of erythrocyte membranes from humans with poorly controlled diabetes [11]. However, other work has emphasized the importance in diabetes of secondary carbonyl molecules, such as methylglyoxal, and the consequent 'carbonyl stress' ([12,13]). Most recently, the oxidative deamination of lysine by a free radical-requiring reaction within the sugar autoxidative pathway has been demonstrated [14]; this reaction can also occur in other radical-mediated pathways.

On the other hand, the free radical chemistry of protein oxidation induced by hydroxyl radicals and Fenton chemistry (involving transition metals and peroxides) has only been adequately understood fairly recently. The pathways of damage induced by radiolytic hydroxyl radicals and transition metal-dependent Fenton chemistry are not identical, partly because of the influence of the site at which the metal is located (e.g. [15]). In each category of attack, most amino acids ultimately can be damaged, and the site of initial radical formation on a protein is not necessarily the site of final damage, indicating that radical centres are mobile (e.g. see [16]). In very early work, Garrison postulated [17,18] a role for some radical intermediates in the oxidation of many amino acids and peptides; the clear, direct detection of those and other radicals followed much later (e.g. [19]). Selective pathways generating chlorinated products (due to hypochlorite and myeloperoxidase) and nitrated products (due to peroxynitrite and congeners) have also been defined.

We were able to propose a pathway for a protein oxidation chain reaction, in which there are two salient features that differentiate it from the chain reaction of lipid oxidation: the importance of alkoxyl radicals in the former pathway, but not the latter; and the relatively low oxygen consumption in the protein pathway [19]. Alkoxyl radicals play a role in polypeptide cleavage reactions [20,21]. The roles for alkoxyl radicals point to the likelihood that mechanisms of chain termination for protein oxidation may well be distinct from those for lipid oxidation, and peroxyl radical-directed chain-breakers such as vitamin E (α-tocopherol) cannot be assumed to be effective against protein oxidation. Consistent with this, for example, in a group of 15 clinically normal humans with relatively high plasma protein carbonyl levels, administration of vitamin E for 18 days had no effect on protein carbonyl values [22]. On the other hand, γ-tocopherol seems to be a selective defence against protein nitration [23], by virtue of specialized chemical reactivities. Hydrophobicity and membrane disposition also influence the accessibility of lipids and proteins to potential antioxidant molecules [24]. While proteins are generally less prone to oxidation from a chemical perspective than lipids, the initiating factors are often in common, for example hydroperoxides and transition metals, and hence prevention of initiation may be similar for lipid and protein oxidation.

Long-lived reactive species on oxidized proteins

We and several other groups have sought to develop assays for oxidized amino acids found on oxidized proteins which could be used on biological materials, and which cumulatively could allow the determination of the relative importance of particular reaction pathways in generating the spectrum of oxidation products observed in a particular physiological or pathological circumstance. Thus, for example, hydroxides of aliphatic amino acids are generated largely by highly reactive species such as hydroxyl radicals, and hardly at all by UV-driven pathways, whereas chloro-tyrosines are formed by pathways involving hypochlorite and/or myeloperoxidase. Di-tyrosines can be formed by both the myeloperoxidase and hydroxyl radical pathways, but in the latter case only when the hydroxyl radical flux is sufficient to allow a high enough instantaneous concentration of tyrosyl radicals so that that their combination to form di-tyrosine reaches a detectable rate. Because of the greater chemical stability of hypochlorite and the enzymic function of myeloperoxidase, even low concentrations can cause the generation of di-tyrosine [25]. We have emphasized [1,2] that caution is necessary in interpreting the literature with regard to protein-carbonyls from complex biological samples. Besides the fact that protein carbonyls are generally measured as a heterogeneous group of entities, rather than a specific product, there are difficulties in distinguishing protein- from other carbonyls, and quantitative data are very inconsistent. However, a valuable approach to identifying individual oxidized proteins after gel electrophoresis exploits the binding of antibodies against 2,4-dinitro-phenylhydrazine (DNPH) to DNPH-derivatized protein-carbonyls. Similar caution is necessary in the interpretation of the more recently popularized advanced oxidation products of proteins (AOPP), which are poorly character-ized as yet (e.g. [26]), even more so than the advanced glycoxidation end-products ('AGEs') after which their name is styled. For example, during intravenous iron supply to haemodialysis patients, enhanced AOPP levels do not correlate with changes in di-tyrosine or in a range of other non-protein oxidative parameters [27].

During the search for suitably stable reaction products, we observed the for-mation of relatively long-lived reactive species, both of an oxidizing and of a reducing nature [28,29]. Hydroperoxides constituted the main oxidizing species generated on proteins during hydroxyl radical attack, whereas the reducing species was significantly but not solely composed of dopa (3,4-dihydroxyphenyl-alanine) formed by the hydroxylation of tyrosine [30]. These reactive intermediates are orders of magnitude more stable than protein radical intermedi-ates of the oxidation chain reaction themselves, and thus can, in principle, diffuse in cells and react with distant biomolecules. We and others have directly demon-strated the reactivity of protein hydroperoxides [31] and protein-bound dopa and their capacity to damage other proteins, lipids and DNA [32,33]. Protein chlor-amines, intermediates in the reactions of hypochlorite, can also damage other biomolecules, in some cases through secondary radical generation [7].

Accumulation of oxidized proteins might be pathogenic

In view of this capacity for secondary damage due to reactive species on oxidized proteins, it can be hypothesized that protein oxidation might be both a symptom of an oxidative affront and potentially a cause of progressive dam-age. The latter might occur through inactivation of protein function due to oxidation (and indeed inactivation is one of the most common early events dur-ing the progressive oxidation of isolated proteins), with consequent cellular failure. For example, oxidation of methionine residues in cellular actin seems to interfere drastically with its function [34]. Even the relatively robust chaperone function of lens α-crystallin can be readily inactivated by arginine modification provoked by some sugar oxidation products [35]. Progressive secondary dam-age might also occur through the capacity of reactive protein moieties to inflict comparably important damage on other biomolecules. Thus there has been a longstanding interest in assessing the occurrence and possible accumulation of oxidized proteins in disease.

Can accumulation of oxidized proteins be demonstrated?

One surprise has been the recurrent observation that even normal human plasma, a fluid containing a dramatic level of antioxidant defences, contains low levels both of cholesterol ester hydroperoxides (cholesterol esters being one of the major pools of oxidizable fatty acids in plasma) and of protein-bound dopa and other aromatic oxidation products [36,37]. Antioxidant defences and proteolytic degradation and excretion thus do not cause complete removal of such products. Low levels are presumably not pathogenic.

However, in several pathologies, such as certain inflammatory diseases, Alzheimer's disease, Parkinson's disease, cataractogenesis in the eye and atherosclerosis in large arteries, more substantial accumulation of oxidation products occurs. Only in a few cases has an attempt been made as yet to determine the spectrum of protein oxidation products that accumulate [36,37]. Evidence for the accumulation of oxidized proteins in aging is equivocal, except for in extracellular connective tissue proteins and possibly skeletal muscle [38,39]. Some evidence exists for intracellular proteins: for example, enhanced protein carbonyls were described recently in the aging canine brain, but this has not been substantiated fully by measurement of specific protein oxidation products [40].

Cataract in the lens of the human eye is one such example. Early observations claimed the occurrence of isolated markers of protein oxidation (e.g. [41,42]). More recently, we systematically investigated the cataract-stage-dependence of the accumulation of a range of markers, including aliphatic amino acid hydroxides, aromatic oxidation products, di-tyrosine and chlorotyrosine [43]. The markers accumulate in a progressive disease-stage-dependent manner, and with a spectrum very similar to that obtained when isolated proteins are exposed to hydroxyl radical attack. Thus, expressed per mol of parent amino acid, dopa predominates, phenylalanine oxidation products are substantial, and aliphatic hydroxides are modest. Levels of di-tyrosine are very low, and chlorotyrosine is also essentially absent. The spectrum observed does not suggest the involvement of hypochlorite/myeloperoxidase-dependent processes. The levels of substitution observed in cataractous lenses, particularly in the nigrescent (black-coloured) samples, are the highest *in vivo* values described in the literature to date. We also showed that the spectrum of products is inconsistent with that generated by direct UV-mediated damage; we envisage that UV may be partly responsible for the supply of co-oxidants, such as hydrogen peroxide, and redox-available transition metals [43].

In the case of atherogenesis, our initial studies, and those of colleagues such as Heinecke, indicated that advanced atherosclerotic plaques contain a more complex spectrum of oxidation products, at lower levels of substitution than observed in the cataract samples [44]. The spectrum is compatible with a hybrid of hydroxyl radical/Fenton chemistry and hypochlorite/myeloperoxidase involvement. Later studies elaborated the stage-dependence of the accumulation, showing that most accumulation occurred in quite late stages of the disease, when assessed in samples of supernatants obtained from total homogenates of plaque [45]. Lipoproteins were present in these samples at significant levels, and they showed many signs of oxidation. Interestingly, some evidence had suggested, in contradiction to ours, that di-tyrosine might accumulate in earlier stages of disease, judged by the determination of total homogenate and of isolated lipoproteins [44]. In subsequent work using a lipoprotein-depleted, protein-rich fraction (R.T. Dean, S. Linton and M. Davies, unpublished work), we obtained evidence indicative of early elevations of dopa (but not di-tyrosine). The possibility remains open that the accumulation of oxidized soluble proteins, although not of oxidized lipids of lipoproteins, might be a very early event in human atherogenesis.

Can protein oxidation be dissociated from pathogenesis?

In a detailed and important series of studies, our colleagues Roland Stocker and his group have demonstrated that, in some animal models of atherogenesis, experimental inhibition of lipid oxidation can be achieved in the vessel wall by dietary administration of appropriate antioxidants, but that this may or may not inhibit atherogenesis, depending on the model in question, and in some cases even depending on the vascular site under study in a given animal model [46–48]. This work has, at the very least, thrown into question the universality of the 'lipoprotein lipid oxidation' hypothesis of atherogenesis, developed by many workers [49] and most clearly promulgated by Steinberg et al. [50]. This hypothesis envisages that the oxidation of LDL is the initial event responsible for vascular injury, cholesterol deposition in the vessel wall, other aspects of atherogenesis and many of the sequelae of atherosclerosis.

However, the dissociation of inhibition of lipid oxidation from atherogenesis in certain models still leaves open the possibility that in others where there is a parallel (i.e. where inhibition of lipid oxidation and atherogenesis are concurrent), lipid oxidation may have an important pathogenic role. When the analogous question is asked of the role of protein oxidation, we as yet have no substantial answer. We are not sure of an appropriate intervention mechanism, since lipid antioxidants are not necessarily effective against protein oxidation, as discussed above. On the other hand, the incipient evidence mentioned above that protein oxidation is at least an early process in atherogenesis makes the need for such an intervention experiment the more pressing.

Such an experiment is even more urgent in the case of cataractogenesis, since protein oxidation leads to the unfolding and aggregation of crystallins, and in the case of S-crystallin at least probably to other modifications such as deamidation [51]. Cumulatively, these processes could be responsible for most of the key features of the disease, such as the progressive opacity and coloration.

Proteolysis of oxidized proteins: model systems

If protein oxidation and oxidized proteins themselves might be pathogenic in certain diseases, what are the important defences against this? We will not discuss further the primary defence, i.e. the avoidance of oxidative stress and damage, to which we have alluded already. A few limited forms of protein oxidation can be repaired; for example, the early steps in the oxidation of methionine are reversible by the methionine suphoxide reductase system, localized primarily in mitochondria [52]. This system may have important antioxidant defence roles [53]. Here, though, we will focus on secondary defences, notably the proteolytic removal of oxidized proteins by their extensive degradation, leading to the potential removal of oxidized amino acids by excretion [54]. There are some acute pathologies in which accentuated protein oxidation seems to occur, and in some of these there is evidence for their accelerated removal. In very-low-birthweight humans, elevated levels of protein carbonyls and lipid oxidation products, but not chlorotyrosine, occur in the cerebrospinal fluid of those individuals prone to cerebral white matter injury [55]. Uraemic muscle also seems to contain elevated levels of protein carbonyls [39]. In sepsis in adult humans, such an elevation precedes accelerated proteolysis *in vivo*, whereby loss of body protein in accompanied by loss of oxidized protein [56]. What is not clear is how general these relationships between protein oxidation and proteolysis may be *in vivo*, and to what degree they are relevant in chronic pathologies such as the two we focus upon.

It has been observed in many systems, from lysates to intact cells, and from intracellular catabolism of endogenous proteins to degradation of exogenous proteins after endocytosis into cells, that modestly oxidized proteins tend to be degraded faster than native proteins, but that extremely oxidized proteins may become resistant [57–59]. Recent examples of this have been described for mitochondrial aconitase in human WI-38 cells [60] and for HepG2 cells after ethanol, peroxide or trichlorobromomethane treatment [61]. Among the resistant forms are aggregates, similar to the amyloid deposits that are formed as inclusions in many diseases [62].

It is clear that endocytosed proteins, including oxidized ones, are generally degraded by the endolysosomal system. In contrast, it has been difficult to elucidate fully the mechanisms responsible for the cellular degradation of endogenous oxidized proteins, partly because of the toxicity commonly associated with oxidative insults needed to foster their generation in intact cells (e.g. [63]). However, recent proteomic approaches have begun to reveal the variety of intracellular proteins that are oxidized during experimental oxidative affronts, such as glyceraldehyde-3-phosphate dehydrogenase in *Saccharomyces cerevisiae* exposed to hydrogen peroxide [64], and a range of iron- and peroxide-metabolizing and structural proteins in iron-induced intracellular oxidation in mammalian cells [65]. Similarly, the cytoskeletal protein ezrin is degraded selectively in clone 9 rat liver cells after exposure to hydrogen peroxide, but corresponding changes in synthesis prevent significant changes in available ezrin [66]; the nature of the chemical changes in ezrin molecules responsible for this are not clear. Protein disulphide isomerase is also degraded and replaced under similar circumstances [67]. In Alzheimer's disease, around 100 selectively oxidized proteins have been noted in two-dimensional gels by DNPH derivatization [68]. Understanding in more detail the nature of the substrates undergoing accelerated proteolysis is of course essential to fully defining the relevant proteolytic mechanisms.

Many studies of the catabolism of oxidized proteins have retreated to the use of isolated organelles, and proteasomes clearly have important capacities for the degradation of oxidized proteins (e.g. reviewed in [69,70]). Recent evidence suggests that this proteasomal role may not require ubiquitinated substrates [71]. Evidence for their importance in intact cells is now strong (for example in HepG2 cells after ethanol or an oxidative affront [61]), although many studies are limited by the imprecision of 'selective' inhibitors. It is extremely difficult to securely quantify the importance of different pathways using approaches based on inhibitors, because of the redundancies built into proteolytic machineries. These redundancies include the fact that lysosomes can catabolize vast quantities of proteins delivered by endocytosis without the need to change the cellular armoury of proteins, yet may normally only be degrading much smaller quantities of endogenous proteins; and that when one proteolytic machinery is inhibited, another may compensate. Proteolytic machineries are present in every part of the cell; for example, the capacity of Lon protease to degrade oxidized mitochondrial aconitase has been addressed recently [60]. One component of the N-end pathway of proteolysis, involving N-terminal arginylation and protein ubiquitination prior to degradation, has recently been shown to have an unexpected possible relevance as 'oxygen' sensor. One of the three N-terminal residues involved is cysteine, and it is apparently oxidized prior to arginylation, with the possible implication that it might also sense oxidized proteins [72]. It is probably fair to say that dogmatic assertions of the dominance of particular pathways have ephemeral followings.

As a new approach to assessing proteolytic mechanisms acting on proteins containing oxidized amino acids, we have developed a method for incorporating oxidized amino acids into intact cells *in vitro* (and *in vivo*) by protein synthesis [73]. Several, but not all, of the oxidized amino acids discussed above can be incorporated in this manner. For example, we have studied in most detail the incorporation of dopa, which is available commercially in radioactive form, but also of hydroxyleucine, hydroxyvaline, and *o*- and *m*-tyrosine. Evidence for the incorporation of these amino acids by protein synthesis includes its inhibition by cycloheximide, and demonstration of the accumulation of the supplied amino acid into cellular proteins, detectable after acid hydrolysis of proteins by the analytical methods we have developed. Incorporation is competitive with the parent amino acid, and in some cases with those amino acids that share the same cellular transporter, but still occurs even when all amino acids are supplied at physiological levels.

This approach has allowed us to control relatively precisely the extent of substitution of normal by oxidized amino acids, and also to focus on populations of proteins of different half-lives, still distinguishing those proteins that contain oxidized amino acids (represented for example by radioactive dopa) from the vast majority of proteins that do not (represented almost as 'cleanly' by labelling with radioactive leucine or other non-oxidized amino acid). We observe the classic biphasic dependence of degradation on the extent of incorporation of oxidized amino acid. Inhibitor studies suggest the involvement of both lysosomes and proteasomes in degradation, with the latter taking a dominant role. However, lysosomal contributions seem to remain modest and unchanging as the extent of substitution of oxidized amino acid increases, while the proteasomal role increases. We have suggested a possible pathway of transfer of incompletely degraded proteins from proteasomes to lysosomes [73,74]. Many aspects of these degradative mechanisms remain obscure and controversial, certainly at a quantitative as opposed to a qualitative level of analysis.

Proteolysis of oxidized proteins as a defence in pathogenesis

In some of the many systems in which oxidized proteins show a biphasic susceptibility to degradation, it is clear that the proteolytic machinery itself is not altered. Thus changes in the nature of the oxidized protein as a substrate, notably unfolding and modified chemistry of the constituent amino acids, can affect susceptibility. However, in some circumstances there are also indications that regulation or decline of proteolytic machineries occurs, or that the oxidized proteins may damage the degradative machinery. For example, there are many indications in the literature that proteasomal function declines in aged tissues, such as the retina [75], and cold exposure has similar effects in some tissues of short-tailed voles [76]. However, such indications are sometimes based on insecure assessments of 'global' proteasomal activity, a determination that is difficult to make [77]. Interestingly, recent evidence has also suggested that phenolic antioxidants, including some that suppress atherogenesis in some models, may themselves down-regulate a group of proteolytic machineries [78].

We have demonstrated that supply of oxidized LDL leads to a biphasic degradative response by isolated mouse macrophages [79,80]. By co-supply of radioactive normal LDL with non-radioactive oxidized LDL, we were able to show that endolysosomal capacities for the degradation of normal proteins were unaffected. However, the possibility remains that the intracellular trafficking of the two populations of substrate molecule might differ, in spite of our evidence to the contrary. Indeed, other workers have argued that there is interference with the degradation of normal proteins under similar circumstances. Similarly, cathepsin B can be inactivated by lipid aldehydes such as are generated in oxidized LDL [81], although the relevance of this in intact cells is still not clear. Thus there are many possible mechanisms by which oxidized proteins might interfere with the catabolism of both normal and oxidized molecules. These range from alterations in protein substrates themselves to alterations in proteolytic enzymes and machineries to altered subcellular trafficking.

To what degree may such changes in protein degradation contribute to the observed accumulation of oxidized proteins in cataract and atherogenesis? Or, to paraphrase the question as stated earlier, why is it that oxidized proteins are not completely removed by the quantitatively powerful proteolytic mechanisms of cells and tissues?

In the case of cataract, we can readily surmise the answer. Proteolysis is almost non-existent in the centre of the lens, and not powerful in the periphery, in complete contrast with most cellular tissues. This, together with the phenomenon of lens coloration, is the reason that cataract has long been a favoured subject for the study of the accumulation of oxidized proteins. The possibilities for intervention in cataractogenesis are thus most readily envisaged at the level of prevention of oxidative damage, rather than removal of oxidative products.

In the case of atherogenesis, we have virtually no information to hand as yet. It is clear that regression of atherogenesis can occur under certain circumstances, but it is not known whether protein oxidation products accumulate in these models, nor whether they decline in parallel with regression. One can now envisage new experimental approaches to the study of the metabolism of oxidized proteins *in vivo*, for example by the supply of precursor oxidized amino acids as we have described. In preliminary experiments we have shown the incorporation of dopa into inflammatory proteins of the mouse peritoneum, and the progressive removal of such protein–dopa.

There are at least two obvious approaches to the use of this method for atherosclerotic sites. One is to supply the oxidized amino acid directly, in the expectation that it will be incorporated into atherogenic sites by cellular metabolism. The other is to create biosynthetically labelled vascular protein components *in vitro* and to supply them to perfused atherosclerotic vessels, or *in vivo* (the latter may be limited by the quantity of incorporation that results). LDL biosynthesized containing dopa will be a particularly interesting substrate for such an experiment.

It is probably only through such quantitative methods that we will begin to gain adequate knowledge of the roles of proteolytic defences against potentially pathogenic oxidized proteins and the reactive species that they can contain.

References

1.   Davies, M. and Dean, R.T. (1997) Radical Mediated Protein Oxidation: From Chemistry to Medicine, Oxford University Press, Oxford
2.   Dean, R.T., Fu, S., Stocker, R. and Davies, M.J. (1997) Biochem. J. **324**, 1–18
3.   Stadtman, E.R. and Levine, R.L. (2000) Ann. N.Y. Acad. Sci. **899**, 191–208
4.   Hensley, K. and Floyd, R.A. (2002) Arch. Biochem. Biophys. **397**, 377–383
5.   Wolff, S.P. and Dean, R.T. (1987) Biochem. J. **245**, 243–250
6.   Dinis, T.C., Santosa, C.L. and Almeida, L.M. (2002) Free Radical Res. **36**, 531–543
7.   Hawkins, C.L., Pattison, D.I. and Davies, M.J. (2002) Biochem. J. **365**, 605–615
8.   Rosen, H., Crowley, J.R. and Heinecke, J.W. (2002) J. Biol. Chem. **277**, 30463–30468
9.   Raftery, M.J. and Geczy, C.L. (2002) J. Am. Soc. Mass Spectrom. **13**, 709–718
10. Gogasyavuz, D., Kucukkaya, B., Ersoz, H.O., Yalcin, A.S., Emerk, K. and Akalin, S. (2002) Int. J. Exp. Diabetes Res. **3**, 145–151
11. Konukoglu, D., Kemerli, G.D., Sabuncu, T. and Hatemi, H.H. (2002) Horm. Metab. Res. **34**, 367–370
12. Eaton, J.W. and Dean, R.T. (2000) in Atherosclerosis: Gene Expression, Cell Interactions and Oxidation (Dean, R.T. and Kelly, D.T., eds), pp. 24–45, Oxford University Press, Oxford
13. Morgan, P.E., Dean, R.T. and Davies, M.J. (2002) Arch. Biochem. Biophys. **403**, 259–269
14. Akagawa, M., Sasaki, T. and Suyama, K. (2002) Eur. J. Biochem. **269**, 5451–5458
15. Schoneich, C. and Williams, T.D. (2002) Chem. Res. Toxicol. **15**, 717–722
16. Edwards, A.M., Ruiz, M., Silva, E. and Lissi, E. (2002) Free Radical Res. **36**, 277–284
17. Garrison, W.M. (1968) Curr. Top. Radiat. Res. **4**, 43–94
18. Garrison, W.M. (1987) Chem. Rev. **87**, 381–398
19. Davies, M.J., Fu, S. and Dean, R.T. (1995) Biochem. J. **305**, 643–649
20. Headlam, H.A. and Davies, M.J. (2002) Free Radical Biol. Med. **32**, 1171–1184
21. Hua, S., Inesi, G., Nomura, H. and Toyoshima, C. (2002) Biochemistry **41**, 11405–11410
22. De, C.R., Rocca, B., Marchioli, R. and Landolfi, R. (2002) Thromb. Haemostasis **87**, 58–67
23. Jiang, Q., Lykkesfeldt, J., Shigenaga, M.K., Shigeno, E.T., Christen, S. and Ames, B.N. (2002) Free Radical Biol. Med. **33**, 1534–1542
24. Kulawiak-Galaska, D., Wozniak, M. and Greci, L. (2002) Acta Biochim. Pol. **49**, 43–49
25. Heinecke, J.W. (2002) Toxicology **177**, 11–22
26. Kaneda, H., Taguchi, J., Ogasawara, K., Aizawa, T. and Ohno, M. (2002) Atherosclerosis **162**, 221–225

27. Tovbin, D., Mazor, D., Vorobiov, M., Chaimovitz, C. and Meyerstein, N. (2002) Am. J. Kidney Dis. **40**, 1005–1012

28. Simpson, J.A., Narita, S., Gieseg, S., Gebicki, S., Gebicki, J.M. and Dean, R.T. (1992) Biochem. J. **282**, 621–624

29. Dean, R.T., Gieseg, S. and Davies, M.J. (1993) Trends Biochem. Sci. **18**, 437–441

30. Gieseg, S.P., Simpson, J.A., Charlton, T.S., Duncan, M.W. and Dean, R.T. (1993) Biochemistry **32**, 4780–4786

31. Wright, A., Bubb, W.A., Hawkins, C.L. and Davies, M.J. (2002) Photochem. Photobiol. **76**, 35–46

32. Luxford, C., Dean, R.T. and Davies, M.J. (2002) Biogerontology **3**, 95–102

33. Ostdal, H., Davies, M.J. and Andersen, H.J. (2002) Free Radical Biol. Med. **33**, 201–209

34. Dalle-Donne, I., Rossi, R., Giustarini, D., Gagliano, N., Di Simplicio, P., Colombo, R. and Milzani, A. (2002) Free Radical Biol. Med. **32**, 927–937

35. Derham, B.K. and Harding, J.J. (2002) Biochem. J. **364**, 711–717

36. Davies, M.J., Fu, S., Wang, H. and Dean, R.T. (1999) Free Radical Biol. Med. **27**, 1151–1163

37. Morin, B., Fu, S., Wang, H., Davies, M.J. and Dean, R.T. (2002) in Oxidative Stress Biomarkers and Antioxidant Protocols (Armstrong, D., ed.), pp. 101–110, The Humana Press, Totowa, NJ

38. Linton, S., Davies, M.J. and Dean, R.T. (2001) Exp. Gerontol. **36**, 1503–1518

39. Lim, P.S., Cheng, Y.M. and Wei, Y.H. (2002) Free Radical Res. **36**, 295–301

40. Head, E., Liu, J., Hagen, T.M., Muggenburg, B.A., Milgram, N.W., Ames, B.N. and Cotman, C.W. (2002) J. Neurochem. **82**, 375–381

41. Truscott, R.J. and Augusteyn, R.C. (1977) Exp Eye Res. **24**, 159–170

42. Truscott, R.J. and Augusteyn, R.C. (1977) Exp Eye Res. **25**, 139–148

43. Fu, S., Dean, R., Southan, M. and Truscott, R. (1998) J. Biol. Chem. **273**, 28603–28609

44. Fu, S., Dean, R.T., Davies, M.J. and Heinecke, J.W. (2000) in Atherosclerosis (Dean, R.T. and Kelly, D., eds), pp. 301–325, Oxford University Press, Oxford

45. Upston, J.M., Niu, X., Brown, A.J., Mashima, R., Wang, H., Senthilmohan, R., Kettle, A.J., Dean, R.T. and Stocker, R. (2002) Am. J. Pathol. **160**, 701–710

46. Witting, P., Pettersson, K., Ostlund-Lindqvist, A.M., Westerlund, C., Wagberg, M. and Stocker, R. (1999) J. Clin. Invest. **104**, 213–220

47. Stocker, R. (1999) Curr. Opin. Lipidol. **10**, 589–597

48. Witting, P.K., Pettersson, K., Letters, J. and Stocker, R. (2000) Arterioscler. Thromb. Vasc. Biol. **20**, E26–E33

49. Dean, R.T. (2000) Redox Rep. **5**, 251–255

50. Steinberg, D., Parthasarathy, S., Carew, T.E., Khoo, J.C. and Witztum, J.L. (1989) N. Engl. J. Med. **320**, 915–924

51. Lapko, V.N., Purkiss, A.G., Smith, D.L. and Smith, J.B. (2002) Biochemistry **41**, 8638–8648

52. Hansel, A., Kuschel, L., Hehl, S., Lemke, C., Agricola, H.J., Hoshi, T. and Heinemann, S.H. (2002) FASEB J. **16**, 911–913

53. Stadtman, E.R., Moskovitz, J., Berlett, B.S. and Levine, R.L. (2002) Mol. Cell. Biochem. **234–235**, 3–9

54. Davies, K.J. (1986) J. Free Radical Biol. Med. **2**, 155–173

55. Inder, T., Mocatta, T., Darlow, B., Spencer, C., Volpe, J.J. and Winterbourn, C. (2002) Pediatr. Res. **52**, 213–218

56. Abu-Zidan, F.M., Plank, L.D. and Windsor, J.A. (2002) Eur. J. Surg. **168**, 119–123

57. Dean, R.T. and Pollak, J.K. (1985) Biochem. Biophys. Res. Commun. **126**, 1082–1089

58. Dean, R.T., Thomas, S.M., Vince, G. and Wolff, S.P. (1986) Biomed. Biochim. Acta **45**, 1563–1573

59. Dean, R.T., Armstrong, S.G., Fu, S. and Jessup, W. (1994) in Free Radicals in the Environment, Medicine and Toxicology (Nohl, H., Esterbauer, H. and Rice-Evans, C., eds), pp. 47–79, Richelieu Press, London

60.   Bota, D.A. and Davies, K.J. (2002) Nat. Cell Biol. **4**, 674–680
61.   Pirlich, M., Muller, C., Sandig, G., Jakstadt, M., Sitte, N., Lochs, H. and Grune, T. (2002) Free Radical Biol. Med. **33**, 283–291
62.   Hatters, D.M. and Howlett, G.J. (2002) Eur. Biophys. J. **31**, 2–8
63.   Duggan, S., Rait, C., Platt, A. and Gieseg, S. (2002) Biochim. Biophys. Acta **1591**, 139–145
64.   Costa, V.M., Amorim, M.A., Quintanilha, A. and Moradas-Ferreira, P. (2002) Free Radical Biol. Med. **33**, 1507–1515
65.   Drake, S.K., Bourdon, E., Wehr, N., Levine, R., Backlund, P., Yergey, A. and Rouault, T.A. (2002) Dev. Neurosci. **24**, 114–124
66.   Grune, T., Reinheckel, T., North, J.A., Li, R., Bescos, P.B., Shringarpure, R. and Davies, K.J. (2002) FASEB J. **16**, 1602–1610
67.   Grune, T., Reinheckel, T., Li, R., North, J.A. and Davies, K.J. (2002) Arch. Biochem. Biophys. **397**, 407–413
68.   Korolainen, M.A., Goldsteins, G., Alafuzoff, I., Koistinaho, J. and Pirttila, T. (2002) Electrophoresis **23**, 3428–3433
69.   Keller, J.N., Gee, J. and Ding, Q. (2002) Ageing Res. Rev. **1**, 279–293
70.   Shringarpure, R. and Davies, K.J. (2002) Free Radical Biol. Med. **32**, 1084–1089
71.   Shringarpure, R., Grune, T., Mehlhase, J. and Davies, K.J. (2003) J. Biol. Chem. **278**, 311–318
72.   Kwon, Y.T., Kashina, A.S., Davydov, I.V., Hu, R.G., An, J.Y., Seo, J.W., Du, F. and Varshavsky, A. (2002) Science **297**, 96–99
73.   Rodgers, K., Wang, H.-J., Fu, S. and Dean, R.T. (2002) Free Radical Biol. Med. **32**, 766–775
74.   Dunlop, R.A., Rodgers, K.J. and Dean, R.T. (2002) Free Radical Biol. Med. **33**, 894–906
75.   Louie, J.L., Kapphahn, R.J. and Ferrington, D.A. (2002) Exp Eye Res. **75**, 271–284
76.   Selman, C., Grune, T., Stolzing, A., Jakstadt, M., McLaren, J.S. and Speakman, J.R. (2002) Free Radical Biol. Med. **33**, 259–265
77.   Rodgers, K.J. and Dean, R.T. (2003) Int. J. Biochem. Cell Biol. **35**, 716–727
78.   Noguchi, N. (2002) Free Radical Biol. Med. **33**, 1480–1489
79.   Jessup, W., Mander, E.L. and Dean, R.T. (1992) Biochim. Biophys. Acta **1126**, 167–177
80.   Mander, E.L., Dean, R.T., Stanley, K.K. and Jessup, W. (1994) Biochim. Biophys. Acta **1212**, 80–92
81.   Crabb, J.W., O'Neil, J., Miyagi, M., West, K. and Hoff, H.F. (2002) Protein Sci. **11**, 831–840

Biochem. Soc. Symp. **70**, 147–161
(Printed in Great Britain)
© 2003 Biochemical Society

13

Proteases as drug targets

**Andy J.P. Docherty[1], Tom Crabbe, James P. O'Connell
and Colin R. Groom**

Celltech R&D, Slough SL1 4EN, U.K.

Abstract

The effective management of AIDS with HIV protease inhibitors, or the use of angiotensin-converting enzyme inhibitors to treat hypertension, indicates that proteases do make good drug targets. On the other hand, matrix metalloproteinase (MMP) inhibitors from several companies have failed in both cancer and rheumatoid arthritis clinical trials. Mindful of the MMP inhibitor experience, this chapter explores how tractable proteases are as drug targets from a chemistry perspective. It examines the recent success of other classes of drug for the treatment of rheumatoid arthritis, and highlights the need to consider where putative targets lie on pathophysiological pathways – regardless of what kind of therapeutic entity would be required to target them. With genome research yielding many possible new drug targets, it explores the likelihood of discovering proteolytic enzymes that are causally responsible for disease processes and that might therefore make better targets, especially if they lead to the development of drugs that can be administered orally. It also considers the impact that biologics are having on drug discovery, and in particular whether biologically derived therapeutics such as antibodies are likely to significantly alter the way we view proteases as targets and the methods used to discover therapeutic inhibitors.

Introduction

Sales of Astra Zeneca's angiotensin-converting enzyme (ACE) inhibitor Zestril in the year 2000 totalled $1228 million. In the same year Roche and Abbott reported sales of $303 million and $210 million respectively for their HIV protease inhibitors Viracept and Kaletra. Clearly, proteolytic enzymes can be profitable drug targets. Furthermore, the speed with which the HIV inhibitors were developed indicates that proteases can be highly tractable for chemists seeking to make orally active inhibitors. Discovery of the HIV pro-

[1]To whom correspondence should be addressed (e-mail adocherty@celltech.co.uk).

tease in 1985 resulted from analysis of the viral nucleotide sequence [1]. Extensive analysis of its catalytic properties followed which, when combined with an early knowledge of its crystal structure, facilitated the rapid development of highly potent and selective inhibitors [2]. By 1997, following demonstration that the inhibitors could reduce viral load in patients, several such inhibitors received approval from the Food and Drug Administration in the U.S.A. for anti-viral use in humans. This example is significant not only because of the brief 12 years that it took to go from target discovery to licensed product, but also because it is an example of the way applied genomics can first reveal the existence of possible new drug targets. One might therefore expect the vast quantity of genome sequence information available to us today to be a very rich hunting ground for further protease drug targets. As Alan Barrett and his colleagues have pointed out, analysis of all the available genome sequences suggests that some 2% encode identifiable peptidase motifs [3]. First, though, let us examine what it is that makes proteolytic enzymes so tractable as drug targets.

Tractability

Proteolytic enzyme active sites are often composed of a deep pocket. This increases the likelihood of small drug-like molecules that mimic the substrate being able to have many possible points of interaction within the pocket. This facilitates tight binding. A variety of methods for discovering short peptide substrates for a particular protease are available, even if the physiological substrate of the protease is a very large protein, or if the true physiological substrate is unknown [4]. The discovery of good substrates is an essential first step, because it allows the development of assays suitable for screening for inhibitors. The susceptibility of such substrates to cleavage, and information derived from the subsequent kinetic analysis, helps define what structures are most readily accommodated within a binding pocket. Early acquisition of this information helps chemists to understand what is required to build enzyme specificity into inhibitor structures – an important consideration when working within large protease families where the therapeutic benefit is expected to derive from inhibiting one or some, but not all, of the family members. Knowledge of substrates derived in this way is an excellent starting point for the rational design of inhibitors that mimic substrates in their binding characteristics, but which have been modified in order to avoid cleavage. This rational approach may reduce the need for large-scale screening for inhibitor leads at the early stages of a drug discovery programme. This is an important point, because the early discovery of a series of leads for which a structure–activity relationship can be derived can be a major hurdle for some other classes of drug target. Examples include receptors for large protein ligands, or kinases where the structure of the various substrates within the active site are all very similar. Another feature that helps make proteases tractable drug targets is the fact that they are very often amenable to structure determination, and crystallographers have a very good track record for solving protease catalytic domain structures at high atomic resolution [5]. Combined with an ability to co-crystallize lead inhibitors, or to 'soak' inhibitors into pre-existing protease crystals, the resulting structural information further facilitates rational inhibitor development. In the case of HIV protease, crystal structures were available very soon after the discovery of the enzyme, and the resulting information impacted on inhibitor design from the very early stages of the project ([6] and references therein). The early acquisition of such information is a trend that is set to increase with the high-throughput crystallography and structure determination methods that are being brought to the fore today [7,8].

Proteolytic enzymes are clearly tractable in terms of lead generation, but the very fact that the leads are often peptide-derived can make their further development a challenge. For ease of use, it is clearly desirable for a drug to be administered orally. This requires that it should be absorbed through the gut and be made bioavailable to the tissues in which it is to act for sufficient periods of time to provide therapeutic benefit. Guidelines for good oral bioavailability have been described by Lipinski et al. [9] and are summarized in Table 1. Generally speaking, compounds that fall within these guidelines compete with ligands or substrates for binding to structurally defined pockets. Proteolytic enzymes usually have such pockets, and they are therefore often referred to as being 'druggable' targets. Unfortunately, inhibitor leads based on peptides tend to be highly susceptible to proteolysis when administered *in vivo*, especially if administered orally. In order to turn a peptide-based inhibitor into an orally bioavailable drug that is not rapidly metabolized and cleared, at the very least it is generally necessary to decrease the M_r and remove most, if not all, of the peptide bonds. This can lead to loss of the potency and selectivity characteristics on the basis of which the compound was first selected. With this in mind, the following section examines just how much effort the pharmaceutical industry has already devoted to protease targets, and the frequency of success.

Success to date

Within chemistry-based drug discovery research, the total number of individual drug targets across the industry (excluding non-human targets, for example in the anti-infectives field) has been estimated to be approx. 480 [10,11]. These targets fall into 130 protein families, of which we estimate that 10% are peptidases [12]. This indicates that a significant amount of the pharmaceutical industry's research effort is devoted to protease targets. Furthermore, this effort appears to be highly productive in terms of yielding compounds that meet the Lipinski guidelines for orally active compounds [9]. For example,

Table 1 Guidelines for oral bioavailability. These guidelines are based on [9]. Note that compounds derived from natural products, or classes that are substrates for biological transporters, will not follow the listed guidelines.

Poor absorption or permeation of a compound is likely when it has:
>5 hydrogen-bond donors (defined as NH or OH)
>10 hydrogen-bond acceptors (defined as O or N atoms)
M_r>500
High lipophilicity: (clogP>5 or MlogP>4.15)

analysis of the industry portfolio of targets reveals that approx. 400 of them have a corresponding drug candidate, as defined by Lipinski et al. [9], that has an IC_{50} in a relevant assay of 10 μM or less [12]. Again, proteases are well represented. For example, if one just takes endopeptidases, we estimate that there are about 14 metallo-, 12 serine, six cysteine and five aspartate endopeptidase drug targets for which a potentially bioavailable 'Lipinski-like' compound exists. This suggests that approx. 10% of all the 'Lipinski-compliant' targets are in fact proteases. However, the sobering fact is that, despite this large amount of research effort, and apparent successful lead generation, only a few of the 120 or so marketed classes of drug target proteases – HIV protease, ACE, thrombin and a few others. Clearly, 'druggable' does not imply 'good drug target'. Other parameters obviously come into play, not least the issue of whether a putative protease target is actually causal in a disease process, and whether its point of activity in a pathway is amenable to therapeutic intervention. These points are considered in the following case study.

Targets in rheumatoid arthritis (RA)

Recent years have seen the introduction of new biological drugs that have dramatically changed the treatment of RA and other inflammatory diseases such as Crohn's disease. The two leading classes of drug both target the inflammatory mediator tumour necrosis factor α (TNFα), either with a neutralizing antibody or with a soluble form of the natural receptor [13]. They are the first true disease-modifying drugs for this indication, which, despite intense research initiatives, had represented a very large unmet medical need. It is worth examining why this particular target has proven to be such an excellent point for therapeutic intervention compared with other targets, including several proteases that, over many years, have received a similar level of attention. The answer is largely due to where TNFα lies on the pathophysiological pathways that lead to RA. As shown in Figure 1, several features of the RA-affected joint represent points of intervention where one would anticipate therapeutic benefit.

Several classes of endopeptidase thought to be responsible for the thinning of cartilage and the erosion of bone have received attention from the pharmaceutical industry. Much attention has been focused on the matrix metalloproteinases (MMPs), and in particular MMP-1 and MMP-3. These enzymes are thought to be responsible for the turnover of collagen and proteoglycan respectively (both of which are essential structural components of the cartilage extracellular matrix), and received much of the early interest [14–16]. In the case of MMP-3, despite strong evidence for its pathophysiological involvement in cartilage breakdown, MMP-3$^{-/-}$ mice proved to be just as susceptible to cartilage loss as wild-type mice [17]. In the case of MMP-1, the potent orally bioavailable inhibitor Trocade (K_i ~3 nM; in vitro IC_{50} ~60 nM), developed by Roche, was found to demonstrate good efficacy in rodent disease models and was well tolerated in human Phase 1 and 2 studies, but was ultimately discontinued because of an 'unfavourable risk/benefit ratio' ([15,18] and Roche Holding AG news release). This is generally interpreted to mean either that no benefit was observed or that administration of the drug made matters worse.

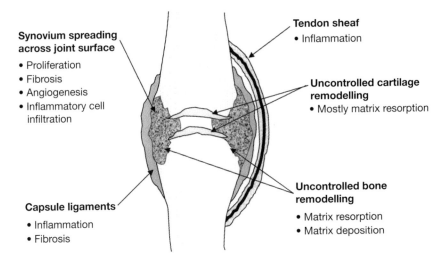

Synovium spreading across joint surface
- Proliferation
- Fibrosis
- Angiogenesis
- Inflammatory cell infiltration

Tendon sheaf
- Inflammation

Uncontrolled cartilage remodelling
- Mostly matrix resorption

Capsule ligaments
- Inflammation
- Fibrosis

Uncontrolled bone remodelling
- Matrix resorption
- Matrix deposition

Figure 1 The RA-affected joint. Points of intervention at which therapeutic benefit might be anticipated are highlighted.

Despite these setbacks, protease inhibitor development continues, focused largely on more recently discovered members of the MMP family such as MMP-13 and various ADAMTS (ADAM with thrombospondin motifs, where ADAM denotes a disintegrin and metalloproteinase) family members that recent studies have shown to be capable of cleaving aggrecan – the polypeptide backbone of proteoglycan [19]. Given the large repertoire of MMPs and ADAMTSs that we now know to be capable of participating in cartilage remodelling, the challenge is to identify the principal contributors in disease and to make appropriately selective inhibitors. Inhibiting the 'wrong' proteases will probably be therapeutically counterproductive, because it has already been shown that knockouts of several MMPs lead to disabilitating phenotypes that one would not wish to recapitulate in humans [20–22]. In addition, there are difficulties with modelling such scenarios in rodent species, because the repertoire of MMPs used by mice, for example, appears to differ from that in humans. On the other hand, redundancy in which proteases are used in particular physiological settings may require that broad-spectrum inhibitors are developed. For example, aggrecan is susceptible to cleavage at various positions by a great many MMPs and ADAMTSs, and it may be that several enzymes contribute to pathological remodelling [19]. However, the use of broad-spectrum inhibitors runs against the apparent need for at least some degree of selectivity to avoid inhibition of normal physiological processes [20,23,24]. Indications that this may present a problem derive from the many broad-spectrum MMP inhibitors that have been evaluated in cancer trials, many of which have suffered from unacceptable side-effect profiles, as well as a lack of efficacy [25].

Taking a step back from the proteolytic enzymes responsible for cartilage and bone resorption, it is apparent that a great many soluble mediators orchestrate the cellular events that occur within the joint in RA. This includes proliferation and vascularization of the synovium, inflammatory cell infiltration, and the consequential activation of bone, cartilage and synovial cells and release of further inflammatory mediators, proteolytic enzymes and their activators. Known as the cytokine network, these mediators include not only the so-called 'destructive' cytokines such as TNFα, interleukin-1β (IL-1β) and IL-17 (among others), but also 'modulatory' cytokines such as IL-10 and growth factors such as transforming growth factor β [26]. Genome mining to find further cytokines, for example IL-10 family members, has been very successful, and has yielded several that are currently being explored for their possible roles in inflammatory disease [27–29]. Several cytokines, such as IL-1β, IL-6, IL-15 and IL-17, are already being pursued as drug targets [30–32]. As extracellular gene products, they are all amenable to evaluation as targets *in vivo* with antibodies that specifically modulate activity by binding to the ligand, or blocking the corresponding receptor. In addition, innovations in the technology surrounding the 'humanization' and manufacturing of antibodies has led to their therapeutic use in a variety of disease settings [33–35]. Innovations whereby cytokine receptors can be manufactured as soluble proteins fused to immunoglobulin Fc domains to give them long circulating half-lives *in vivo* has similarly led to their application as both proof-of-concept reagents and therapeutics for use in humans [36]. As a consequence, although TNFα is not in itself tractable as a traditional chemistry target, it has proven to be an excellent target for biologics.

The two leading products are Remicade, an anti-TNFα antibody from Centocor that is licensed for use in RA and Crohn's disease, and Enbrel, a p75-TNF receptor–Fc from Immunex, which is licensed for the treatment of RA and psoriatic RA. Sales of these drugs reported for the first three-quarters of 2002 were $918 million and $567 million respectively. The established pharmaceutical industry therefore cannot ignore the impact of biologics in large, chronic indications where orally absorbed chemically derived products were previously thought to be the only kind of product that would meet with patient compliance. In fact, one might have expected that the cost and intravenous administration required for Remicade and Enbrel might mitigate against them being highly successful drugs. It turns out, however, that patients given anti-TNF treatment benefit from an immediate anti-inflammatory effect, as well as longer-term disease-modifying benefits [37]. The success of Remicade and Enbrel is therefore due not only to the lack of any previously existing disease-modifying treatments for RA, but also to the fact that they deliver a clinically demonstrable benefit to patients very soon after their first administration. Although there are good reasons to believe that inhibitors of some of the proteolytic enzymes described above may have helped to maintain bone and cartilage integrity, it is unlikely that they would have the same profound anti-inflammatory properties of the anti-TNFα treatments. In consequence, the clinical trial design for Trocade required much more complex read-outs for efficacy, involving the use of MRI (magnetic resonance imaging) to measure changes to bone and cartilage integrity over long periods of time. TNFα therefore turns out to be a far superior target for therapeutic intervention because of its upstream position in the disease process and the fact that (probably) small changes in levels of its activity are sufficient to change the balance of cytokines, leading to a readily measurable clinical benefit (Figure 2). Nonetheless, this approach has required the development of brand new classes of therapeutic agent.

One obvious outcome of the success with anti-TNFα agents is that cytokine-mediated pathways, in particular the TNFα pathway, are now considered to be validated in terms of therapeutic intervention for inflammatory disease in humans. The whole pathway has therefore been examined for possible additional points of intervention, and proteolytic enzymes feature highly [13]. Instead of focusing on the matrix-degrading proteases induced downstream by TNFα and other inflammatory mediators such as IL-1β, a search for chemistry targets has focused on the proteolytic events that give rise to these mediators. Interleukin-1β-converting enzyme (ICE) is a cysteine endopeptidase belonging to the caspase family, and is responsible for the intracellular processing of an IL-1β precursor prior to secretion of the cytokine in active form [38]. TNFα-converting enzyme (TACE) is a member of the adamalysin family of metallo-endopeptidases, and is responsible for the cleavage of transmembrane-domain-anchored TNFα, leading to its rapid shedding from inflammatory cells following acute inflammatory

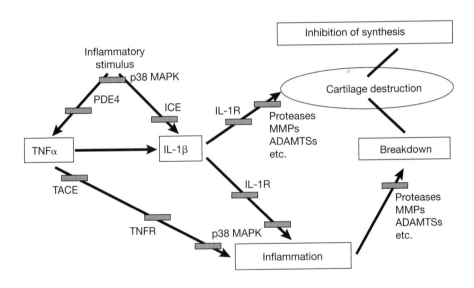

Figure 2 Targets within the TNFα pathway. The clinical success of antibodies or soluble receptors that modulate the activity of TNFα highlights several possible protease targets on the same pathway. The inhibition of downstream targets such as proteases involved in cartilage destruction may have less impact than the inhibition of targets that are additionally involved in inflammatory processes. ICE and TACE are the subject of inhibitor development programmes, but currently inhibitors aimed at the non-protease targets type IV phosphodiesterase (PDE4) and p38 MAP kinase (p38 MAPK) appear to be more advanced. Points of intervention more readily targeted by biologics are also apparent, and include blockers of the receptors for TNF (TNFR) or IL-1β (IL-1R), or IL-1β itself, for which a drug (Anakinra, a recombinant IL-1β receptor antagonist) has recently been launched [74].

challenge [39]. Since being cloned by a number of pharmaceutical companies, both ICE and TACE have been pursued as drug targets. Several groups have reported ICE inhibitors, but only Vertex Pharmaceuticals appear to have a compound in clinical development for RA ([40] and company news release). TACE inhibitors have emerged from the various series of compounds developed originally for the MMPs (structural similarities place TACE and the MMPs within the metzincin family), but to date none appear to have entered clinical development [41]. As targets for synthetic orally bioavailable chemistry programmes, both ICE and TACE present significant challenges in terms of selectivity if inappropriate inhibition of related family members is to be avoided. Furthermore, the relative contributions in disease of cell-bound compared with shed TNFα are not entirely clear, and what could be achieved therapeutically with a TACE inhibitor that only prevents shedding is hard to predict. In the case of ICE, it is not clear how critical its activity is to the occurrence of IL-1β in the RA joint. However, it does seem likely that the therapeutic use of neutralizing antibodies, natural antagonists or soluble receptors significantly lowers the levels of cytokine in the inflamed joint. Whether a similar decrease can be achieved with ICE or TACE inhibitors is an issue that has yet to be addressed.

In fact, the most promising targets on the TNFα pathway for chemically derived drugs appear not to be proteases (Figure 2). Upstream on the pathway, orally active inhibitors of type IV phosphodiesterase are potent inhibitors of TNFα release from lipopolysaccharide-stimulated human macrophages, and a number of companies have entered compounds into clinical development for the treatment of asthma [42]. Elsewhere on the pathway, orally active inhibitors of p38 mitogen-activated protein kinase (MAP kinase) are looking promising in a variety of inflammation models [43,44]. Such inhibitors appear to act upstream to prevent synthesis of TNFα (and IL-1β) following inflammatory cell stimulation. They also have some downstream activity on p38 MAP kinase-dependent signalling pathways that operate when TNFα or IL-1β bind to their receptors, although other pathways also exist at this point [45]. Many pharmaceutical companies are very active in this area, including Vertex Pharmaceuticals, Boehringer Ingelheim Pharmaceuticals and Scios Inc., who all have compounds in clinical development for RA (company news releases).

The following section considers how one might apply what has been learnt about 'good' targets in the RA field to a better understanding of how we should mine genome sequence information for new proteolytic enzyme targets.

Genome mining

The human genome is estimated to encode around 30000 gene products [46,47]. If we assume that where the protein corresponding to one member of a gene family has been proven to bind a drug-like molecule, other members will also be able to do so, we can estimate the number of 'druggable' targets. Through homology searches and analysis of gene family sizes, it has been estimated that, using this definition, some 3000 gene products are likely to be druggable (Figure 3) [12]. However, only those that play causal roles in disease processes, or which when modulated provide symptomatic benefit, can become viable drug targets. Estimates of the number of disease-related genes vary widely depending on whether the number is based solely on expression data in disease tissue or on more substantial data such as evidence from knockout experiments. Estimates range between 3000 and 10000 [48–50]. How many causally involved gene products fall into the druggable gene product camp is even less clear, but one estimate based on the number of anti-fungal targets arising from the yeast genome puts this as low as 2–5%, which is in broad agreement with data derived from knockout experiments [12,51] (Figure 3). This indicates that the number of useful new drug targets, as defined by the traditional requirements of chemistry programmes aimed at orally bioavailable drugs, may be a rather modest several hundred.

However, this overlooks the fact that the 30000 or so predicted gene products do not represent all the possible targets. Splice variation, post-translational modification, variation in glycosylation, functional synapses composed of multimeric protein complexes, and non-catalytic domains in the case of proteases, very often exist. If these are considered to be potential molecular targets, the total number of such targets becomes greater than the number of gene products (Figure 3). Most of these targets are likely to be intractable to traditional chemistry strategies. However, a number of them may make excellent starting points for the development of biologics in the form of therapeutic proteins, soluble receptor-based drugs, antibody therapeutics, or components of therapeutic strategies based on gene therapy, vaccination or antisense technology, etc. The success of Remicade and Enbrel tells us that at least some of these therapeutic approaches are eminently feasible.

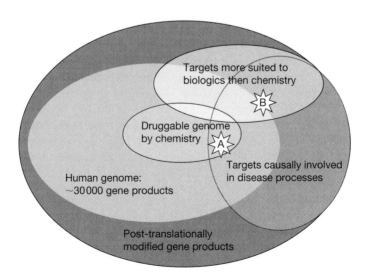

Figure 3 How many new drug targets will genomics generate? Intersection A indicates those targets that may be exploited by chemistry, while intersection B indicates those targets that may be exploited by biologics. The latter category makes available targets that have not traditionally been recognized to be suitable for chemistry.

One feature of genome research is the speed and scale at which it identifies potential new drug targets. For example, following the discovery that TACE, the protease responsible for shedding TNFα, was an adamalysin and consisted of several domains, including a disintegrin and metallo-endopeptidase (metzincin) domain, intense genome mining rapidly extended the known members of this family to 30 or more [52–54]. Further genome mining subsequently uncovered the ADAMTS family [55]. Rapidly increasing in size, this family includes some enzymes that cleave aggrecan [56]. The striking point is that genome analysis revealed the existence of many new family members, and expression was demonstrated in a variety of tissues, long before most of them (with a few exceptions [57,58]) had been evaluated in 'wet' experiments as proteases. Indeed genetics, for example through the use of gene knockouts, may indicate which of the candidate proteases is responsible for a phenotype before physiologically meaningful enzyme assay results are available [51]. In fact, in the case of ADAM-33, genetics revealed an association with asthma in humans only months after the gene was first reported [59,60]. Such technological innovations are clearly very powerful. Nonetheless, it is often the case that the significance of a particular phenotype in a knockout mouse is hard to relate to disease in the human adult. Furthermore, expression profiling to reveal the existence of a new protease in disease tissues only demonstrates a disease association. It does not demonstrate that the protease is causal in the disease process, nor that the protease is tractable for drug discovery, or that its activity represents a feasible point for therapeutic intervention. There is, therefore, a need for powerful *in vivo* target validation strategies. If such strategies are to be dependent on selective bioavailable compounds, it would be desirable to have the means to generate compounds that modulate the activity of a relatively novel protease before detailed knowledge of its biochemical properties exists. Furthermore, given the size of most proteolytic enzyme families and the structural similarities within active sites, the need for inhibitor selectivity cannot be ignored, and the means to obtain such selectivity needs to be sought. These are difficult challenges for early chemistry programmes. In contrast, antibodies can make excellent target validation reagents, although not for cellular targets, where RNA-mediated interference technology may have a big role to play [61]. Furthermore, thanks to great progress in the ways in which antibodies can be engineered and manufactured cost-effectively (for example to remove inessential Fc domains or to create small monovalent Fabs expressed in *Escherichia coli* [62–64]), there are encouraging signs that they are likely to be useful for the long-term treatment of chronic diseases [65]. Could proteolytic enzymes be targets for inhibitory monovalent Fabs?

Proteolytic enzymes as antibody drug targets?

A clear limitation in the use of antibodies is the requirement for extracellular targets. With that exception, they bring potential opportunities for inhibition that do not arise readily from the use of chemically derived inhibitors. For example, the high-affinity, tight-binding and exquisite specificity properties that an antibody can possess arise from multiple interactions over a relatively large surface area [66,67]. They do not share a dependency on the kind of deep binding pocket that would typically be recognized as necessary for low-molecular-mass inhibitor compounds. The fact that it has been possible to isolate an inhibitory single-domain camelid antibody with a K_i towards α-amylase of 10 nM which, although binding to the catalytic site, did not contact the catalytic residues, suggests that useful inhibitory antibodies directed against proteases should be possible [68]. Many proteolytic enzymes consist of multi-domain structures in which the catalytic domain is just one of many. Several examples exist where it has been shown that specificity for the natural (generally, in this case, high-molecular-mass) substrate is dependent on the involvement of an additional domain to the one that contributes the catalytic mechanism. Examples include the C-terminal domain of MMP-1 required for collagen cleavage, and the thrombospondin-like motif in ADAMTS-4 required for aggrecanase cleavage [69,70]. It is probable that antibodies directed against these domains would act as functional inhibitors of enzyme activity. In fact, recent new high-throughput ways in which antibodies are isolated mean that it is now feasible to consider screening for inhibitory antibodies from the entire immune repertoire of an immunized animal. The selected lymphocyte antibody method (SLAM) means that all that is required is an assay for detecting such functional antibodies [71]. The immune repertoire of an immunized animal, in the form of B cells plated out in microtitre wells, is screened and, following the identification of wells containing the required inhibitory activity, DNA encoding the corresponding antibody variable regions are recovered by PCR and expressed recombinantly. An important feature of this approach is that a precise knowledge of which part of a novel protease contributes to the proteolytic property one wishes to inhibit is not essential. In fact, the resulting antibodies can be used subsequently as powerful reagents to define the molecular components of the protease that contribute to its function both *in vitro* and *in vivo*.

New chemistry strategies for inhibitor discovery

In addition to the breakthroughs in antibody technology, developments in chemistry also promise to enhance the drug discovery process. As with SLAM, developments in what has been called 'chemical genomics' also offer the prospect of discovering protease inhibitor leads without a great deal of biochemical characterization of the target, or a bias towards a conventional substrate-binding pocket. In a strategy called the automated ligand identification system (ALIS), pools of combinatorially synthesized compounds are screened for their ability to bind to a target protein of interest [72]. Compounds that bind tightly (depending on the stringency of the conditions used) are separated from the rest of the pool by size-exclusion chromatography and subsequently identified by mass spectroscopy. Tight-binding compounds are only then tested for their inhibitory and selectivity properties. Another strategy, based on identifying individual compounds among libraries arrayed on to a solid surface on the basis of their binding to a protein of interest, has also been described [73]. If applied to protease targets, this approach opens up the prospect of discovering novel leads without a detailed knowledge of the enzyme's substrate specificity pocket. Furthermore, it offers the possibility of identifying leads that act through binding to alternative sites, such as allosteric sites or non-catalytic exosites. New classes of inhibitor that act though such sites may provide routes to more selective inhibitors than has sometimes been possible by targeting the substrate-binding pocket with competitive antagonists.

Conclusions

Experience shows that proteolytic enzymes are tractable targets for lead discovery and are 'druggable', in that they can give rise to orally active drugs. Thanks to readily identifiable peptidase motifs, a multitude of potential protease targets have arisen from genome mining. Some of the questions that we would ask of such a new target are summarized in Table 2. Clearly, generating data to show whether a particular protease is truly causal in a disease process is not easy. In some cases it is apparent that their presence in disease tissue is a consequence of the disease rather than a significant cause. This may be the case with MMPs found in the presence of the stroma around proliferating tumour cells in cancer tissues. In inflammatory diseases such as RA, targets other than matrix-degrading proteases have clearly been shown to represent superior points for intervention. This in turn has required the development of non-chemically derived biologics as the therapeutic agent of choice – in this case for neutralizing the activity of TNFα. In the near future, a significant number of new 'post-genomic' targets may not in fact be tractable for chemistry-based programmes, and will instead be reliant on the application of biologics technology for target validation and, very likely, for drug development. Close scrutiny of how well such drugs work, and the pathways in which they act, should help highlight better chemistry targets, including proteases. Armed with some of the new methods for identifying novel leads, such proteolytic enzymes should be suitable targets for orally administered drugs.

Table 2 Proteolytic enzyme target checklist.

Is it a druggable target?	
Is it at a critical point on a pathophysiological pathway?	
Is it of a proven therapeutic class?	
Peptide hormone release	✓
Viral processing enzymes	✓
Matrix remodelling	X
Amyloid processing	?
Cell surface shedding or activation events	?
Intracellular processing events	?
Are there other potential drug/antibody binding sites?	
Would non-chemical intervention be feasible?	
Does its cleavage product represent a surrogate for therapeutic efficacy?	

We thank Alan Barrett at the Wellcome Trust Sanger Institute for the MEROPS database – a resource that has facilitated many of our protease mining expeditions. We also gratefully acknowledge our colleagues within the Chemistry, Biology, Pharmacology and Clinical Departments at Celltech, Gill Murphy at the Cambridge Institute for Medical Research, Vera Knäuper at the University of York and Vim Van den Berg at the University Medical Centre in Nijmegen for the many fruitful discussions that culminated in this chapter.

References

1. Ratner, L., Haseltine, W., Patarca, R., Livak, K.J., Starcich, B., Josephs, S.F., Doran, E.R., Rafalski, J.A., Whitehorn, E.A., Baumeister, K. et al. (1985) Nature (London) **313**, 277–284
2. Dunn, B.M. (1998) in Handbook of Proteolytic Enzymes (Barrett, A.J., Rawlings, N.D. and Woessner, J.F., eds), pp. 919–928, Academic Press, London and San Diego
3. Barrett, A.J., Rawlings, N.D. and Woessner, J. F. (eds) (1998) Handbook of Proteolytic Enzymes, Academic Press, London and San Diego
4. Nixon, A.E. (2002) Curr. Pharm. Biotechnol. **3**, 1–12
5. Bode, W., Fernandez-Catalan, C., Tschesche, H., Grams, F., Nagase, H. and Maskos, K. (1999) Cell. Mol. Life Sci. **55**, 639–652
6. Blundell, T.L., Lapatto, R., Wilderspin, A.F., Hemmings, A.M., Hobart, P.M., Danley, D.E. and Whittle, P.J. (1990) Trends Biochem. Sci. **15**, 425–430
7. Buchanan, S.G., Sauder, J.M. and Harris, T.J.R. (2002) Curr. Pharm. Des. **8**, 1173–1188
8. Blundell, T.L., Jhoti, H. and Abell, C. (2002) Nat. Rev. Drug Discovery **1**, 45–54
9. Lipinski, C., Lombardo, F., Dominy, B. and Feeney, P. (1997) Adv. Drug Delivery Rev. **23**, 3–25
10. Drews, J. (1996) Nat. Biotechnol. **14**, 1516–1518
11. Drews, J. and Ryser, S. (1997) Nat. Biotechnol. **15**, 1318–1319
12. Hopkins, A.L. and Groom, C.R. (2002) Nat. Rev. Drug Discovery **1**, 727–730
13. Higgs, G.A. and Henderson, B. (eds) (2000) Novel Cytokine Inhibitors, Birkhauser, Basel, Boston and Berlin
14. Clark, I.M., Rowan, A.D. and Cawston, T.E. (2000) Curr. Opin. Anti-inflammatory Immunomodulatory Invest. Drugs **2**, 16–25
15. Montana, J. and Baxter, A. (2000) Curr. Opin. Drug Discovery Dev. **3**, 353–361
16. Brinckerhoff, C.E. and Matrisian, L.M. (2002) Nat. Rev. Mol. Cell Biol. **3**, 207–213
17. Mudgett, J.S., Hutchinson, N.I., Chartrain, N.A., Forsyth, A.J., McDonnell, J., Singer, I.I., Bayne, E.K., Flanagan, J., Kawka, D., Shen, C.F. et al. (1998) Arthritis Rheum. **41**, 110–121
18. Lorenz, H-M. (2000) Curr. Opin. Anti-inflammatory Immunomodulatory Invest. Drugs **2**, 47–52
19. Mort, J.S. and Billington, C.J. (2001) Arthritis Res. **3**, 337–341
20. Vu, T.H. and Werb, Z. (2000) Genes Dev. **14**, 2123–2133
21. Martignetti, J.A., Al Aqeel, A., Sewairi, W.A., Boumah, C.E., Kambouris, M., Al Mayouf, S., Sheth, K.V., Al Eid, W., Dowling, O., Harris, J. et al. (2001) Nat. Genet. **28**, 261–265
22. Corry, D.B., Rishi, K., Kanellis, J., Kiss, A., Song, L.Z., Xu, J., Feng, L., Werb, Z. and Kheradmand, F. (2002) Nat. Immunol. **3**, 347–353
23. Chang, C. and Werb, Z. (2001) Trends Cell Biol. **11**, s37–s43
24. Egeblad, M. and Werb, Z. (2002) Nat. Rev. Cancer **2**, 161–174
25. Coussens, L.M., Fingleton, B. and Matrisian, L.M. (2002) Science **295**, 2387–2392
26. Arend, W.P. (2001) Arthritis Care Res. **45**, 101–106
27. Blumberg, H., Conklin, D., Xu, W.F., Grossmann, A., Brender, T., Carollo, S., Eagan, M., Foster, D., Haldeman, B.A., Hammond, A. et al. (2001) Cell **104**, 9–19

28. Fickenscher, H., Hor, S., Kupers, H., Knappe, A., Wittmann, S. and Sticht, H. (2002) Trends Immunol. **23**, 89–96
29. Wolk, K., Kunz, S., Asadullah, K. and Sabat, R. (2002) J. Immunol. **168**, 5397–5402
30. Miossec, P. (2001) Cell. Mol. Biol. **47**, 675–678
31. Aggarwal, S. and Gurney, A.L. (2002) J. Leukocyte Biol. **71**, 1–8
32. Van den Berg, W.B. (2002) Curr. Rheumatol. Rep. **4**, 232–239
33. King, D.L. (1998) Applications and Engineering of Monoclonal Antibodies, Taylor Francis, London
34. Dijk, M.A. and van de Winkel, J.G.J. (2001) Curr. Opin. Chem. Biol. **5**, 368–374
35. Reichert, J.M. (2002) Curr. Opin. Mol. Therap. **4**, 110–118
36. Meager, A. (2000) in Novel Cytokine Inhibitors (Higgs, G.A. and Henderson, B., eds), pp. 157–176, Birkhauser, Basel, Boston and Berlin
37. Maini, R.N. (2002) in Novel Cytokine Inhibitors (Higgs, G.A. and Henderson, B., eds), pp. 145–156, Birkhauser, Basel, Boston and Berlin
38. Miller, D.K., Calaycay, J.R., Howard, A.D., Kostura, M.J., Molineaux, S.M. and Thornberry, N.A. (1993) Ann. N.Y. Acad. Sci. **696**, 133–148
39. Black, R.A. and White, J.M. (1998) Curr. Opin. Cell Biol. **10**, 654–659
40. Ku, G., Faust, T., Lauffer, L.L., Livingston, D.J. and Harding, M.W. (1996) Cytokine **8**, 377–386
41. Nelson, F.C. and Zask, A. (1999) Exp. Opin. Invest. Drugs **8**, 383–392
42. Griswold, D.E., Webb, E.F., Badger, A.M., Gorycki, P.D., Levandoski, P.A., Barnette, M.A., Grous, M., Christensen, S. and Torphy, T.J. (1998) J. Pharmacol. Exp. Ther. **287**, 705–711
43. Underwood, D.C., Osborn, R.R., Bochnowicz, S., Webb, E.F., Rieman, D.J., Lee, J.C., Romanic, A.M., Adams, J.L., Hay, D.W. and Griswold, D.E. (2000) Am. J. Physiol. Lung Cell Mol. Physiol. **279**, L895–L902
44. English, J.M. and Cobb, M.H. (2002) Trends Pharmacol. Sci. **23**, 40–45
45. Kumar, S., Blake, S.M. and Emery, J.G. (2001) Curr. Opin. Pharmacol. **1**, 307–313
46. Lander, E.S., Linton, L.M., Birren B., Nusbaum, C., Zody, M.C., Baldwin, J., Devon, K., Dewar, K., Doyle, M., FitzHugh, W. et al. (2001) Nature (London) **409**, 860–921
47. Venter, J.C., Adams, M.D., Myers, E.W., Li, P.W., Mural, R.J., Sutton, G.G., Smith, H.O., Yandell, M., Evans, C.A., Holt, R.A. et al. (2001) Science **291**, 1304–1351
48. Drews, J. (2000) Science **287**, 1960–1964
49. Claverie, J.M. (2001) Science **291**, 1255–1257
50. Walke, D.W., Han, C., Shaw, J., Wann, E., Zambrowicz, B. and Sands, A. (2001) Curr. Opin. Biotechnol. **12**, 626–631
51. Zambrowicz, B.P. and Sands, A.T. (2003) Nat. Rev. Drug Discovery **2**, 38–51
52. Killar, L., White, J., Black, R. and Peschon, J. (1999) Ann. N.Y. Acad. Sci. **878**, 442–452
53. Schlondorff, J. and Blobel, C.P. (1999) J. Cell Sci. **112**, 3603–3617
54. Stone, A.L., Kroeger, M. and Sang, Q.X. (1999) J. Protein Chem. **18**, 447–465
55. Tang, B.L. (2001) Int. J. Biochem. Cell Biol. **33**, 33–44
56. Arner, E.C. (2002) Curr. Opin. Pharmacol. **2**, 322–329
57. Jia, L.G., Shimokawa, K., Bjarnason, J.B. and Fox, J.W. (1996) Toxicon **34**, 1269–1276
58. Howard, L., Lu. X., Mitchell, S., Griffiths, S. and Glynn, P. (1996) Biochem. J. **317**, 45–50
59. Yoshinaka, T., Nishii, K., Yamada, K., Sawada, H., Nishiwaki, E., Smith, K., Yoshino, K., Ishiguro, H. and Higashiyama, S. (2002) Gene **282**, 227–236
60. Van Eerdewegh, P., Little, R.D., Dupuis, J., Del Mastro, R.G., Falls, K., Simon, J., Torrey, D., Pandit, S., McKenny, J., Braunschweiger, K. et al. (2002) Nature (London) **418**, 426–430
61. Shi, Y. (2003) Trends Genet. **19**, 9–12
62. Gilliland, L.K., Walsh, L.A., Frewin, M.R., Wise, M.P., Tone, M., Hale, G., Kioussis, D. and Waldmann, H. (1999) J. Immunol. **162**, 3663–3671
63. Chapman, A.P. (2002) Adv. Drug Delivery Rev. **54**, 531–545

64. Humphreys, D.P. (2003) Curr. Opin. Drug Discovery Dev. **6**, 188–196
65. Reichert, J.M. (2001) Nat. Biotechnol. **19**, 819–822
66. Bentley, G.A., Alzari, P.M., Amit, A.G., Boulot, G., Guillon-Chitarra, V., Fischmann, T., Lascombe, M.B., Mariuzza, R.A., Poljak, R.J., Riottot, M.M. et al. (1989) Philos. Trans. R. Soc. London B Biol. Sci. **323**, 487–494
67. Amit, A.G., Mariuzza, R.A., Phillips, S.E. and Poljak, R.J. (1986) Science **233**, 747–753
68. Desmyter, A., Spinelli, S., Payan, F., Lauwereys, M., Wyns, L., Muyldermans, S. and Cambillau, C. (2002) J. Biol. Chem. **277**, 23645–23650
69. Murphy, G., Allan, J.A., Willenbrock, F., Cockett, M.I., O'Connell, J.P. and Docherty, A.J. (1992) J. Biol. Chem. **267**, 9612–9618
70. Tortorella, M., Pratta, M., Liu, R.Q., Abbaszade, I., Ross, H., Burn, T. and Arner, E. (2000) J. Biol. Chem. **275**, 25791–25797
71. Babcook, J.S., Leslie, K.B., Olsen, O.A., Salmon, R.A. and Schrader, J.W. (1996) Proc. Natl. Acad. Sci. U.S.A. **93**, 7843–7848
72. Falb, D. and Jindal, S. (2002) Curr. Opin. Drug Discovery Dev. **5**, 532–539
73. Freundlieb, S. and Gamer, J. (2002) New Drugs **3**, 54–60

Biochem. Soc. Symp. **70**, 163–178
(Printed in Great Britain)
© 2003 Biochemical Society

14

How serpins change their fold for better and for worse

Robin W. Carrell[1] and James A. Huntington

University of Cambridge, Department of Haematology, Wellcome Trust/MRC Building, Hills Road, Cambridge CB2 2XY, U.K.

Abstract

The serpins differ from the many other families of serine protease inhibitors in that they undergo a profound change in topology in order to entrap their target protease in an irreversible complex. The solving of the structure of this complex has now provided a video depiction of the changes involved. Cleavage of the exposed reactive centre of the serpin triggers an opening of the five-stranded A-sheet of the molecule, with insertion of the cleaved reactive loop as an additional strand in the centre of the sheet. The drastic displacement of the acyl-linked protease grossly disrupts its active site and gives an overall loss of 40% of ordered structure. This ability to provide effectively irreversible inhibition explains the selection of the serpins to control the proteolytic cascades of higher organisms. The conformational mechanism provides another advantage in its potential to modulate activity. Sequential crystallographic structures now provide clear depictions of the way antithrombin is activated on binding to the heparans of the microcirculation, and how evolution has utilized this mobile mechanism for subtle variations in activity. The complexity of these modulatory mechanisms is exemplified by heparin cofactor II, where the change in fold is seen to trigger multiple allosteric effects. The downside of the mobile mechanism of the serpins is their vulnerability to aberrant intermolecular β-linkages, resulting in various disorders from cirrhosis to thrombosis. These provide a well defined structural prototype for the new entity of the conformational diseases, including the common dementias, as confirmed by the recent identification of the familial neuroserpin dementias.

 Time-lapse movies of results presented in this chapter can be viewed at http://www.portlandpress.com/bssymp/070/bss0700163add.htm

[1]To whom correspondence should be addressed (e-mail rwc1000@cam.ac.uk).

Introduction

The serpins were the first proteins to be identified with a mechanism requiring not just a change in their shape, but also a profound change in their topology [1–8]. The puzzle that accompanied the recognition of this mobile mechanism was that it is clearly unique to the serpins, although they are just one of some 20 families of serine protease inhibitors [9,10]. The other families of protease inhibitors are all small proteins with tightly configured tertiary structures designed to hold an exposed reactive-centre peptide loop in a fixed conformation. However, although the core protein structures differ in each of the families, all of their exposed reactive loops have convergently evolved an identical conformation that precisely matches the active site of the chymotrypsin family of serine proteases [11]. Thus almost all of the serine protease inhibitors function by a simple lock-and-key blocking mechanism (Figure 1a). The serpins too, although they are much larger proteins, have a flexible reactive loop that readily takes up the same canonical conformation (Figure 1b). However, it is clear that this mobile mechanism must provide additional selective advantages, as the serpins have become the predominant serine protease inhibitors in higher organisms. This predominance is readily apparent in human plasma, where serpins such as antithrombin control the coagulation cascade by inhibiting thrombin and factor Xa, α_1-antitrypsin modulates inflammation by inhibiting the elastase released by neutrophils, and plasminogen activator inhibitor-1 (PAI-1) controls fibrin degradation. The advantage provided by the mobile mechanism of inhibition must indeed be considerable, as it was recognized early on that this mobility also brings with it a vulnerability to mutations, with a wide range of diseases being shown to arise from mutations affecting the mobile domains of the serpin molecule [12,13]. What is the advantage of the mobile mechanism of the serpins? Why are the serpins more effective inhibitors? Why have they been selected as the predominant protease inhibitors in higher organisms?

The challenge of these questions was set some 15 years ago in the context of the review by Barrett and Salvesen of proteinase inhibitors [9]. It was apparent at that time that a definitive answer required the determination of the sequence of structural changes involved in the serpin mechanism. In effect, this requires a video depiction formed of a sequence of frames, each frame being a crystallographic structure at a different stage in the conformational movements. It seemed a daunting task. However, there was good reason to believe that it was achievable. Evolution and serendipity provided some early frames for the video – serendipity in the crystallization of cleaved α_1-antitrypsin [14] and evolution in the spontaneous insertion of the reactive loop into the A β-sheet of the molecule in PAI-1 [15] and in the freezing of the reactive loop of ovalbumin in its exposed form [16]. The need then was to fill in the intervening crystallographic frames, a task that took more than a decade, culminating just 2 years ago with the final frame of the inhibitory mechanism [8]. The completed sequence of structures explains in a most satisfying way not only the distinct advantage of the serpin inhibitory mechanism, but also how the mobility of this mechanism has been adapted to allow the subtle modulations of

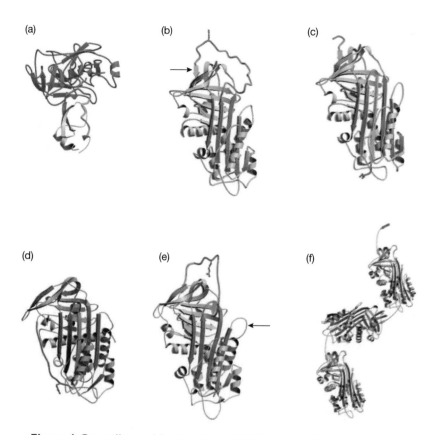

Figure 1 Crystallographic structures. (a) The canonical conformation of the reactive-centre loop (green) of bovine pancreatic trypsin inhibitor precisely matches the active site of trypsin (magenta). (b) The reactive centre of the active form of the serpin α_1-antitrypsin can adopt the same canonical conformation [24]. (c) The cleaved reactive loop of serpins (yellow) immediately inserts as an extra middle strand in the five-stranded A β-sheet (red) [14]. (d) The same transition to the hyperstable six-stranded A-sheet occurs in latent antithrombin, with release of strand 1 of the C-sheet [arrowed in (b)] and insertion of the intact reactive loop into the A-sheet [46]. (e) Antithrombin circulates in a relatively inactive form, with partial insertion of the reactive loop into the A-sheet and internal orientation of the reactive-centre arginine (green). Activation by heparin causes transition to the active conformation as in (b). The attachment site of the missing oligosaccharide in β-antithrombin is arrowed [26,27,29,46]. (f) Aberrant opening of the A-sheet allows the formation of intermolecular β-strand linkages to give loop–sheet polymers, which are commonly disordered but illustrated here with an ordered crystallographic example [63].

activity required, from tissue to tissue, in higher organisms. The bonus has been not only a clearer understanding of the way that mutations affecting these mechanisms cause disease, but also an opening up of prospects for the prevention and treatment of these diseases [17,18].

The videos: cassette crystallization

The serpins have a strongly conserved template structure formed by nine helices (A–I) and three β-pleated sheets (A–C), with the dominant feature of the structure being the five-stranded A-sheet [2]. The inhibitory function of the serpins is dependent on a triggered opening of the A-sheet to allow the insertion of the cleaved reactive-centre loop as an additional strand in the centre of the sheet (Figure 1c). The driving force for this change is a shift from the metastable active form, with a 'melting' temperature of 60°C, to the hyperstable inactive form with a melting temperature typically of 120°C or more. The conservation of this complex mechanism has resulted in a superimposable identity of backbone structures within the serpins. and thus validates the use of structures from different serpins as frames for the video depictions [19–27]. A breakthrough in the determination of a range of conformations resulted from the development of a cassette technology allowing the ready crystallization of diverse conformations of antithrombin.

The development of the cassette approach arose from a proposal in 1991 that the serpins in general, and antithrombin in particular, could, if perturbed, incorporate their intact reactive loop into the A-sheet to give a hyperstable inactive form [3]. The prediction that this transition to a six-stranded A-sheet occurred in the latent form of PAI-1 was subsequently confirmed crystallographically [15], but there was much scepticism as to whether antithrombin could undergo a similar transition. Experimental evidence for the formation of latent antithrombin was frustratingly inconsistent. The reason is now apparent [28]. Antithrombin does indeed readily undergo an irreversible transition to the latent form both *in vivo* and *in vitro* (Figure 1d). However, as soon as a latent molecule forms, it β-links to a molecule of active antithrombin to give a heterodimer, which on native gel electrophoresis runs close to the normal active component. The structural evidence for this dimer came from the coincidental crystallization of antithrombin in two different centres [26,27]. In both instances, crystallization trials had been set up with purified active antithrombin, but it was the heterodimer of active and latent antithrombin that crystallized, with linkage of the reactive loop of the active molecule to the vacated strand 1 position in the C-sheet of the latent molecule. Subsequently, Lei Jin and colleagues utilized this finding to successfully determine the long-sought-after structure of the antithrombin–heparin complex [29]. They prepared latent antithrombin and then added to it active antithrombin in the presence of the core heparin pentasaccharide, to give ready and reproducible crystallization of the heterodimer of latent and heparin-bound active antithrombin.

Subsequently this cassette approach, using latent antithrombin as a starting template, has been utilized by us and colleagues to give some 20 different conformational forms of antithrombin. These structures allowed the completion of a series of videos, including the transition from the active to the latent conformation, the detailed atomic-level interactions at the point of opening of the A-sheet [30,31], and the binding and activation of antithrombin by heparin. Although these and other completed video sequences provided a coherent outline of the sequence of movements involved in the major conformational transitions in the serpins, there was a critical absentee. Many attempts by a number of groups over a period of 15 years had failed to crystallize the inhibitory serpin–protease complex.

Solving the serpin–protease complex

The elusive crystal structure of the final complex formed by a serpin with its target protease became labelled as the 'Holy Grail' of the field [32]. The challenges for crystallization were caused by the nature of the serpin mechanism itself. Since the reaction pathway is bifurcated, it is never possible to get a perfect 1:1 complex. Very small amounts of active protease are sufficient to selectively degrade molecules of the complex, giving heterogeneous products that counter attempts at crystallization. In addition, the serpin–protease complex undergoes a slow deacylation, resulting in cleaved serpin and active protease. To overcome these problems, we chose reaction conditions containing a large excess of the inhibitor α_1-antitrypsin over the protease trypsin, followed by precise separation of the complex and then its crystallization at refrigerated temperature. The crystals obtained gave X-ray diffraction at a resolution that showed interactions at the atomic level, as well as revealing the overall structure of the complex [8]. The result (Figure 2) immediately answers questions relating to the function of the serpins, and fills in the missing frames that now complete a crystallographic video of the way the serpins entrap their target proteases (http://www.portlandpress.com/bssymp/070/bss0700163add.htm).

The consecutive structures of the video show how the protease takes the bait of the exposed reactive loop and initially forms a Michaelis complex [33,34] identical to that formed by the other families of serine protease inhibitors (Figure 1a). In all of these families, including the serpins, the initial mechanism is the same, with the formation of a tight non-covalent complex followed by nucleophilic attack of the reactive centre (P1) carbonyl carbon of the inhibitor by the catalytic serine of the protease (Figure 2b). In the families of small protease inhibitors, this process halts at various stages of the catalytic cycle, prior to the separation of the cleaved ends of the loop. Inherently, though, this complex is reversible, which results ultimately in the release of active protease and intact inhibitor. Thus, mechanistically, the small-protein protease inhibitors are reversible, tight-fitting inhibitors, whose inherent structural rigidity prohibits proteolysis. The serpin mechanism is distinct, in that it relies on the success of proteolysis and on the structural mobility of the serpin. It is best to think of a serpin as a substrate whose rate-limiting step is deacylation, since left long enough the complex dissociates to yield the products of proteolysis. Thus the reaction of the protease with the serpin is typical of that with any substrate, proceeding through the Michaelis complex and the first tetrahedral intermediate to yield the acyl–enzyme intermediate. At this stage, an ester bond exists between the active-site serine of the protease and the carbonyl carbon of the P1 residue of the serpin. It is here that the special property of the serpins becomes evident. The release of the cleaved C-terminal peptide fragment from the active site is accompanied in the serpins by a dramatic conformational rearrangement, with insertion of the reactive-centre loop into the A-sheet and a flinging displacement of the attached protease to the other end of the molecule. As the structure of the

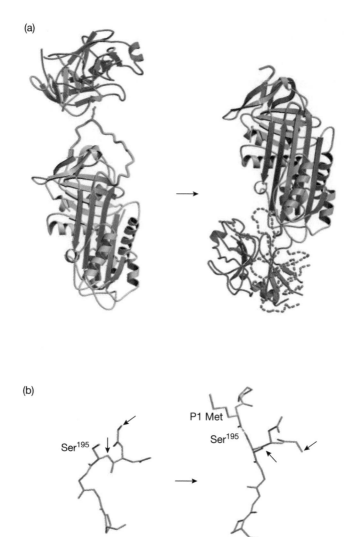

Figure 2 Serpin inhibitory mechanism. (a) The 70 Å displacement of the protease from one pole to another of the serpin results in a plucking-distortion of the active site of the protease, with a loss of ~40% of the ordered structure of the protease (magenta). (b) The disruption of the active site of the protease (left) is seen (right) to result from the ester linkage of the reactive-centre P1 methionine to the active Ser[195] of the protease. The serine is plucked away from the catalytic site, with consequent loss of the oxanion hole formed by Ser[195] and Gly[193] (arrowed). Further details are given in [8]. A time-lapse movie of this mechanism can be viewed at http://www.portlandpress.com/bssymp/070/bss0700163add.htm

complex shows (Figure 2a), this forced displacement distorts the protease, causing it to lose 40% of its ordered structure.

The effectiveness of the serpins is due not just to the entrapment and gross distortion of the protease, but rather to the lethal disruption of the active site of the protease that accompanies this distortion. The active serine of the protease, tethered by its ester linkage to the displaced loop of the serpin, is effectively pulled away from its position within the catalytic triad (Figure 2b). The displacement of the catalytic serine results in the distension of the catalytic loop to which it is attached, causing the destruction of the oxyanion hole. The role of the oxyanion hole is to stabilize the tetrahedral intermediate, where a negative charge exists on the carbonyl oxygen. Deacylation occurs through the nucleophilic attack of water, which proceeds through a tetrahedral intermediate. This step is effectively prevented by the removal of the ester bond from the catalytic diad responsible for potentiating the water oxygen as a nucleophile, and by increasing the energy of the tetrahedral intermediate by distending the oxyanion hole. The critical factor in this lethal plucking action is the short length of the cleaved reactive loop, limited in all serpins to 17 amino acids [35]. The consequence is an overlap of the two structures, which contributes to the crushing distortion of the protease together with a plucking of the active serine away from its catalytic site (Figure 2). Moreover, the overlap of structures in the final complex, together with the plucking action of the displaced loop, perturb the structure of the protease to an extent that makes it susceptible to proteolytic cleavage. Normally, functional proteases are folded in such a way as to be highly resistant to proteolytic attack. However, it had been observed that proteases in complex with a serpin were readily cleaved at multiple sites in their structure [36–38]. We can now see that each of these cleavages occurs in the portion of the protease molecule that is disordered in the complex.

This disorder of the protease is also affected by a breaking of the principal bond responsible for the conformational integrity of the active protease. Serine proteases are initially synthesized and secreted in an inactive and only partially folded zymogen form. Their activation depends on an N-terminal cleavage, with formation of a salt bridge between the new N-terminal amine on Ile^{16} and Asp^{194}, to give a 20% increase in ordered structure [39]. However, this process is reversed by the disruption of the catalytic triad of the protease that occurs on complexing with the serpin. This does not by itself ensure complete irreversibility, although it does slow the release of active protease to an extent that the half-life of the complex is measured in years rather than seconds. However, the serpin mechanism ensures that, well before such release occurs, the unfolded protease will be irreversibly destroyed by incidental proteolysis.

Heparin and the modulation of inhibition

The irreversibility of binding of the protease enables the serpins to provide the complete inhibition required for the control of proteolytic pathways such as coagulation. These pathways consist of series of different proteases, each one capable of amplifying the production of the next protease in the series [40]. Thus the release of just one or two molecules of the coagulation protease Factor Xa can rapidly result in the activation of tens of thousands of molecules of thrombin, with the consequent onset of potentially fatal thrombosis. To counter this, antithrombin must rapidly and irreversibly inhibit any Factor Xa released. However, if there was total efficiency in inhibition, then coagulation could never take place and we would all bleed to death.

Evolution has met this problem by utilizing the structural mobility of antithrombin to allow the initiation of coagulation, but not its progression to thrombosis. There are two different requirements of a natural anticoagulant: in the large vessels it must function slowly, so as to allow the initial formation of the fibrin plug, but peripherally it must be active immediately to protect the vulnerable microvasculature. The elegant mechanism evolved by antithrombin to meet these needs is seen in a video sequence (http://www.portlandpress.com/bssymp/070/bss0700163add.htm) formed from a series of crystallographic structures determined with our colleagues (PDB codes 1ANT, 1AZX, 1BR8, 1BTH, 1DZ, 1DZH, 1EO3, 1EO4, 1EO5, 2ANT, 1JVQ, 1LK, 1LK6, 1NQ9 and 1OYH). Antithrombin circulates with its reactive-centre peptide loop partially inserted into the A-sheet [26,27], such that the reactive-centre arginine is internally oriented and is only slowly accessible to proteolytic attack (Figure 1e). In the peripheral circulation, however, it binds to, and is activated by, the glycosaminoglycan heparans that line the microvasculature [41]. Antithrombin contains a basic site on its D-helix that binds specifically to a defined pentasaccharide sequence present both in heparans and in the derivatized therapeutic agent heparin [42,43]. Binding by the pentasaccharide induces an overall change in the conformation of antithrombin [29], with the formation of a new helix, the extension of the D-helix, the closure of the five-stranded A-sheet and the expulsion of the reactive centre into an exposed active position, as shown in Figure 1(b). The essential requirement for this ability to switch inhibitory activity on and off was dramatically demonstrated 20 years ago, with the identification of a natural mutation that replaces the reactive-centre methionine of α_1-antitrypsin with an arginine [44]. This change of the reactive-centre P1 residue altered the specificity of inhibition of the α_1-antitrypsin from being an inhibitor of elastase to being an inhibitor of thrombin and Factor Xa. In effect, the mutation converted the α_1-antitrypsin not just into an antithrombin, but into a permanently activated antithrombin. The consequence was a life-threatening bleeding disorder.

The other aspect of the conformational change revealed by the series of structures is the rearrangement of the pentasaccharide-binding site that accompanies the transition to the six-stranded A-sheet during complex formation. This results in both the release of the antithrombin–protease complex into the circulation for catabolic destruction and the freeing of the heparin or heparan to enable it to bind and activate another molecule of antithrombin. Nor is this the only way that the mobile mechanism has been utilized to provide modulation of activity. The most accessible link between the heparin-induced changes in the binding site on the D-helix and the shift in the reactive loop is an exposed peptide loop that connects the top of the D-helix to strand 2 in the A-sheet (arrowed in Figure 1e). This hD–s2A loop contains an oligosaccharide attachment site that differs from other such sites in antithrombin in having the sequence Asn-Xaa-Ser rather than Asn-Xaa-Thr. The presence of the serine

leads to only 80% glycosylation at this point. The remaining 20%, β-antithrombin [45], constitutes the most active inhibitory component in the blood. This is
because the absence of the bulky oligosaccharide on the mobile hD–s2A loop
allows a more ready transition to the fully activated conformation [46]. Thus β-
antithrombin binds preferentially to the heparans of the microvasculature,
where it provides a first line of defence against thrombosis [47]. The combined
effects of these modulatory adaptations of antithrombin provide the necessary
subtle gradation of anticoagulant activity from tissue to tissue.

The complexity of the modulation of antithrombin is, however, overshad-
owed by that of another glycosaminoglycan-activated serpin, heparin cofactor II
(HCII). This is an enigmatic inhibitor, which is present in human plasma at high
concentrations and with a targeted specificity for thrombin, but with, as yet, no
defined physiological role. Earlier work [48] had shown how the targeting of
HCII against thrombin is dependent on its unique possession of an extended 80-
residue N-terminal tail. The tail contains an acidic domain with a sequence closely
similar to that of the leech anticoagulant hirudin. Recent crystallographic structures of HCII and of its complex with thrombin [34] show how this hirudin-like
domain binds to an exosite on thrombin and is then sandwiched against the body
of the serpin so as to form the initial Michaelis complex. HCII resembles
antithrombin in that it circulates with a partially inserted reactive loop, and has a
virtually identical heparin-binding site. Thus the two inhibitors have similar
mechanisms of activation by glycosaminoglycans, but with the activation of HCII
resulting in the release of its N-terminal tail as well as a reorientation of its reactive-centre loop. The elegance of the mechanism and of its allosteric control is
evident in the full video depiction available from the combined series of structures,
as summarized and referenced in the legend to Figure 3 [34]. The extraordinary
feature is the way in which the tail can extend, like a fisherman's line, to 'catch' the
thrombin and then guide it to optimal inhibitory docking with the serpin.
Conversely, the subsequent displacement of the thrombin to the other pole of the
serpin results in the disruption of the binding exosite on thrombin, with release of
the tail and then its cleavage to give a chemotactically active peptide [49].

Evolution in progress

It is surprising to find that antithrombin and HCII have independently
evolved the same mechanism of partial loop insertion, with each having identical
heparin-binding sites linked to the opening and closing of the A-sheet. How
does the evolution of such intricate mechanisms take place? An insight into how
this may occur, mutation-by-mutation, is seen in the current British population.
Within recent times, a main risk to survival in the fertile age group was the frequent occurrence of severe haemorrhage after childbirth. One response to this,
apparent in the European, is the emergence of a mutation in coagulation Factor
V, which predisposes to thrombosis [50]. However, another response, of particular interest here, is the appearance of a mutation in antithrombin that achieves
the same effect by subtly modifying the serpin inhibitory mechanism. A survey
of British blood donors [51] showed the relatively high frequency of a mutation
in the hinge of the reactive loop – the portion that first enters the A-sheet on

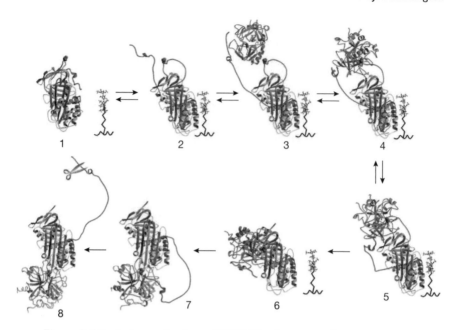

Figure 3 Allosteric mechanism of HCII. The glycosaminoglycan activation of serpins is shown as video frames based on the crystallographic structures of HCII and antithrombin–heparin complexes [29,34]. First, HCII binds to endothelial glycosaminoglycans, with expulsion of the partially inserted reactive loop (yellow) and release of the N-terminal tail (magenta) (steps 1 and 2). The hirudin-like domain (magenta) then binds to an exosite on thrombin and guides it to form the Michaelis complex (steps 3–5). Next, the cleaved reactive loop of HCII inserts into the A-sheet, with release of the glycosaminoglycan and displacement of the thrombin to the other pole of HCII (steps 6 and 7). Finally (step 8), the consequent distortion of the protease (orange) disrupts the binding exosite. This frees the N-terminal tail, allowing its cleavage by resident plasma proteases, with release of a chemotactic peptide. A time-lapse movie of this mechanism can be viewed at http://www.portlandpress.com/bssymp/070/bss0700163add.htm

cleavage of the reactive centre (Figures 1b and 2). The mutation codes for the replacement of a conserved alanine, 10 residues prior to the reactive centre (P10), by a serine [52]. This change, from the small apolar side chain of alanine to the somewhat larger polar side chain of serine, will predictably decrease slightly the rate of entry of the cleaved reactive loop into the A-sheet. This rate is critical to the inhibitory mechanism, which has been likened to a race between insertion of the loop (and hence inhibition) and deacylation of the serpin–protease linkage to give substrate-like cleavage of the serpin. Studies of the mutant antithrombin by Mushunje [53] show that the only significant change in activity is in the inhibition of Factor Xa and thrombin in the presence of heparin. The physiological consequence is that the mutant antithrombin will continue to be an effective inhibitor of coagulation proteases in the main circulation. However, it will be a less effective inhibitor when bound to the heparans of the capillaries and sinusoidal vessels that are the source of blood loss after childbirth. Thus the effect of the mutation will predictably be an increase in maternal survival rates, with the downside being an increased rate of thrombosis in travelling professors – but this last is unlikely to threaten the survival of our species!

Conformational disease and dementia

A feature of the serpins is the number of diseases that result from their malfunction [12]. In part this is due to the critical contribution that they make to essential pathways, but the diseases arising from mutations in the serpins also show an unusual diversity in their nature. Mutations in most proteins commonly result in a readily understandable loss of function of the protein, but in the serpins they often result in gain-of-function diseases [18]. Thus the mutations in α_1-antitrypsin commonly present in people of European descent not only result in a loss of protection for the lungs against attack by elastases, but are also associated with the slow onset of liver cirrhosis [17,54]. Similarly, mutations in antithrombin may cause little apparent loss in inhibitory activity, and then quite suddenly result in a catastrophic loss of activity to give the onset of life-threatening thrombosis [55,56]. We now know that disease manifestations such as these result from amino acid substitutions that affect the mobile domains of the molecule. Moreover, we can see in crystallographic detail the structural consequences of such changes in the serpins and the sequence of conformational events that leads to them [57,58]. The end result in almost all cases is an increase in β-structure. What was unexpected was the diversity of ways in which this could occur, with the changes being monomeric, dimeric or polymeric [59–61], and with polymeric changes (Figure 1f) resulting in either disordered or highly ordered aggregation [62–65].

It was also clear that the diverse manifestations that accompany these changes in the serpins mirror the perplexing clinical findings in other single-protein diseases, and most notably in the dementias of Alzheimer's disease, Parkinson's disease and the prion encephalopathies [66–68]. These neurodegenerative diseases, just as with the cirrhosis arising from serpin abnormalities, develop over a period of years and usually only became apparent in old age. In particular, the bizarre features of the prion encephalopathies, with the infective propagation of conformational changes and the isoform specificity of this propagation, are also seen to occur, in a more benign way, with antithrombin [57]. As with the serpins, each of the neurodegenerative diseases arises from a change in the fold of an underlying protein, with β-strand or β-sheet linkages leading to aggregation and accumulation of the protein. However, a survey of the medical literature shows the serpinopathies and the dementias to be just part of a much wider entity, the conformational diseases [69], arising from protein instability and typically resulting in late-onset and sporadic disorders. Conformational diseases occur when an individual protein undergoes a change in fold, with resultant self-association and tissue deposition. Previously such diseases, and in particular the common dementias, had been classified as amyloidoses [70]. This reflects the frequent finding in these diseases of deposits of amyloid, a highly ordered β-fibrillar aggregate [71]. Amyloid specifically binds the dye Congo Red and has a typical birefringence and fibre diffraction pattern. However, amyloid is not always present in the conformational diseases, and represents just one of a number of end-point structural rearrangements. In the dementias there is increasing evidence that the pathological species are the earlier oligomers that precede amyloid formation, and not the amyloid itself [72,73]. Nevertheless, there was an understandable reluctance by researchers in this field to accept the relevance of studies of α_1-antitrypsin and cirrhosis, in which amyloid does not feature, to the pathogenesis of dementia. The events that broke through this reluctance commenced with the insistence of a general practitioner in upstate New York that an autopsy be carried out on a middle-aged patient who had developed an atypical 'Alzheimers' dementia [74].

To put this breakthrough into context, some 2 or 3 years earlier James Whisstock had prepared the first crystallographic video of the overall conformational change in the serpins [30]. He used this video to examine in detail the shifts in amino acid side chains that occurred as the A-sheet of the serpins opened, with a focus on a particular serine (Ser[53]), a mutation of which was known to cause the polymerization of α_1-antitrypsin. In analysing the video, he pointed out that two other amino acids, Ser[56] and His[334], played an even more critical role in the opening of the A-sheet (Figure 4a). The relevance of these deductions was highlighted by the results of the autopsy and its subsequent follow-up in Syracuse, NY [13,75]. The autopsy showed the presence of numerous particulate aggregates in the neurons of the brain. These intracellular inclusions had an appearance and staining properties identical with those of the inclusions found in liver cells in α_1-antitrypsin cirrhosis. Furthermore, the neuronal inclusions were formed of polymers of a brain-specific serpin, neuroserpin, containing a mutation at the same site (Ser[53]) that resulted in the liver inclusions of α_1-antitrypsin. The age of onset of dementia in this index case and of other affected family members was close to 50 years. Subsequently a second family was identified with similar findings, but with a mutation at Ser[56] and an age of onset of dementia around 30 years [76]. The full validation of the predictions made from the video analysis of Whisstock and colleagues [77] came with the identification of a third individual with the onset of dementia at age 15 years and a mutation in neuroserpin at His[334]. Taken together, the findings with the mutants of neuroserpin and their correlation with other serpinopathies provides convincing evidence that protein aggregation is in itself a sufficient cause of late-onset dementia. There is, over our lifetime, a significant and steady loss of both hepatocytes and neurons. For most people the loss of neurons does not, even in later years, affect the essential social and housekeeping functions of life, although the safety margin is small. However, if this loss of neurons is even slightly accelerated by cellular protein aggregation, as in the families with the neuroserpin mutations, then there will be the inability to cope that we label clinically as dementia. The evidence for this correlation between protein aggregation and the rate of neurodegeneration is clear in Figure 4(b), in which the severity of conformational instability is seen to be related to the magnitude of inclusion body formation and hence to the age of onset of disease.

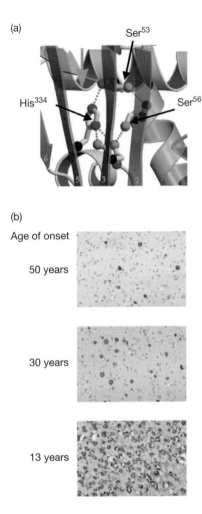

Figure 4 Mutations in neuroserpin causing dementia. (a) Video depiction of the side-chain movements involved in entry of the reactive loop
(arrowed) into the A-sheet of serpins, showing the necessary movement of
Ser[53]. However, Ser[56] was predicted to be even more critically involved, with
the hydrogen-bond network formed by His[334] affecting the interactions of both
of these serines [30,77]. These predictions were borne out by the subsequent
recognition [13,75] of mutations in neuroserpin of Ser[53] (onset of neurodegeneration at 50 years), Ser[56] (onset at 30 years) and His[334] (onset at 15 years*).
(b) The correlation of the age of onset of dementia with the degree of conformational instability is seen in the stained brain sections showing intracellular
aggregation of polymerized neuroserpin. Template numbering [2] is used
throughout this paper; *the '13 years' section is from a further case [75] with a
mutation causing overall disruption of the region shown in a.

The fourth dimension

Many currently working in the field of proteases and inhibitors will have seen the transition from the single dimension of protein sequences to the three dimensions of crystallography. This transition involved changes in thinking and perception, but these were minor compared with the changes required for the study of proteins that undergo functional changes in fold. As biochemists we have become accustomed to dealing with molecules whose properties and behaviour in any given buffer and temperature is predictable and reproducible. However, in dealing with the flexibility of fold seen with the serpins, a new dimension has been introduced, that of time. Measurements carried out on stored antithrombin will give different results over a period of days and weeks, no matter how carefully the sample has been stored. In the meantime, some of the antithrombin will have been converted into the latent form and then dimerized to give a heterodimer. The changes are compounded by even small differences in pH or salt concentration. We can see clearly now the reasons for the frustrations and uncertainties that affected earlier work with the serpins. Because mobile mechanisms require consideration in four dimensions, new approaches to the handling and presentation of data had to be developed. Just as crystallography requires stereo views, so the study of conformational transitions requires video depictions. True videos, made frame-by-frame from a sequence of individual structures, should not be confused with cartoons, any more than a crystallographic structure should be confused with a molecular model. A true molecular video can be used to examine conformational changes at all levels, from outline skeletal movements down to atomic detail. As indicated in Figure 4(a), individual hydrogen bonds can be seen to form, stretch, break and then be re-formed with therapeutic blocking agents [78]. We can be confident of the fidelity of what we see because, as shown in Figure 4(b), the predictions made have been confirmed subsequently by the most testing of all transgenics – the experiments of Nature in humans.

This work was supported by the Wellcome Trust, the Medical Research Council (U.K.) and the National Institutes of Health (U.S.A.). We are grateful to Dr Richard Davis (SUNY at Syracuse, NY, U.S.A.) for help with Figure 4.

References

1. Carrell, R.W. and Boswell, D.R. (1986) in Proteinase Inhibitors (Barrett, A. and Salvesen, G., eds), pp. 403–420, Elsevier, Amsterdam
2. Huber, R. and Carrell, R.W. (1989) Biochemistry **28**, 8951–8966
3. Carrell, R.W., Evans, D.L. and Stein, P.E. (1991) Nature (London) **353**, 576–578
4. Carrell, R. and Travis, J. (1985) Trends Biochem. Sci. **10**, 20–24
5. Gettins, P.G.W., Patston, P.A. and Olson, S.T. (1996) Serpins: Structure, Function and Biology, R.G. Landes Company, Austin, TX
6. Irving, J.A., Pike, R.N., Lesk, A.M. and Whisstock, J.C. (2000) Genome Res. **10**, 1845–1864
7. Silverman, G.A., Bird, P.I., Carrell, R.W., Church, F.C., Coughlin, P.B., Gettins, P.G., Irving, J.A., Lomas, D.A., Luke, C.J., Moyer, R.W. et al. (2001) J. Biol. Chem. **276**, 33293–33296
8. Huntington, J.A., Read, R.J. and Carrell, R.W. (2000) Nature (London) **407**, 923–926
9. Barrett, A.J. and Salvesen, G. (eds) (1986) Proteinase Inhibitors, Elsevier, Amsterdam
10. Laskowski, M. and Qasim, M.A. (2000) Biochim. Biophys. Acta **1477**, 324–337

11. Bode, W. and Huber, R. (2000) Biochim. Biophys. Acta **1477**, 241–252
12. Stein, P.E. and Carrell, R.W. (1995) Nat. Struct. Biol. **2**, 96–113
13. Davis, R.L., Shrimpton, A.E., Holohan, P.D., Bradshaw, C., Feiglin, D., Sonderegger, P., Kinter, J., Becker, L.M., Lacbawan, F., Krasnewich, D. et al. (1999) Nature (London) **401**, 376–379
14. Loebermann, H., Tokuoka, R., Deisenhofer, J. and Huber, R. (1984) J. Mol. Biol. **177**, 531–556
15. Mottonen, J., Strand, A., Symersky, J., Sweet, R.M., Danley, D.E., Geoghegan, K.F., Gerard, R.D. and Goldsmith, E.J. (1992) Nature (London) **355**, 270–273
16. Stein, P.E., Leslie, A.G.W., Finch, J.T., Turnell, W.G., McLaughlin, P.J. and Carrell, R.W. (1990) Nature (London) **347**, 99–102
17. Carrell, R.W. and Lomas, D.A. (2002) N. Engl. J. Med. **346**, 45–53
18. Lomas, D.A. and Carrell, R.W. (2002) Nat. Rev. Genet. **3**, 759–768
19. Wei, A., Rubin, H., Cooperman, B.S. and Christianson, D.W. (1994) Nat. Struct. Biol. **1**, 251–258
20. Gooptu, B., Hazes, B., Chang, W.S., Dafforn, T.R., Carrell, R.W., Read, R.J. and Lomas, D.A. (2000) Proc. Natl. Acad. Sci. U.S.A. **97**, 67–72
21. Xue, Y., Bjorquist, P., Inghardt, T., Linschoten, M., Musil, D., Sjolin, L. and Deinum, J. (1998) Structure **6**, 627–636
22. Sharp, A.M., Stein, P.E., Pannu, N.S., Carrell, R.W., Berkenpas, M.B., Ginsburg, D., Lawrence, D.A. and Read, R.J. (1999) Structure **7**, 111–118
23. Elliott, P.R., Lomas, D.A., Carrell, R.W. and Abrahams, J.P. (1996) Nat. Struct. Biol. **3**, 676–681
24. Elliott, P.R., Abrahams, J.-P. and Lomas, D.A. (1998) J. Mol. Biol. **275**, 419–425
25. Song, H.K., Lee, K.N., Kwon, K.-S., Yu, M.-H. and Suh, S.W. (1995) FEBS Lett. **377**, 150–154
26. Schreuder, H.A., de Boer, B., Dijkema, R., Mulders, J., Theunissen, J.H., Grootenhis, P.D. and Hol, W.G. (1994) Nat. Struct. Biol. **1**, 48–54
27. Carrell, R.W., Stein, P.E., Fermi, G. and Wardell, M.R. (1994) Structure **2**, 257–270
28. Zhou, A., Huntington, J.A. and Carrell, R.W. (1999) Blood **94**, 3388–3396
29. Jin, L., Abrahams, J.P., Skinner, R., Petitou, M., Pike, R.N. and Carrell, R.W. (1997) Proc. Natl. Acad. Sci. U.S.A. **94**, 14683–14688
30. Whisstock, J. (1996) in Haematology, p. 314, Ph.D. Thesis, University of Cambridge
31. Carrell, R.W. and Stein, P.E. (1996) Biol. Chem. Hoppe-Seyler **377**, 1–17
32. Stratikos, E. and Gettins, P.G. (1999) Proc. Natl. Acad. Sci. U.S.A. **96**, 4808–4813
33. Ye, S., Cech, A.L., Belmares, R., Bergstrom, R.C., Tong, Y., Corey, D.R., Kanost, M.R. and Goldsmith, E.J. (2001) Nat. Struct. Biol. **8**, 979–983
34. Baglin, T.P., Carrell, R.W., Church, F.C., Esmon, C.T. and Huntington, J.A. (2002) Proc. Natl. Acad. Sci. U.S.A. **99**, 11079–11084
35. Zhou, A., Carrell, R.W. and Huntington, J.A. (2001) J. Biol. Chem. **276**, 27541–27547
36. Kaslik, G., Patthy, A., Balint, M. and Graf, L. (1995) FEBS Lett. **370**, 179–83
37. Stavridi, E.S., O'Malley, K., Lukacs, C.M., Moore, W.T., Lambris, J.D., Christianson, D.W., Rubin, H. and Cooperman, B.S. (1996) Biochemistry **35**, 10608–10615
38. Egelund, R., Petersen, T.E. and Andreasen, P.A. (2001) Eur. J. Biochem. **268**, 673–685
39. Bode, W. and Huber, R. (1978) FEBS Lett. **90**, 265–269
40. Olson, S.T. and Björk, I. (1992) in Thrombin: Structure and Function (Berliner, L.J., ed.), pp. 159–217, Plenum Press, New York
41. Marcum, J.A., McKenney, J.B. and Rosenberg, R.D. (1984) J. Clin. Invest. **74**, 341–350
42. Choay, J., Petitou, M., Lormeau, J.C., Sinay, P., Casu, B. and Gatti, G. (1983) Biochem. Biophys. Res. Commun. **116**, 492–499
43. Lindahl, U., Thunberg, L., Backstrom, G., Riesenfeld, J., Nordling, K. and Bjork, I. (1984) J. Biol. Chem. **259**, 12368–12376
44. Owen, M.C., Brennan, S.O., Lewis, J.H. and Carrell, R.W. (1983) N. Eng. J. Med. **309**, 694–698
45. Peterson, C.B. and Blackburn, M.N. (1985) J. Biol. Chem. **260**, 610–615

46. McCoy, A.J., Pei, X.Y., Skinner, R., Abrahams, J.-P. and Carrell, R.W. (2003) J. Mol. Biol. **326**, 823–833

47. Fribelius, S., Isaksson, S. and Swedenberg, J. (1996) Arteroscler. Thromb. Vasc. Biol. **16**, 1292–1297

48. Tollefsen, D.M. (1997) Adv. Exp. Med. Biol. **425**, 35–44

49. Church, F.C., Pratt, C.W. and Hoffman, M. (1991) J. Biol. Chem. **266**, 704–709

50. Dahlback, B. (1995) Thromb. Haemostasis **74**, 139–148

51. Tait, R.C., Walker, I.D., Perry, D.J., Islam, S.I.A., Daly, M.E., McCall, F., Conkie, J.A. and Carrell, R.W. (1994) Br. J. Haematol. **87**, 106–112

52. Perry, D.J., Daly, M.E., Tait, R.C., Walker, I.D., Brown, K., Beauchamp, N.J., Preston, F.E., Gyde, H., Harper, P.L. and Carrell, R.W. (1998) Thromb. Haemostasis **79**, 249–253

53. Mushunje, A. (2002) Blood Coagulation Fibrinolysis **13**, Ab

54. Laurell, C.-B. and Eriksson, S. (1963) Scand. J. Clin. Lab. Invest. **15**, 132–140

55. Bruce, D., Perry, D.J., Borg, J.-Y., Carrell, R.W. and Wardell, M.R. (1994) J. Clin. Invest. **94**, 2265–2274

56. Beauchamp, N.J., Pike, R.N., Daly, M., Butler, L., Makris, M., Dafforn, T.R., Zhou, A., Fitton, H.L., Preston, F.E., Peake, I.R. and Carrell, R.W. (1998) Blood **92**, 2696–2706

57. Carrell, R.W. and Gooptu, B. (1998) Curr. Opin. Struct. Biol. **8**, 799–809

58. Carrell, R.W., Huntington, J.A., Mushunje, A. and Zhou, A. (2001) Thromb. Haemostasis **86**, 14–22

59. Lomas, D.A., Evans, D.L., Finch, J.T. and Carrell, R.W. (1992) Nature (London) **357**, 605–607

60. Aulak, K.S., Eldering, E., Hack, C.E., Lubbers, Y.P.T., Harrison, R.A., Mast, A., Cicardi, M. and Davis, III, A.E. (1993) J. Biol. Chem. **268**, 18088–18094

61. Lindo, V.S., Kakkar, V.V., Learmonth, M., Melissari, E., Zappacosta, F., Panico, M. and Morris, H.R. (1995) Br. J. Haematol. **89**, 589–601

62. Schulze, A.J., Huber, R., Degryse, E., Speck, D. and Bischoff, R. (1991) Eur. J. Biochem. **202**, 1147–1155

63. Huntington, J.A., Pannu, N.S., Hazes, B., Read, R., Lomas, D.A. and Carrell, R.W. (1999) J. Mol. Biol. **293**, 449–455

64. Dunstone, M.A., Dai, W., Whisstock, J.C., Rossjohn, J., Pike, R.N., Feil, S.C., Le Bonniec, B.F., Parker, M.W. and Bottomley, S.P. (2000) Protein Sci. **9**, 417–420

65. Janciauskiene, S., Carlemalm, E. and Eriksson, S. (1995) Biol. Chem. Hoppe-Seyler **375**, 103–109

66. Selkoe, D.J. (2001) Physiol. Rev. **81**, 741–766

67. Polymeropoulos, M.H., Lavedan, C., Leroy, E., Ide, S.E., Dehejia, A., Dutra, A., Pike, B., Root, H., Rubenstein, J., Boyer, R. et al. (1997) Science **276**, 2045–2047

68. Prusiner, S.B., Scott, M.R., DeArmond, S.J. and Cohen, F.E. (1998) Cell **93**, 337–348

69. Carrell, R.W. and Lomas, D.A. (1997) Lancet **350**, 134–138

70. Tan, S.Y. and Pepys, M.B. (1994) Histopathology **25**, 403–414

71. Sunde, M., Serpell, L.C., Bartlam, M., Fraser, P.E., Pepys, M.B. and Blake, C.C. (1997) J. Mol. Biol. **273**, 729–739

72. Conway, K.A., Lee, S.J., Rochet, J.C., Ding, T.T., Williamson, R.E. and Lansbury, P.T.J. (2000) Proc. Natl. Acad. Sci. U.S.A. **97**, 571–576

73. Walsh, D.M., Klyubin, I., Fadeeva, J.V., Cullen, W.K., Anwyl, R., Wolfe, M.S., Rowan, M.J. and Selkoe, D.J. (2002) Nature (London) **416**, 535–539

74. Davis, R.L., Holohan, P.D., Shrimpton, A.E., Tatum, A.H., Daucher, J., Collins, G.H., Todd, R., Bradshaw, C., Kent, P., Feiglin, D. et al. (1999) Am. J. Pathol. **155**, 1901–1913

75. Davis, R.L., Shrimpton, A.E., Carrell, R.W., Lomas, D.A., Gerhard, L., Baumann, B., Lawrence, D.A., Yepes, M., Kim, T.S., Ghetti, B. et al. (2002) Lancet **359**, 2242–2247

76. Yerby, M.S., Shaw, C.M. and Watson, J.M. (1986) Neurology **36**, 68–71

77. Whisstock, J.C., Skinner, R., Carrell, R.W. and Lesk, A.M. (2000) J. Mol. Biol. **295**, 651–665

78. Zhou, A., Stein, P.A., Huntingdon, J.A. and Carrell, R.W. (2003) J. Biol. Chem. **278**, 15116–15122

Biochem. Soc. Symp. **70**, 179–199
(Printed in Great Britain)
© 2003 Biochemical Society

15

Cystatins

Magnus Abrahamson[1], Marcia Alvarez-Fernandez and Carl-Michael Nathanson

Department of Clinical Chemistry, Institute of Laboratory Medicine, University of Lund, University Hospital, S-221 85 Lund, Sweden

Abstract

Chicken egg white cystatin was first described in the late 1960s. Since then, our knowledge about a superfamily of similar proteins present in mammals, birds, fish, insects, plants and some protozoa has expanded, and their properties as potent peptidase inhibitors have been firmly established. Today, 12 functional chicken cystatin relatives are known in humans, but a few evolutionarily related gene products still remain to be characterized. The type 1 cystatins (A and B) are mainly intracellular, the type 2 cystatins (C, D, E/M, F, G, S, SN and SA) are extracellular, and the type 3 cystatins (L- and H-kininogens) are intravascular proteins. All true cystatins inhibit cysteine peptidases of the papain (C1) family, and some also inhibit legumain (C13) family enzymes. These peptidases play key roles in physiological processes, such as intracellular protein degradation (cathepsins B, H and L), are pivotal in the remodelling of bone (cathepsin K), and may be important in the control of antigen presentation (cathepsin S, mammalian legumain). Moreover, the activities of such peptidases are increased in pathophysiological conditions, such as cancer metastasis and inflammation. Additionally, such peptidases are essential for several pathogenic parasites and bacteria. Thus cystatins not only have capacity to regulate normal body processes and perhaps cause disease when down-regulated, but may also participate in the defence against microbial infections. In this chapter, we have aimed to summarize our present knowledge about the human cystatins.

The cystatin superfamily

The cystatins constitute a protein superfamily of enzyme inhibitors. Their main function is probably to ensure protection of cells and tissues against the proteolytic activity of lysosomal peptidases that are released either occa-

---

[1]To whom correspondence should be addressed
(e-mail Magnus.Abrahamson@klinkem.lu.se)

sionally at normal cell death, or 'intentionally' by proliferating cancer cells or by invading organisms, such as parasites [1].

The first cystatin to be isolated was a papain and ficin inhibitor character-ized from chicken egg white [2,3]. The name 'cystatin' was proposed for the first time by Barrett [4], and was later used to describe homologous proteins in the same superfamily [5]. The first cystatin sequence determined was that of a protein isolated from human urine, which was called 'γ-trace' due to its elec-trophoretic mobility [6]. A few years later the function of γ-trace as a cystatin was assigned and it was renamed cystatin C [7]. Today we know that cystatin C is one of a dozen human cystatins, all with different properties as cysteine pep-tidase inhibitors and each with unique expression and distribution patterns. The present review will focus primarily on a summary of those human mem-bers of the cystatin superfamily.

Different cystatins can be grouped in three major protein families or 'types' (Figure 1), defined by similarity analysis of mammalian inhibitors. The original denomination of a family was based on the premise that those proteins included should display ≥50% amino acid identity at sequence alignments [8]. Nowadays, we have switched to use the term 'type' according to the criteria for grouping cystatins proposed by Rawlings and Barrett [9]. In this way, a protein belongs to a certain cystatin type, not only on the basis of sequence similarities with homologues, but also depending on "how many cystatin-like segments are present" and "whether the protein presents disulphide bonds". A new clas-sification scheme for peptidase inhibitors has been presented very recently by

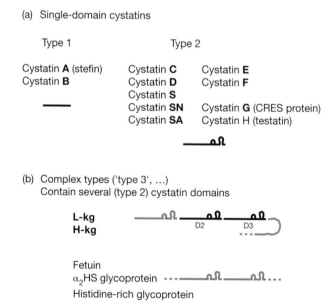

Figure 1 **Human cystatins.** The human cystatin superfamily members with proven function as cysteine peptidase inhibitors are indicated in **bold**. kg, kininogen; HS, Heremans–Schmid; CRES, cystatin-related epididymal and spermatogenic.

Barrett and colleagues, applying criteria developed for the fundamental classification of peptidases in the MEROPS database (http://merops.sanger.ac.uk).
Although the first release has been stressed to be a preliminary version, this
database is a milestone achievement that will facilitate peptidase inhibitor
research, and the classification used is likely to be a main reference for many
years to come. In the MEROPS classification, the cystatins all belong to the
same family, on the basis of statistically significant similarities in amino acid
sequence. Different inhibitor families that are thought to share evolutionary
origin can and will then be grouped into a clan. The cystatins entered into the
classification scheme to date all belong to MEROPS Family I25 of Clan IH.
The strict sequence-defined classification of peptidase inhibitors is problematic
in some cases, however. Some cystatins, and many other larger proteins with
peptidase inhibitory function, contain several inhibitor domains. These
domains are treated as single inhibitor units in the MEROPS classification, but
the additional level of organization of multi-domain inhibitors and the relationship of such multi-domain proteins is not always obvious from the
sequence-based classification. We therefore feel that the grouping of cystatins
in types still adds value to a review on cystatins, especially when the possible
biological functions of the inhibitors are discussed.

Type 1 cystatins

The type 1 cystatins are intracellular cystatins that are present in the
cytosol of many cell types, but can also appear in body fluids at significant concentrations [10]. They are single-chain polypeptides of ~100 amino acid
residues, and present neither disulphide bonds nor carbohydrate side chains.
There are two human representatives, cystatins A and B, which will be
described in detail below. Homologues have been found in mouse, rat, cow and
pig (reviewed in [11]). The type 1 cystatins have been called 'stefins' to stress
their difference from other cystatin superfamily members. However, they have
a general structure very similar to the 'cystatin fold' of other cystatins and, consequently, similar activity as cysteine peptidase inhibitors. In a similarity tree
(Figure 2a) or evolutionary tree (Figure 2b) calculated from human cystatin

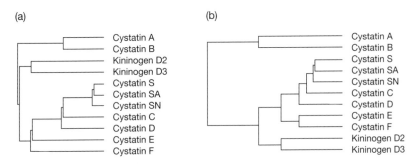

Figure 2 Relationships between human cystatins. Human cystatin
sequences were analysed using algorithms in the GCG package, to illustrate
their similarity (**a**) and evolutionary relationships (**b**).

sequences, it is obvious that the two type 1 cystatins are quite closely related, and constitute a distinct subgroup. In the MEROPS classification, the type 1 cystatins form Subfamily A of Family I25.

Type 2 cystatins

The type 2 cystatins are mainly extracellular, secreted proteins. They are synthesized with 19–28-residue signal peptides, leading to transport of the mature protein across the cell membrane into the extracellular space. They are broadly distributed and found in most body fluids [10]. Like type 1 cystatins, they are single-chain polypeptides, but larger (about 120 residues long). Also like type 1 cystatins, they contain a conserved Gln-Xaa-Val-Xaa-Gly segment known as the 'cystatin motif' in the central part of their sequences. Mammalian and avian type 2 cystatins all present two conserved disulphide bridges in the C-terminal end of the sequence, with 10 and 20 residues in between the respective cysteines, and a characteristic Pro-Trp pair in their C-terminal segments. They may be phosphorylated [12]. The avian type 2 cystatin, chicken egg white (CEW) cystatin, has had its tertiary structure solved by both X-ray crystallography and NMR spectroscopy [13,14], which has defined the 'cystatin fold' (see below). The 'classic' human type 2 cystatins, cystatins C, D, S, SA and SN, are >50% identical at the sequence comparison level. The more recently elucidated cystatins E/M and F show sequence identities of 30–35% compared with the classic cystatins, but should still be seen as type 2 cystatins due to overall similarity, and the presence of a signal peptide and conserved disulphide bridges (Figure 2). This is also the case for the as yet uncharacterized gene products of the cystatin multigene locus (see below), if they prove to be functional inhibitors, as their deduced sequences both contain signal peptides and conserved disulphide-forming Cys residues. The human type 2 cystatins are grouped in Subfamily B of MEROPS Family I25.

Type 3 cystatins

These are multi-domain proteins. The mammalian cystatins belonging to this type are the kininogens [15], which were first described as kinin precursor proteins. These proteins are high-molecular-mass cystatins (60–120 kDa) presenting three tandemly repeated type 2-like cystatin domains, with a total of eight disulphide bridges (six conserved and two additional at the beginning of cystatin domains D2 and D3). They are glycosylated proteins, but the carbohydrate attachments sites are not present in the cystatin domains. Only the second and third cystatin domains (D2 and D3) present cysteine peptidase inhibitory activity [16]. These domains show a quite high degree of sequence identity when aligned with mammalian type 2 cystatin sequences (Figure 2). There are three different kininogens in mammals: H- (high-molecular-mass) and L- (low-molecular-mass) kininogens, both of which are found in several mammals, and T-kininogen, found only in the rat [17]. Kininogens are intravascular in localization; they are found in blood plasma and, as a result of diffusion, in synovial and amniotic fluids [10]. Like the type 2 cystatins, cystatin domains D2 and D3 with inhibitory activity in human kininogens (Figure 1) are grouped in Subfamily B of MEROPS Family I25.

Other cystatin-related proteins

Fetuin in mammals, sarcocystatin and *Drosophila* cystatin from insects, puff adder venom cystatin from a reptile, *Bm*-CPI-1 and *Bm*-CPI-2 from a parasite, and the plant 'phyto-cystatin', oryzacystatin (the structure of which has been determined by NMR [18]), are examples of a wide array of cystatin-related proteins included in the cystatin superfamily (for individual references, see [9,19]), illustrating that cystatins are old proteins in evolutionary terms and, hence, are present in very different organisms. Several of these cystatin-related proteins show quite low sequence identity with the three types mentioned above, and some of them have been shown to be devoid of inhibitory activity. Others are potent cysteine peptidase inhibitors from species quite distant from mammals, and are synthesized with signal peptides, but have no or only one disulphide bridge, which makes it difficult to group them into the types defined for mammalian cystatins. Yet others are functional inhibitors that are multi-domain proteins (such as the potato multicystatin). Although these sequences have not yet been entered in the MEROPS classification and database, all of them will probably be found to belong to Family I25, due to evolutionary relationships of amino acid sequences. It is too early to speculate on whether they will all be found in Subfamily B, or if yet another subfamily (or perhaps several additional subfamilies) may be recognized.

Human cystatins

Some basic characteristics of the human cystatins, which are useful for the identification of different cystatins appearing at purification of cysteine peptidase inhibitors [20] or when protein array data are analysed, are summarized in Table 1. Individual protein, cDNA or gene sequences can easily be retrieved through MEROPS, and several useful protein sequence alignments have been published [9,21,22].

Cystatin A

Cystatin A (also called 'epidermal SH-protease inhibitor', 'epidermal TPI', 'ACPI' or 'CPI-A' at original characterization of the inhibitor; 'stefin' in the original amino acid sequence report; and 'stefin A' in some recent papers) is a type 1 cystatin (MEROPS id I25.001). Human cystatin A is composed of 98 amino acid residues [23]. A cDNA encoding cystatin A was first identified as corresponding to an mRNA species that is up-regulated following UV irradiation of skin [24]. Its gene was originally mapped to chromosome 3cen-q21 by amplifying cystatin A fragments by PCR from hamster–human hybrid DNA [25]. The expression pattern is quite restricted, with especially high levels of the inhibitor in skin and some blood cells. The tertiary structure of human cystatin A has been elucidated by NMR spectroscopy [26].

Cystatin B

Cystatin B (also called 'NCPI', 'liver TPI' or 'CPI-B' at original characterization of the inhibitor; and 'stefin B' in the three-dimensional structure report) is a type 1 cystatin (MEROPS id I25.003). Human cystatin B is com-

Table 1 Some molecular characteristics of human low-molecular-mass cystatins. N-terminal sequences were aligned on the evolutionarily conserved Gly residue present in the N-terminal segments of all functional cystatins [78]. ac, acetyl.

Cystatin	Type	Amino acids (n)	Molecular mass (Da)	N-terminal sequence	pI	Preprotein
A	1	98	11006	MIPG-	4.5–4.7	No
B	1	98	11175	acMMCG-	5.6–6.3	No
C	2	120	13343*	SSPGKPPRLVG-	9.3	146 aa
D	2	122	13885†	GSASAQSRTLAG-	6.8–7.0	142 aa
S	2	121	14189‡	SSSKEENRIIPG-	4.4–4.6	141 aa
SN	2	121	14316	WSPKEEDRIIPG-	6.6–6.8	141 aa
SA	2	121	14351	WSPQEEDRIIEG-	4.6	141 aa
E/M	2	121	13652	RPQERMVG-	7-8	149 aa
F	2	126	14534	GPSPDTCSQDLNSRVKPG-	6-7	145 aa

*Cystatin C from urine contains a hydroxylated Pro residue, Hyp[3], giving a molecular mass of 13359 Da [6].
†Cystatin D also exists in a form with Arg replacing Cys[26], giving a molecular mass of 13938 Da [44].
‡Phosphorylated forms of cystatin S, with a higher molecular mass, are present in saliva [60].

posed of 98 amino acid residues [27]. A cDNA encoding cystatin B was deposited in the GenBank database in 1992 (accession number L03558). The gene was later cloned and mapped to chromosome 21 [28]. The expression pattern is broad in human cells and tissues, and the inhibitor can be seen as a general cytosolic inhibitor in mammalian cells, where it probably serves to protect the cell against leakage of lysosomal enzymes. Cystatin B is one of the three human cystatins for which genetic evidence points to an important biological function. Loss-of-function alterations in the cystatin B gene (either through a multiplied repeat unit in the promoter or through point mutations in the structural gene) are present in both alleles of the cystatin B gene in patients with a form of progressive myoclonus epilepsy [28]. Seemingly, this causes disease due to increased apoptosis in the brain, resulting in loss of critical neurons at an early age [29]. The tertiary structure of human cystatin B has been elucidated by X-ray crystallography [30].

Cystatin C

Cystatin C (also called 'post-γ-globulin' in early characterization of the protein, and 'γ-trace' in the original protein sequence report) is a type 2 cystatin (MEROPS id I25.004) and is the most thoroughly studied of the mammalian cystatins. The primary structure of the protein was determined by Grubb and Löfberg in 1982 [6]. Later, two groups independently determined the sequence of CEW cystatin [31,32] and showed that it was highly similar to the primary structure of γ-trace/cystatin C. Mature human cystatin C is composed of 120 amino acid residues [6]. The cDNA sequence [33] revealed that it is synthesized as a preprotein with a 26-residue signal peptide. The cystatin C gene is located in the cystatin multigene locus on chromosome 20 [34–36] (Table 2) and is of the housekeeping gene type [37]. Consequently, studies of cystatin C expression show that it is produced in a wide variety of human tissues and cell lines [37,38], and it is the type 2 cystatin that shows the most widespread body distribution. It is found in all body fluids at significant concentrations, and in particularly high levels in seminal plasma (~50 mg/l or 3.7 μM) and cerebrospinal fluid (~5.8 mg/l or 430 nM) [10,37]. Cystatin C has strikingly good inhibitory properties for all papain-like (family C1) peptidases investigated. Thus the inhibitor can be seen as a major, general extracellular cysteine peptidase inhibitor in mammals. The concentrations of cystatin C are sufficiently high to allow the inhibitor to efficiently inhibit any lysosomal cysteine peptidase that is released into a body fluid [10]. This is necessary for rapid trapping and neutralization of the peptidase activity in an 'emergency' inhibition, as described by Turk et al. [39].

A point mutation in the cystatin C gene, resulting in the substitution Leu^{68}→Gln, is responsible for the dominantly inherited Icelandic type of amyloidosis, hereditary cystatin C amyloid angiopathy (HCCAA). Patients with HCCAA suffer from paralysis and dementia as successive brain haemorrhages take place from a very young age, typically concluding in death at the age of around 30 years (reviewed in [40]). HCCAA is primarily to be seen as a conformational disease, in which the abnormal protein aggregates formed as such are responsible for the pathology. The possibility cannot be ruled out, how-

Table 2 Type 2 cystatin genes. The cystatin genes at the locus on human chromosome 20 and the cystatin E/M gene on chromosome 11 are sorted according to gene name. *CST1*–*CST8* all encode functional cystatins. The inhibitory properties of the products of the *CST9L*, *CST11* and *CSTL1* genes have not yet been determined. The main tissue of expression of the genes and the locus ID in the GenBank database are indicated. n.d., not determined.

Gene name	Common name	Chromosome	Tissue	Locus ID
CST1	Cystatin SN	20p11.21	Salivary gland	1469
CST2	Cystatin SA	20p11.21	Salivary gland	1470
CST3	Cystatin C	20p11.21	Ubiquitous	1471
CST4	Cystatin S	20p11.21	Salivary gland	1472
CST5	Cystatin D	20p11.21	Salivary gland, tear duct	1473
CST6	Cystatin E/M	11q13	Epithelium (skin)	1474
CST7	Cystatin F	20p11.21	Haematopoietic cells	8530
CST8	Cystatin G/CRES	20p11.21	Testis, epididymis	10047
CST9L	Testatin	20p11.21	Testis	128821
CST11	Cystatin 11	20p11.21	Testis	140880
CSTL1	Cystatin-like 1	20p11.21	Testis	128817
CSTP1	Cystatin pseudogene 1	20p11.21	n.d.	1480
CSTP2	Cystatin pseudogene 2	20p11.21	n.d.	1481
BC024006	Cystatin I pseudogene	20p11.21	Testis	164380

ever, that the partial loss of cysteine peptidase inhibitory function in the heterozygous patients, resulting when cystatin C concentrations in the brain are reduced to approximately half of normal levels due to the aggregation, also has an effect on the disease phenotype.

Cystatin D

Cystatin D (MEROPS id I25.005) was first identified as the product of a gene with resemblance to the cystatin C gene [41]. It contains all the classical features of a type 2 cystatin. In the 122-amino-acid polypeptide chain, the four disulphide-forming Cys residues as well as the central 'cystatin motif' and the C-terminal Pro-Trp motif are perfectly conserved [41,42]. Two alleles of the cystatin D gene are known, coding for either Cys or Arg as residue 26 [43]. It has been shown that this variation has no effect on the enzyme-binding properties of the inhibitor [44]. The cystatin D gene is located in the cystatin multigene locus on chromosome 20 [45] (Table 2) and was originally shown to be expressed in the parotid gland [41]. The protein is found in saliva, and also in small amounts in tears. However, cystatin D could not be detected in seminal or blood plasma, milk or cerebrospinal fluid [42].

Cystatin E/M

Cystatin E/M (MEROPS id I25.006) was first identified as a cDNA sequence. In a search for new members of the cystatin superfamily in human EST (expressed sequence tag) libraries, it was discovered as a rare mRNA species, originally detected in only two out of >650 cDNA libraries (called cystatin E [46]). These libraries were made from amniotic cells and foetal skin, and it was concluded that the expression of the cystatin E/M gene was quite restricted to epithelial cells [46]. Independently, cystatin E/M was discovered as the product of a gene that was down-regulated in mammary tumour cells compared with surrounding cells (called cystatin M [47]).

Cystatin E/M is a 121-amino-acid mature polypeptide with a 28-residue theoretical signal peptide. The amino acid sequence identity of the mature polypeptide is approx. 34% with the cystatin C sequence and 26–30% with cystatins D, S, SA and SN. The polypeptide chain contains some unusual features, including a five-residue insertion between amino acids 76 and 77 and a deletion of residue 91 (cystatin C numbering). Despite these differences, the typical type 2 cystatin features are perfectly conserved. In contrast with the classical human cystatins, cystatin E/M contains an N-glycosylation site at residue 108, and recombinant cystatin E/M expressed in a baculovirus expression system is indeed glycosylated [46]. Unlike all other human type 2 cystatin genes, the cystatin E/M gene is not located on chromosome 20, but on chromosome 11 [48] (Table 2). Northern blot experiments on tissue mRNA have shown somewhat conflicting results, although it has not been debated that, at the cellular level, various epithelial cells are the main producers of cystatin E/M. In one of the original studies, it was reported that the gene was expressed in all major human tissues, with the strongest signals seen in liver, ovary and pancreas, and small amounts of cystatin E/M could be isolated from urine using an affinity column [46]. However, other Northern blot studies indicated that cystatin E/M is not expressed in liver or brain, and in very low amounts in heart [47]. Recently, Zeeuwen and co-workers have shown results indicating that the expression of cystatin E/M is quite restricted to human epidermal keratinocytes and sweat glands [49], but appears to be less restricted in mouse tissues [50]. Cystatin E/M is the third cystatin for which genetic evidence points to a particular biological function. A null mutation in the cystatin E/M gene causes juvenile lethality and defective epidermal cornification in a particular mouse strain, a disease phenotype resembling human type 2 harlequin ichthyosis [50].

Cystatin F

Cystatin F (also known as 'leukocystatin' or 'CMAP'; MEROPS id I25.007) was independently discovered as a cDNA sequence by three different lines of research. In one, cystatin F was found through sequence similarity searches of ESTs in a database of human cDNA library sequences [51]. A total of 54 clones were found in 20 different libraries. The positive clones were all derived from cDNA libraries of different haematopoeitic cells and cell lines. Northern blot experiments showed expression in a variety of tissues, with particularly high expression in peripheral blood leucocytes, thymus and spleen [51]. Independently, Halfon et al. [52] screened a wide variety of blood cells, blood cell lines and foetal tissues in humans, and found cystatin F (called leukocystatin) primarily in resting T cells, pre-monocytic cells, activated stem cell-derived dendritic cells and some natural killer cell clones. Later, a third identification of mouse cystatin F [called cystatin-related metastasis-associated protein (CMAP)] was reported, as an mRNA species that was up-regulated in the metastasizing tumour cell line IMC-HM [53]. Human cancer cell lines were subsequently screened for cystatin F transcripts, expression being demonstrated in some of these cell lines [54].

The human and mouse cystatin F genes each contain four exons instead of the usual three in type 2 cystatin genes [52,53,55]. The human gene is located in the cystatin multigene locus on chromsome 20 (Table 2). The expression of the gene is readily regulated, as has recently been demonstrated in human U937 cells [55]. In this premyeloid cell line, differentiation towards a granulocytic pathway by all-*trans*-retinoic acid or towards a monocytic pathway through stimulation with phorbol ester both resulted in marked down-regulation of cystatin F expression. Also, unusually for type 2 cystatins, a large fraction of the inhibitor is localized intracellularly in lysosome-like granules in U937 cells, indicating that cystatin F may have an intracellular function, e.g. to regulate enzymes involved in antigen presentation [55]. Cystatin F protein levels in human tissues and body fluids are generally low, but it has been reported that normal blood serum contains detectable amounts of the inhibitor (~1 μg/l or 7 pM) [51], and pleural fluid from patients with inflammatory lung disorders may contain concentrations of cystatin F up to ~3 μg/l (190 pM) [56]. Cystatin F presents the conserved 'cystatin' and Pro-Trp motifs as well as the four Cys residues forming disulphide bridges typically found in classical type 2 cystatins. Two additional Cys residues are found in the N-terminal part of the polypeptide (residues 1 and 37; cystatin C numbering), which appear to form a third disulphide bridge, making cystatin F unique among the single-domain cystatins [51]. Like cystatin E/M, cystatin F is a glycoprotein. Two N-glycosylation sites are found in the polypeptide sequence, one N-terminally at position 36 and one C-terminally at position 88. Both possible sites are glycosylated in insect cell-produced or human U937 cell-derived cystatin F [51,55].

Cystatins S, SA and SN

Cystatin S (MEROPS id I25.008), cystatin SA (I25.009) and cystatin SN (I25.010) display amino acid sequence identities of 90% at pairwise comparisons, with an overall identity with the cystatin C sequence of ~50%. These cystatins were named after the fluid from which they were first isolated, i.e. saliva [57–59]. Later it was shown that these inhibitors can also be found in tears, urine and seminal plasma [10]. They are 121-amino-acid non-glycosylated proteins that are synthesized with signal peptides, and their mature sequences contain the typical type 2 cystatin motifs. Cystatin S appears in different forms in saliva due to partial phosporylation at two sites [60]. The cystatin S, SA and SN genes [61,62] are all located in the cystatin multigene locus on chromosome 20 [35] (Table 2). The expression patterns for these genes are quite restricted, to salivary, tear and some of the male sex glands. Thus cystatins S, SA and SN could, like cystatin D, be seen as specialized glandular cystatins, and possibly function in secretions as defence inhibitors of exogenous cysteine peptidases.

Additional functional type 2 cystatins

Several other functional type 2 cystatins may be found among the as yet not fully characterized gene products at the cystatin multigene locus (Table 2). In 1992, Cornwall et al. [63] described cystatin-related epididymal-specific (CRES) protein, the deduced product of a transcript isolated from a mouse epididymis cDNA library. Later studies showed that the gene was also expressed in spermatids, and the name of the protein was thereafter modified slightly to cystatin-related epididymal and spermatogenic (CRES) protein [64]. The human homologue of this protein has recently been identified, and our preliminary results demonstrate that it is indeed a functional cystatin (C.-M. Nathanson and M. Abrahamson, unpublished work). We have tentatively called it cystatin G, as it appears to be a twelfth functional member of the cystatin superfamily, at least in humans.

A second cystatin-like protein thought to be specific for the male reproductive organ was called testatin when originally cloned from a mouse foetal testis cDNA library [65–67]. It is still not known if this gene product is a functional cystatin. Yet another testis-specific type 2 cystatin-like mRNA was very recently reported by Hamil et al. [68]. The theoretical protein product has tentatively been called cystatin 11. Its expression in humans has not been widely studied, and only three clones are found in the public EST databases. It is still not known if this gene product is a functional cystatin. In the mouse, an mRNA for 'cystatin T' has been discovered in a testis cDNA library [69]. This gene product may be a functional cystatin, but the most similar of the human cystatin genes contains inactivating mutations, rendering it a pseudogene (Table 2).

Kininogens

L-kininogen (also known as α_2-CPI or α_2-TPI) and H-kininogen (α_1-CPI, α_1-TPI) both contain the functional cystatin domains D2 and D3 (MEROPS id I25.016–017). The kininogen molecule is divided into three regions: the 'heavy chain' N-terminal region, the kinin region in the core and the 'light chain' region towards the C-terminus. The large and miniscule light chains of H- and L-kininogen respectively are what distinguish between the two molecules. A cDNA for α_2-cysteine peptidase inhibitor was cloned in 1984 [70], showing surprisingly high identity with the bovine H- and L-kininogens sequenced the year before [71,72]. Only one kininogen gene is present in the human genome, on chromosome 3q26-qter, which encodes both H- and L-kininogens [73,74]. The gene covers ~27 kb and comprises 11 exons. The first nine of these encode sequences found in both L- and H-kininogens. Exon 10 is spliced in two different ways to result in two mRNAs. L-kininogen mRNA contains only a short portion of the 5′ part of exon 10, the bradykinin sequence, plus the whole of exon 11. H-kininogen mRNA, however, contains the whole of exon 10, but excludes exon 11. Thus the divergent C-terminal sequences of the two proteins originate from the splicing differences of exons 10 and 11.

The main site of kininogen expression is undoubtedly the liver, as is the case for most intravascular proteins. L- and H-kininogens are found in approximately equimolar amounts in blood, at a total concentration of ~8 μM [10]. Thus the kininogens are a major source of inhibitory capacity in the circulation, for systemic protection against leaking lysosomal cysteine peptidases or enzymes derived from invading micro-organisms. Smaller amounts of the kininogens are found in all other body fluids tested except saliva. However, in secretions and the extracellular space in tissues, the single-domain cystatins are the predominant cysteine peptidase inhibitors [10].

Structure of cystatins

True, functional cystatins inhibit papain-like (family C1) cysteine peptidases through tight (K_i in the nanomolar to picomolar range) and reversible binding in a substrate-competing mechanism, to form an equimolar complex with the peptidase [75–79]. As multi-domain proteins, kininogens are the exception to this rule, and can bind two peptidase molecules simultaneously [80].

Clues as to which parts of the cystatin molecule are involved in binding to C1 peptidases were first obtained from functional studies. Controlled enzymic cleavage, site-directed mutagenesis and fluorescence analysis of human cystatin C, CEW cystatin and kininogens [16,78,81–87] indicated three main peptidase-interacting segments. The first is the N-terminal region, where the well conserved Gly[11] residue (human cystatin C numbering) is positioned. The second segment is located in the central part of the sequence, displaying the Gln-Xaa-Val-Xaa-Gly 'cystatin motif' at positions 55–59. The third segment includes the conserved Pro[105]-Trp[106] pair in type 2 and 3 cystatins (type 1 cystatins present Pro-His/Gly in the corresponding positions).

The three-dimensional structures of types 1 and 2 cystatins and of the plant cystatin, oryzacystatin [13,14,19,27,30,88], confirmed the biochemical results and gave us an insight into how cystatins inhibit C1 peptidases. The structure models showed a very conserved tertiary structure, the so-called 'cystatin fold' (Figure 3), which comprises a five-stranded antiparallel β-pleated sheet wrapped around a 5-turn α-helix. The major difference between type 1 and 2 cystatins is the nearly 20-residue-long 'appendix loop' between strands 3 and 4 of the β-sheet in type 2 cystatins, which is missing from type 1 cystatins. The three parts implicated in involvement in target enzyme binding by functional studies are localized in the N-terminal segment (Gly[11]), the first β-hairpin loop (L1) between strands 2 and 3 (Gln[55]-Xaa-Val-Xaa-Gly[59]) and the second β-hairpin loop (L2) between strands 4 and 5 (Pro[105]-Trp[106]). Together, they form a wedge-shaped edge [13].

Bode et al. [89] proposed an intermolecular interaction mechanism based on docking experiments performed for the individual tertiary structures of papain and N-terminally truncated CEW cystatin [13,90], and results suggesting involvement of the N-terminal segment in inhibition [78]. In this model, the tripartite edge enters the active-site cleft of the target enzyme without making any covalent interactions with it, blocking its entrance in such a way that a substrate cannot be hydrolysed by the enzyme. According to this model, the binding site in cystatins is complementary to the peptidase active-site cleft. In this way, the S3–S2–S1 subsite pockets in the peptidase should accommodate the side chains of residues in positions 9, 10 and 11 (cystatin C numbering) of the N-terminal segment of the cystatin, and the first and second hairpin loops should interact with the residues forming the S1′ and S2′ subsites respectively in the enzyme. No cleavage of the inhibitor is possible, as residue 12 in the cystatin is unable to fit into the S1′ pocket. This 'key-in-the-hole' fit does not require substantial conformational changes in either the inhibitor or the enzyme. This agrees with the fact that cystatins inhibit papain and several cathepsins in a one-step reaction [83]. The X-ray structure of cystatin B in complex with papain [30] confirmed most aspects of this model of inhibition mechanism. For the inhibition of cathepsin B, however, there is evidence for a two-step kinetic reaction [91]. The first step is regulated by the anchoring of the N-terminal segment in the narrower side of the cleft (the non-primed S subsites), followed by the displacement of the occluding loop of the enzyme. In this way, the other two segments of the cystatin are free to bind to the broader part of the active cleft (the S′ subsites) in the second step of the reaction.

In addition to the X-ray structures of CEW cystatin and cystatin B, NMR studies have elucidated the structures of cystatin A and the plant cystatin oryzacystatin [18] (Figure 3), Moreover, Janowski et al. [88] recently determined the structure of human cystatin C in its dimeric form. The structure is that of a perfectly symmetrical dimer, as observed earlier in NMR studies [92]. In the dimer, two cystatin C molecules interact with each other to form two identical domains. This is possible due to subdomain swapping of the two cystatin molecules involved, where each molecule by itself loses its tertiary structure but maintains its secondary elements. In this way, each domain is constituted by elements of both molecules, retaining the conserved 'cystatin fold', as other monomeric cystatins do. The subdomain swapping occurs in the first hairpin loop of the monomeric molecules, resulting in destruction of the C1 peptidase binding site. In this domain switch region, or 'open interface', the two unstable 'molten globule-like' molecules establish main-chain–main-chain

M. Abrahamson et al.

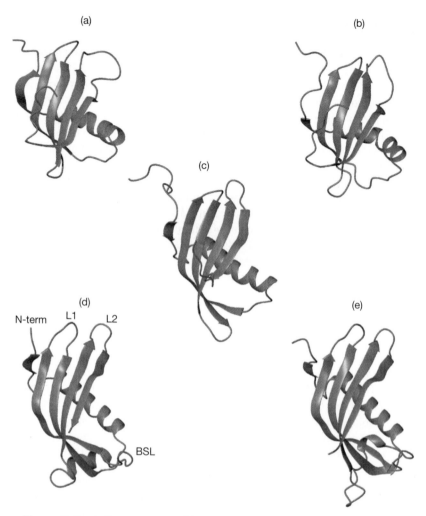

Figure 3 Cystatin structures. The cystatin structures known from X-ray crystallography or NMR spectroscopy studies are illustrated. (**a**) Human cystatin A (PDB code 1dvd) [21]; (**b**) human cystatin B (1STF) [30]; (**c**) oryzacystatin from rice (1EQG) [18]; (**d**) CEW cystatin (1CEW) [13]; (**e**) theoretical cystatin C monomer, obtained by cutting the structure of the human cystatin C dimer (1G96) [88] in half and modelling the L1 loop. The cystatin parts responsible for C1 peptidase binding and inhibition are indicated: the N-terminal segment (N-term), the first (L1) and second (L2) hairpin loops. Also indicated is the BSL involved in the inhibition of leguman-like (family C13) cysteine peptidases.

interactions that are not present in the monomeric form. These additional interactions ensure dimer stabilization. In spite of substantial loss of conformation to form the open interface, each domain still resembles the C1 peptidase-binding region of the previously solved monomeric cystatin structures, strongly indicating that the binding site in cystatin C should be very similar to that in its homologues (Figure 3).

Function of cystatins as inhibitors of family C1 peptidases

All true cystatins are tight-binding inhibitors of papain and other C1 peptidases. Nevertheless, some cystatins show more restricted inhibition profiles than others, indicating their evolution towards more specific biological functions. Others, such as cystatin C, appear to have broad reactive properties as general cysteine peptidase inhibitors (Table 3). A partial explanation of the different inhibitory properties of cystatins seems to be the sequence of their N-terminal segments. This has been elaborated by analysis of cystatin variants obtained by site-directed mutagenesis and hybrid cystatin constructs, with results showing that the N-terminal segment structure is an important determinant of the inhibitory properties of cystatins [86,87,93]. However, other parts of the C1 peptidase-binding region of cystatins naturally also affect inhibition, but there is no straightforward key specificity-determining residue, as in many serine peptidase inhibitors.

Cystatins F and D are examples of cystatins with more specialized biological functions implied from their inhibitory properties. Neither of them can inhibit cathepsin B. Cystatin F is unusual structurally in having positively charged residues in the usually hydrophobic C1 peptidase-binding region of cystatins. The presence of Lys^9 (in the N-terminal segment), Lys^{58} (in the first hairpin loop) and even His^{109} (at the end of the second hairpin loop) are probably the consequence of an evolutionary step to generate a more specific inhibitor of an as yet unknown target peptidase. In addition, the two extra Cys residues that probably form a disulphide bridge in the N-terminal segment of cystatin F (Cys^1 and Cys^{37}) should affect the overall look of the C1 peptidase-binding site. This disulphide bridge might fix the seven-residue elongated N-terminus of cystatin F to the body of the protein, which probably means that the N-terminal segment would lose the flexibility that seems to be an important property for the inhibitory activity of cystatins in general, and may even adopt a secondary structure stabilized by the neighbouring α-helix. Like cystatin F, cystatin D also presents charged residues in the L2 loop, i.e. Glu^{107}, Asp^{108} and Lys^{109}. In the case of cystatin D, however, the sequence of its flexible N-terminal segment seems to provide a major explanation of its inhibitory properties [93]. Thus, although cystatin structures generally look very much alike (Figure 3), more fine-tuned structural differences in different cystatins are certainly important for their inhibitory function.

Function of cystatins as inhibitors of family C13 peptidases

Legumains are quite unique cysteine endopeptidases in that they are strictly specific for hydrolysing the peptide bond after an asparagine residue, and form a distinct enzyme family defined by amino acid sequence (family C13). Their proteolytic mechanism is, for evolutionary, structural and functional reasons, very unlike that of family C1 peptidases [94,95]. Even so, pig kidney legumain was found to be inhibited by human cystatin C and CEW cystatin [94]. Further studies showed that cystatin C binds to pig legumain in an equimolar ratio, without getting cleaved, in a tight, reversible and substrate-

Table 3 Equilibrium constants for dissociation (K_i) of complexes between human cystatins and family C1 peptidases. n.d., not determined.

Cystatin	K_i (nM)					Refs
	Papain	Cathepsin B	Cathepsin H	Cathepsin L	Cathepsin S	
A	0.019	8.2	0.31	1.3	0.05	[76,98]
B	0.12	73	0.58	0.23	0.07	[76,98]
C	0.00001	0.27	0.28	<0.005	0.008	[7,98,99]
D	1.2	>1000	7.5	18		
0.27	[42,44]					
E/M	0.39	32	n.d.	n.d.	n.d.	[46]
F	1.1	>1000	n.d.	0.31	n.d.	[51]
S	108	n.d.	n.d.	n.d.	n.d.	[58]
SA	0.32	n.d.	n.d.	n.d.	n.d.	[59]
SN	0.016	19	n.d.	n.d.	n.d.	[10]
L-kininogen	0.015	600	0.72	0.017	n.d.	[16,17]

competitive way [96]. Cystatin C variants with alterations in the tripartite wedge forming the C1 peptidase-binding region were equally as effective as wild-type cystatin C in inhibiting legumain, as was the dimeric form of human cystatin C with a completely disrupted C1 peptidase-binding site (see above). This strongly indicated that the legumain-inhibitory site is not the same as that for the inhibition of C1 peptidases. Moreover, legumain activity could be titrated with cystatin C complexed to papain and papain activity could be titrated with cystatin C complexed with legumain, proving that the binding sites for the two different peptidases are independent from each other. Finally, a ternary papain–cystatin C–legumain complex could be formed, demonstrating that the cystatin binding sites for C1 and C13 peptidases do not overlap [96].

When the array of human cystatins was investigated, it was observed that only some of them could inhibit mammalian legumain [96]. Only three type 2 cystatins, i.e. cystatins C, E/M and F, showed inhibitory activity (K_i values of 0.20, 0.0016 and 10 nM respectively). Sequence alignments and analysis of the structure of CEW cystatin indicated that the back-side loop (BSL), located between the main α-helix and the second strand of the β-sheet, possibly in conjunction with the appendix loop between the third and fourth strands of the β-sheet, was the probable location of the binding site for legumain. The BSL at the C-terminal end of the α-helix appeared the most promising candidate, as a quite conserved sequence pattern around residue Asn^{39} (cystatin C numbering) could be seen in human cystatins C, E/M and F, as well as in CEW cystatin [96]. To verify the involvement of Asn^{39} in legumain inhibition, a variant with this residue substituted by lysine was produced. This variant was found to be completely inactive against legumain (>5000-fold decrease in affinity compared with the wild-type protein), but was still active against cathepsin B. This result clearly indicated that the BSL where Asn^{39} is located is responsible for the legumain-inhibitory activity of some cystatins.

Based on all the biochemical evidence summarized above, we conclude that the human type 2 cystatins C, E/M and F, as well as CEW cystatin and the parasite cystatin Bm-CPI-2 [97], may bind to mammalian legumain in a substrate-like manner, with the Asn^{39} side chain interacting with the S1 subsite of the enzyme, but with no possibility of the enzyme hydrolysing the subsequent peptide bond. In addition to Asn^{39}, other residues in the BSL are well conserved in those type 2 cystatins that display legumain-inhibitory ability, forming what could be part of a second, legumain-binding, 'cystatin motif': $Ser(Thr)^{38}-Asn^{39}-Asp(Ser)^{40}$.

We thank Dr Alan Barrett for many stimulating discussions and for a fruitful collaboration lasting since M.A. had the pleasure to visit Strangeways Research Laboratory in 1985. This work was supported by grants from the Swedish Medical Research Council (no. 09915), the Medical Faculty at the University of Lund, the A. Österlund Foundation and the Crafoord Foundation.

References

1.   Turk, B., Turk, V. and Turk, D. (1997) Biol. Chem. **378**, 141–150
2.   Fossum, K. and Whitaker, J.R. (1968) Arch. Biochem. Biophys. **125**, 367–375

3.   Sen, L.C. and Whitaker, J.R. (1973) Arch. Biochem. Biophys. **158**, 623–632
4.   Barrett, A.J. (1981) Methods Enzymol. **80**, 771–778
5.   Barrett, A.J., Fritz, H., Grubb, A., Isemura, I., Järvinen, M., Katunuma, N., Machleidt, W., Müller-Esterl, W., Sasaki, M. and Turk, V. (1986) Biochem. J. **236**, 312
6.   Grubb, A. and Löfberg, H. (1982) Proc. Natl. Acad. Sci. U.S.A. **79**, 3024–3027
7.   Barrett, A.J., Davies, M.E. and Grubb, A. (1984) Biochem. Biophys. Res. Commun. **120**, 631–636
8.   Dayhoff, M.O., Barker, W.C. and Hunt, L.T. (1978) in Atlas of Protein Sequence and Structure (Dayhoff, M.O., ed.), vol. 5, pp. 9–20, National Biomedical Research Foundation, Washington, DC
9.   Rawlings, N.D. and Barrett, A.J. (1990) J. Mol. Evol. **30**, 60–71
10. Abrahamson, M., Barrett, A.J., Salvesen, G. and Grubb, A. (1986) J. Biol. Chem. **261**, 11282–11289
11. Barrett, A.J., Rawlings, N.D., Davies, M.E., Machleidt, W., Salvesen, G. and Turk, V. (1986) in Proteinase Inhibitors (Barrett, A.J. and Salvesen, G., eds), vol. 12, pp. 515–569, Elsevier, New York
12. Laber, B., Krieglstein, K., Henschen, A., Kos, J., Turk, V., Huber, R. and Bode, W. (1989) FEBS Lett. **248**, 162–168
13. Bode, W., Engh, R., Musil, D., Thiele, U., Huber, R., Karshikov, A., Brzin, J., Kos, J. and Turk, V. (1988) EMBO J. **7**, 2593–2599
14. Dieckmann, T., Mitschang, L., Hofmann, M., Kos, J., Turk, V., Auerswald, E.A., Jaenicke, R. and Oschkinat, H. (1993) J. Mol. Biol. **234**, 1048–1059
15. Ohkubo, I., Kurachi, K., Takasawa, T., Shiokawa, H. and Sasaki, M. (1984) Biochemistry **23**, 5691–5697
16. Salvesen, G., Parkes, C., Abrahamson, M., Grubb, A. and Barrett, A.J. (1986) Biochem. J. **234**, 429–434
17. Müller-Esterl, W., Iwanaga, S. and Nakanishi, S. (1986) Trends Biochem. Sci. **11**, 336–339
18. Nagata, K., Kudo, N., Abe, K., Arai, S. and Tanokura, M. (2000) Biochemistry **39**, 14753–14760
19. Margis, R., Reis, E.M. and Villeret, V. (1998) Arch. Biochem. Biophys. **359**, 24–30
20. Abrahamson, M. (1994) Methods Enzymol. **244**, 685–700
21. Turk, V. and Bode, W. (1991) FEBS Lett. **285**, 213–219
22. Ni, J., Alvarez-Fernandez, M., Danielsson, L., Chillakuru, R.A., Zhang, J., Grubb, A., Su, J., Gentz, R. and Abrahamson, M. (1998) J. Biol. Chem. **273**, 24797–24804
23. Machleidt, W., Borchart, U., Fritz, H., Brzin, J., Ritonja, A. and Turk, V. (1983) Hoppe-Seylers Z. Physiol. Chem. **364**, 1481–1486
24. Kartasova, T., Cornelissen, B.J., Belt, P. and van de Putte, P. (1987) Nucleic Acids Res. **15**, 5945–5962
25. Hsieh, W.T., Fong, D., Sloane, B.F., Golembieski, W. and Smith, D.I. (1991) Genomics **9**, 207–209
26. Martin, J.R., Craven, C.J., Jerala, R., Kroon-Zitko, L., Zerovnik, E., Turk, V. and Waltho, J.P. (1995) J. Mol. Biol. **246**, 331–343
27. Ritonja, A., Machleidt, W. and Barrett, A.J. (1985) Biochem. Biophys. Res. Commun. **131**, 1187–1192
28. Pennacchio, L.A., Bouley, D.M., Higgins, K.M., Scott, M.P., Noebels, J.L. and Myers, R.M. (1998) Nat. Genet. **20**, 251–258
29. Shannon, P., Pennacchio, L.A., Houseweart, M.K., Minassian, B.A. and Myers, R.M. (2002) J. Neuropathol. Exp. Neurol. **61**, 1085–1091
30. Stubbs, M.T., Laber, B., Bode, W., Huber, R., Jerala, R., Lenarcic, B. and Turk, V. (1990) EMBO J. **9**, 1939–1947
31. Turk, V., Brzin, J., Longer, M., Ritonja, A., Eropkin, M., Borchart, U. and Machleidt, W. (1983) Hoppe-Seylers Z. Physiol. Chem. **364**, 1487–1496

32. Schwabe, C., Anastasi, A., Crow, H., McDonald, J.K. and Barrett, A.J. (1984) Biochem. J. 234, 429–434
33. Abrahamson, M., Grubb, A., Olafsson, I. and Lundwall, Å. (1987) FEBS Lett. 216, 229–233
34. Abrahamson, M., Islam, M.Q., Szpirer, J., Szpirer, C. and Levan, G. (1989) Hum. Genet. 82, 223–226
35. Saitoh, E., Sabatini, L.M., Eddy, R.L., Shows, T.B., Azen, E.A., Isemura, S. and Sanada, K. (1989) Biochem. Biophys. Res. Commun. 162, 1324–1331
36. Schnittger, S., Gopal Rao, V.V.N., Abrahamson, M. and Hansmann, I. (1993) Genomics 16. 50–55
37. Abrahamson, M., Olafsson, I., Palsdottir, A., Ulvsbäck, M., Lundwall, Å., Jensson, O. and Grubb, A. (1990) Biochem. J. 268, 287–294
38. Thomas, T., Schreiber, G. and Jaworowski, A. (1989) Dev. Biol. 134, 38–47
39. Turk, B., Turk, D. and Salvesen, G.S. (2002) Curr. Pharm. Des. 8, 1623–1637
40. Olafsson, I. and Grubb, A. (2000) Int. J. Exp. Clin. Invest. 7, 70–79
41. Freije, J.P., Abrahamson, M., Olafsson, I., Velasco, G., Grubb, A. and López-Otín, C. (1991) J. Biol. Chem. 266, 20538–20543
42. Freije, J.P., Balbín, M., Abrahamson, M., Velasco, G., Dalbøge, H., Grubb, A. and López-Otín, C. (1993) J. Biol. Chem. 268, 15737–15744
43. Balbín, M., Freije, J.P., Abrahamson, M., Velasco, G., Grubb, A. and López-Otín, C. (1993) Hum. Genet. 90, 668–669
44. Balbín, M., Hall, A., Grubb, A., Mason, R.W., López-Otín, C. and Abrahamson, M. (1994) J. Biol. Chem. 269, 23156–23162
45. Freije, J.P., Pendás, A.M., Velasco, G., Roca, A., Abrahamson, M. and López-Otín, C. (1993) Cytogenet. Cell. Genet. 62, 29–31
46. Ni, J., Abrahamson, M., Zhang, M., Alvarez-Fernandez, M., Grubb, A., Su, J., Yu, G.-L., Li, Y., Parmelee, D., Xing, L. et al. (1997) J. Biol. Chem. 272, 10853–10858
47. Sotiropoulou, G., Anisowicz, A. and Sager, R. (1997) J. Biol. Chem. 272, 903–910
48. Stenman, G., Åstrom, A.K., Roijer, E., Sotiropoulou, G., Zhang, M. and Sager, R. (1997) Cytogenet. Cell. Genet. 76, 45–46
49. Zeeuwen, P.L., van Vlijmen-Willems, I.M., Jansen, B.J., Sotiropoulou, G., Curfs, J.H., Meis, J.F., Janssen, J.J., van Ruissen, F. and Schalkwijk, J. (2001) J. Invest. Dermatol. 116, 693–701
50. Zeeuwen, P.L., van Vlijmen-Willems, I.M., Hendriks, W., Merkx, G.F. and Schalkwijk, J. (2002) Hum. Mol. Genet. 11, 2867–2875
51. Ni, J., Alvarez-Fernandez, M., Danielsson, L., Chillakuru, R.A., Zhang, J., Grubb, A., Su, J., Gentz, R. and Abrahamson, M. (1998) J. Biol. Chem. 273, 24797–24804
52. Halfon, S., Ford, J., Foster, J., Dowling, L., Lucian, L., Sterling, M., Xu, Y., Weiss, M., Ikeda, M., Liggett, D. et al. (1998) J. Biol. Chem. 273, 16400–16408
53. Morita, M., Yoshiuchi, N., Arakawa, H. and Nishimura, S. (1999) Cancer Res. 59, 151–158
54. Morita, M., Hara, Y., Tamai, Y., Arakawa, H and Nishimura, S. (2000) Genomics 67, 87–91
55. Nathanson, C.-M., Wassélius, J., Wallin, H. and Abrahamson, M. (2002) Eur. J. Biochem. 269, 5502–5511
56. Werle, B., Sauckel, K., Nathanson, C.-M., Bjarnadottir, M., Spiess, E., Ebert, W. and Abrahamson, M. (2003) Biol. Chem, 384, 281–287
57. Isemura, S., Saitoh, E. and Sanada, K. (1984) J. Biochem. (Tokyo) 96, 489–498
58. Isemura, S., Saitoh, E. and Sanada, K. (1986) FEBS Lett. 198, 145–149
59. Isemura, S., Saitoh, E. and Sanada, K. (1987) J. Biochem. (Tokyo) 102, 693–704
60. Isemura, S., Saitoh, E., Sanada, K. and Minakata, K. (1991) J. Biochem. (Tokyo) 110, 648–654
61. Saitoh., E., Kim, H.S., Smithies, O. and Maeda, N. (1987) Gene 61, 329–338
62. Saitoh., E., Isemura, S., Sanada, K. and Ohnishi, K. (1992) Agents Actions Suppl. 38, 340–348
63. Cornwall, G.A., Orgebin-Crist, M.C. and Hann, S.R. (1992) Mol. Endocrinol. 6, 1653–1664

64. Cornwall, G.A. and Hann, S.R. (1995) Mol. Reprod. Dev. **41**, 37–46
65. Tohonen, V., Österlund, C. and Nordqvist, K. (1998) Proc. Natl. Acad. Sci. U.S.A. **95**, 14208–14213
66. Kanno, Y., Tamura, M., Chuma, S., Sakura, T., Machida, T. and Nakatsuji, N. (1999) Int. J. Dev. Biol. **43**, 777–784
67. Eriksson, A., Tohonen, V., Wedell, A. and Nordqvist, K. (2002) Mol. Hum. Reprod. **8**, 8–15
68. Hamil, K.G., Liu, Q., Sivashanmugam, P., Yenugu, S., Soundararajan, R., Grossman, G., Richardson, R.T., Zhang, Y.L., O'Rand, M.G., Petrusz, P. et al. (2002) Endocrinology **143**, 2787–2796
69. Shoemaker, K., Holloway, J.L., Whitmore, T.E., Maurer, M. and Feldhaus, A.L. (2000) Gene **245**, 103–108
70. Ohkubo, I., Kurachi, K., Takasawa, T., Shiokawa, H. and Sasaki, M. (1984) Biochemistry **23**, 5691–5697
71. Kitamura, N., Takagaki, Y., Furuto, S., Tanaka, T., Nawa, H. and Nakanishi, S. (1983) Nature (London) **305**, 545–549
72. Nawa, H., Kitamura, N., Hirose, T., Asai, M., Inayama, S. and Nakanishi, S. (1983) Proc. Natl. Acad. Sci. U.S.A. **80**, 90–94
73. Kitamura, N., Kitagawa, H., Fukushima, D., Takagaki, Y., Miyata, T. and Nakanishi, S. (1985) J. Biol. Chem. **260**, 8610–8617
74. Fong, D., Smith, D.I. and Hsieh, W.T. (1991) Hum. Genet. **87**, 189–192
75. Anastasi, A., Brown, M.A., Kembhavi, A.A., Nicklin, M.J., Sayers, C.A., Sunter, D.C. and Barrett A.J. (1983) Biochem. J. **211**, 129–138
76. Green, G.D., Kembhavi, A.A., Davies, M.E. and Barrett, A.J. (1984) Biochem. J. **218**, 939–946
77. Nicklin, M.J. and Barrett, A.J. (1984) Biochem. J. **223**, 245–253
78. Abrahamson, M., Ritonja, A., Brown, M.A., Grubb, A., Machleidt, W. and Barrett, A.J. (1987) J. Biol. Chem. **262**, 9688–9694
79. Björk, I. and Ylinenjarvi, K. (1990) Biochemistry **29**, 1770–1776
80. Higashiyama, S., Ishiguro, H., Ohkubo I., Fujimoto, S., Matsuda, T. and Sasaki, M. (1986) Life Sci. **39**, 1639–1644
81. Machleidt, W., Thiele, U., Laber, B., Assfalg-Machleidt, I., Esterl, A., Wiegand, G., Kos, J., Turk, V. and Bode, W. (1989) FEBS Lett. **243**, 234–238
82. Nycander, M. and Björk, I. (1990) Biochem. J. **271**, 281–284
83. Lindahl, P., Nycander, M., Ylinenjarvi, K., Pol, E. and Björk, I. (1992) Biochem. J. **286**, 165–171
84. Auerswald, E.A., Genenger, G., Assfalg-Machleidt, I., Machleidt, W., Engh, R.A. and Fritz, H. (1992) Eur. J. Biochem. **209**, 837–845
85. Abrahamson, M., Mason, R.W., Hansson, H., Buttle, D.J., Grubb, A. and Ohlsson, K. (1991) Biochem. J. **273**, 621–626
86. Hall, A., Dalbøge, H., Grubb, A. and Abrahamson, M. (1993) Biochem. J. **291**, 123–129
87. Hall, A., Håkansson, K., Mason, R.W., Grubb, A. and Abrahamson, M. (1995) J. Biol. Chem. **270**, 5115–5121
88. Janowski, R., Kozak, M., Jankowska, E., Grzonka, Z., Grubb, A., Abrahamson, M. and Jaskolski, M. (2001) Nat. Struct. Biol. **8**, 316–320
89. Bode, W., Engh, R., Musil, D., Laber, B., Stubbs, M., Huber, R. and Turk, V. (1990) Biol. Chem. Hoppe-Seyler **371**, 111–118
90. Kamphuis, I.G., Kalk, K.H., Swarte, M.B. and Drenth, J. (1984) J. Mol. Biol. **179**, 233–256
91. Nycander, M., Estrada, S., Mort, J.S., Abrahamson, M. and Björk, I. (1998) FEBS Lett. **422**, 61–64
92. Ekiel, I., Abrahamson, M., Fulton, D.B., Lindahl, P., Storer, A.C., Levadoux, W., Lafrance, M., Labelle, S., Pomerleau, Y., Groleau, D. et al. (1997) J. Mol. Biol. **271**, 266–277

93. Hall, A., Ekiel, I., Mason, R.W., Kasprzykowski, F., Grubb, A. and Abrahamson, M. (1998) Biochemistry **37**, 4071–4079

94. Chen, J.-M., Dando, P.M., Rawlings, N.D., Brown, M.A., Young, N.E., Stevens, R.A., Hewitt, E., Watts, C. and Barrett, A.J. (1997) J. Biol. Chem. **272**, 8090–8098

95. Chen, J.-M., Rawlings, N.D., Stevens, R.A.W. and Barrett, A.J. (1998) FEBS Lett. **441**, 361–365

96. Alvarez-Fernandez, M., Barrett, A.J., Gerhartz, B., Dando, P.M., Ni, J. and Abrahamson, M. (1999) J. Biol. Chem. **274**, 19195–19203

97. Manoury, B., Gregory, W.F., Maizels, R.M. and Watts, C. (2001) Curr. Biol. **11**, 447–451

98. Brömme, D., Rinne, R. and Kirschke, H. (1991) Biomed. Biochim. Acta **50**, 631–635

99. Lindahl, P., Abrahamson, M. and Björk, I. (1992) Biochem. J. **281**, 49–55

Biochem. Soc. Symp. **70**, 201–212
(Printed in Great Britain)
© 2003 Biochemical Society

16

Designing TIMP (tissue inhibitor of metalloproteinases) variants that are selective metalloproteinase inhibitors

Hideaki Nagase*[1] and Keith Brew†

*Kennedy Institute of Rheumatology Division, Imperial College London, London W6 8LH, U.K., and †Department of Biomedical Sciences, Florida Atlantic University, Boca Raton, FL 33431, U.S.A.

Abstract

The tissue inhibitors of metalloproteinases (TIMPs) are endogenous inhibitors of the matrix metalloproteinases (MMPs), enzymes that play central roles in the degradation of extracellular matrix components. The balance between MMPs and TIMPs is important in the maintenance of tissues, and its disruption affects tissue homoeostasis. Four related TIMPs (TIMP-1 to TIMP-4) can each form a complex with MMPs in a 1:1 stoichiometry with high affinity, but their inhibitory activities towards different MMPs are not particularly selective. The three-dimensional structures of TIMP–MMP complexes reveal that TIMPs have an extended ridge structure that slots into the active site of MMPs. Mutation of three separate residues in the ridge, at positions 2, 4 and 68 in the amino acid sequence of the N-terminal inhibitory domain of TIMP-1 (N-TIMP-1), separately and in combination has produced N-TIMP-1 variants with higher binding affinity and specificity for individual MMPs. TIMP-3 is unique in that it inhibits not only MMPs, but also several ADAM (a disintegrin and metalloproteinase) and ADAMTS (ADAM with thrombospondin motifs) metalloproteinases. Inhibition of the latter groups of metalloproteinases, as exemplified with ADAMTS-4 (aggrecanase 1), requires additional structural elements in TIMP-3 that have not yet been identified. Knowledge of the structural basis of the inhibitory action of TIMPs will facilitate the design of selective TIMP variants for investigating the biological roles of specific MMPs and for developing therapeutic interventions for MMP-associated diseases.

[1]To whom correspondence should be addressed (e-mail h.nagase@imperial.ac.uk).

Introduction

The extracellular matrix (ECM) in multicellular organisms holds cells and tissues together, and creates appropriate cellular environments by forming organized molecular scaffolds. The spatial and temporal turnover of ECM molecules is therefore an integral part of many biological processes, such as embryo development, organ morphogenesis, tissue remodelling and repair, cell differentiation, migration, growth and apoptosis [1,2]. Various types of proteinases are involved in the catabolism of the ECM, but the major ECM-degrading enzymes are the matrix metalloproteinases (MMPs) or matrixins [3,4]. More than 30 matrixins have been found, including those in hydra [5], sea urchins [6] and plants [7]. In humans, 23 matrixin genes have been identified. Recent studies have also indicated that members of the ADAM (a disintegrin and metalloproteinases) and ADAMTS (ADAM with thrombospondin motifs) families participate in ECM processing and degradation [8]. The activities of these metalloproteinases must be highly regulated under physiological conditions. Indeed, the expression levels of many matrixins in normal steady-state tissues are very low or negligible, but they are transcriptionally regulated by growth factors, hormones, inflammatory cytokines, cell–cell and cell–matrix interactions, and cellular transformations [1,2], suggesting that they function upon demand. The activities of matrixins are also controlled through zymogen activation and inhibition by endogenous protein inhibitors. The disruption of this balance may result in diseases, such as arthritis, atheroma, cancer, nephritis, encephalomyelitis, fibrosis, etc. [9].

The major endogenous inhibitors of matrixins are the tissue inhibitors of metalloproteinases (TIMPs) and a plasma glycoprotein, α_2-macroglobulin (α_2M), and related molecules. Human α_2M consists of four identical subunits of molecular mass 180 kDa, and it inhibits most endopeptidases, regardless of their substrate specificity and catalytic mechanism, by the unique 'trap' mechanism. This mechanism was first proposed by Alan Barrett and Phyllis Starkey in 1973 [10], based on their observations that most endopeptidases form a complex with α_2M, but that this interaction is triggered by cleavage of α_2M subunits, and that inhibition of the bound proteinase is observed with a large protein substrate, but not with a small peptidic substrate. These results indicate that α_2M does not block the active site of the proteinase, but it blocks enzymic action on a large protein substrate by steric hindrance. Matrixins bind reasonably rapidly to α_2M. Collagenase 1 (MMP-1) preferentially binds to α_2M when equal concentrations of α_2M and TIMP-1 are present in solution [11]. However, α_2M is considered to inhibit matrixins in the fluid phase, whereas TIMPs regulate them in the tissue.

General biochemical properties of TIMPs

The vertebrate TIMPs comprise a family of four members (TIMP-1 to TIMP-4). Each TIMP consists of 184–195 amino acids, and the four TIMPs show approx. 40% sequence identity [12]. The expression of these TIMPs in different tissues and cell types is also controlled in order to maintain balanced

ECM catabolism during tissue remodelling under physiological conditions [13,14]. Vertebrate TIMPs have 12 conserved cysteine residues and contain N-terminal and C-terminal domains each stabilized by three disulphide bonds, whereas invertebrate TIMPs have different structural features, although they are related to vertebrate TIMPs. Two TIMPs in *Caenorhabditis elegans* lack the C-terminal domain, and *Drosophila* TIMP lacks two cysteines corresponding to Cys^{13} and Cys^{120} of human TIMP-1, but has two additional cysteines in its C-terminal domain. *Drosophila* TIMP probably has two disulphide bonds in the N-terminal domain and four disulphide bonds in the C-terminal domain [12].

Vertebrate TIMPs interact with MMPs in a 1:1 stoichiometry with high affinity. They exhibit some differences in specificity. Thus TIMP-1 inhibits most MMPs with K_i values of around 0.1–3.0 nM [15], but is a poor inhibitor of MMP-14 (membrane-type 1-MMP) [16] and MMP-19 [17]. TIMPs-2, -3 and -4 inhibit all MMPs tested so far, but TIMP-2 and TIMP-4 bind MMP-2 very tightly [18–20]. Hutton et al. [18] reported that the overall dissociation constant of the complex between TIMP-2 and MMP-2 was 0.6 fM, indicating essentially irreversible complex formation. This tight binding is due largely to the interaction between the C-terminal domain of TIMP-2 and the C-terminal haemopexin domain of MMP-2 [18,21]. TIMP-3 inhibits MMP-1 and MMP-2 with K_is in the low nanomolar range, but its affinity for MMP-3 is lower [22]. TIMP-3, on the other hand, is unique in that that it inhibits several members of the ADAM and ADAMTS families. Target proteases include ADAM-10 [23], ADAM-12 [24], ADAM-17/tumour necrosis factor α (TNFα)-converting enzyme [25], ADAMTS-1 [26], ADAMTS-4 [22,27] and ADAMTS-5 [22], whereas ADAM-8 and ADAM-9 are not inhibited by TIMPs [28]. However, the structural bases of these differential inhibitory features among TIMPs are not well understood.

In addition to interacting with active MMPs, some TIMPs bind to the zymogens of MMPs (proMMPs). ProMMP-2 (gelatinase A) binds to TIMP-2, -3 or -4 through interactions between its C-terminal haemopexin domain and the C-terminal domain of the TIMP [20,29–31]. Also, proMMP-9 (progelatinase B) binds to TIMP-1 or TIMP-3 via their C-terminal domains [31,32]. These binary complexes inhibit mature MMPs, as the N-terminal inhibitory domain of the TIMP is exposed to the solvent. Ternary complex formation between proMMP-2, TIMP-2 and MMP-14 (membrane-type 1-MMP) is essential for the activation of proMMP-2 by MMP-14 on the cell surface. As discussed in Chapters 1, 6 and 21 in this volume, the TIMP-2-bound MMP-14 acts as a 'receptor' for proMMP-2, and the adjacent MMP-14, free of TIMP-2, acts as an 'activator'.

Mechanism of inhibition of MMPs by TIMPs

TIMP-1 was first isolated as a collagenase inhibitor from the culture medium of connective tissues [33,34]. Later it was shown to inhibit other MMPs [35]. The first three-dimensional structure of a TIMP was the NMR structure of the N-terminal domain of TIMP-2 (N-TIMP-2), determined by Williamson et al. [36]. It revealed a five-stranded β-barrel with a Greek key topology and two α-helices, a structure known as an OB fold, but the study did not identify the site that interacts with MMPs. Proteins with an OB fold include a group of oligonucleotide- and oligosaccharide-binding (OB) proteins, such as staphylo-coccal nuclease, bacterial enterotoxins and some tRNA synthases [37].

To investigate the potential interaction site of TIMPs for MMPs, we used the differential proteolytic susceptibility of TIMP-1 and the TIMP-1–MMP-3 complex [38]. The study suggested that the region near Val^{69} and Cys^{70} of TIMP-1, which is disulphide-bonded to Cys^1, is involved in the interaction with MMPs. This proposal was based on the observation that neutrophil elastase cleaved the Val^{69}–Cys^{70} bond of TIMP-1 and that this single proteolytic event inactivated the inhibitor, but TIMP-1 was protected from inactivation by elastase when it was complexed with MMP-3 (stromelysin 1). The inhibitory activity of TIMP-1 was recovered from the elastase-treated complex after dissociation. The NMR solution structure of N-TIMP-2 shows that the region around the Cys^1–Cys^{70} disulphide bond, comprising Glu^{67}-Ser-Val-Cys^{70} and Cys^1-Thr-Cys-Val^4 (residue numbering from TIMP-1), forms a contiguous ridge, suggesting that this region might represent the reactive site of the inhibitor. We also conducted an extensive mutational study of N-TIMP-1 [39]. One N-TIMP-1 mutant with a substitution of Ala for Thr^2 (T2A) exhibited strikingly differential effects on three MMPs: compared with the wild type, the affinities for MMP-1, MMP-2 and MMP-3 decreased by approx. 1600-, 30- and 250-fold respectively (Table 1). Mutations at other sites changed TIMP-1 activity, but the changes were relatively modest and, unlike the T2A mutant, the proportional change was similar for all three MMPs [39]. Mutation of either Cys^1 or Cys^{70}, which are disulphide-bonded in native TIMP-1, had large disruptive effects on TIMP-1 inhibitory activity [39,40]. Again these results suggest that the reactive site of TIMPs is located in or close to the ridge region.

The mechanism of MMP inhibition by TIMP was elucidated by the crystal structure of the complex of TIMP-1 and the catalytic domain of MMP-3 lacking the C-terminal haemopexin domain ([41]; see Figure 1a), and subsequently the structure of the TIMP-2–MMP-14 complex [42], both determined by Bode and colleagues. The structure of the TIMP-1–MMP-3 complex showed that TIMP-1 is a 'wedge-shaped' molecule, and that the ridge described above was inserted into the active site of the enzyme. In this complex, approx. 75% of the protein–protein contacts in TIMP-1 are from the N-terminal stretch of Cys^1 to Val^4 and the loop region of Met^{66} to Val^{69} that are linked by the Cys^1–Cys^{70} disulphide bond. Residues 1–4 of TIMP-1 bind to the enzyme in an analogous way as the P1-P1'-P2'-P3' residues of a peptide substrate (proteolytic cleavage occurs between P1 and P1'), and the side chain of Thr^2 extends into the large S1' specificity pocket (the substrate-binding pocket that accepts the P1' residue of the substrate) of MMP-3. Ser^{68} and Val^{69} occupy part of the active site (S2 and S3 subsites respectively), but they are in a nearly opposite orientation to that of a bound peptide substrate (Figures 1a and 1b). The active-site Zn^{2+} of the enzyme is bidentately chelated by the α-amino acid and the carbonyl group of the N-terminal Cys^1. The OH group of Thr^2 interacts with Glu^{202} of MMP-3 and displaces a water molecule from the active site that is essential for peptide hydrolysis. Figure 1(c) highlights the residues of TIMP-1 within 4 Å of MMP-3 to identify the reactive ridge.

Table 1 K_i values for wild-type **N-TIMP-1, N-TIMP-2 and N-TIMP-3, and for various N-TIMP-1 mutants.** Selected mutants of N-TIMP-1 had substitutions of Thr[2], Val[4] and/or Ser[68]. The data are from Meng et al. [44] and Wei et al. [46].

	K_i (nM)		
TIMP	MMP-1	MMP-2	MMP-3
N-TIMP-1			
Wild type	0.4	0.4	0.2
T2S	10	0.8	0.2
T2G	7000	>40000	560
T2A	836	15	50
T2L	37	0.4	1.3
T2I	105	2.2	81.6
T2V	0.6	108	1.2
T2M	4.4	0.3	0.3
T2F	17	6.8	5.2
T2N	788	6.4	18
T2Q	348	4.8	12
T2D	3250	500	440
T2E	2290	173	187
T2K	668	12	28
T2R	~2000	4.8	11
V4A	0.4	0.5	0.2
V4S	2.1	0.4	1.6
V4I	6.4	0.3	0.2
V4K	8.8	1.1	2.6
S68Y	61	2.7	0.3
S68A	14	5.2	32
S86E	16	3.2	14
S68R	48	18	49
T2L/V4S	140	0.3	6
T2L/V4S/S68A	~3000	4.0	14
T2S/V4A	3.3	0.4	0.1
T2S/V4A/S68Y	15	0.1	0.05
T2R/V4I	Inactive	10	4.0
N-TIMP-2	0.6	0.5	9.3
N-TIMP-3	1.2	4.3	67

Altered specificity of TIMP-1 due to mutagenesis of residue 2

The S1′ specificity pockets of the various MMPs differ in size, and therefore substrate specificity is greatly influenced by the nature of this pocket [43]. TIMPs have Thr or Ser in position 2. We therefore speculated that TIMP variants with a larger side chain at this position may be more selective for different MMPs. We used N-TIMP-1 rather than the full-length TIMP-1, as it is readily folded from the inclusion bodies of *Escherichia coli* when overexpressed. The His[2] mutant did not fold *in vitro*, but other mutants were compatible with the native fold, as judged by near-UV CD spectra. The position 2 variants were tested against MMP-1, -2 and -3 prototypes of collagenase, gelatinase and

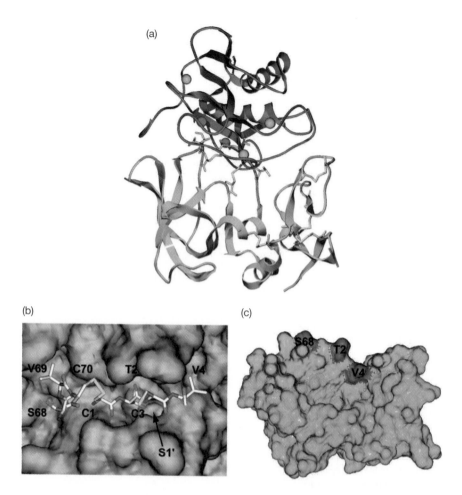

Figure 1 Structures of TIMP-1 and the catalytic domain of MMP-3 in complex. (a) Ribbon diagram of TIMP-1 bound to the catalytic domain of MMP-3. TIMP-1 is shown in green and MMP-3 is in blue. The regions of TIMP-1 that have atoms lying within 4 Å of MMP-3 are depicted in red. Zinc and calcium atoms are shown in purple and orange respectively. Disulphide-bonded cystines are shown in yellow. (b) Surface of the MMP catalytic domain with contact residues from TIMP-1. The binding region of TIMP-1 adjacent to the Cys[1]–Cys[70] disulphide bond is shown in stick representation, with nitrogen in blue, oxygen in red, sulphur in yellow and carbon in white. TIMP-1 residues are labelled in the single-letter code, and the S1′ pocket is indicated by an arrow. (c) Surface structure of TIMP-1. The reactive ridge region represented by atoms lying within 4 Å of MMP-3 is shown in pink, and residues Thr[2], Val[4] and Ser[68] are highlighted in blue. The rest of the TIMP-1 residues are shown in green. The image was prepared from the Brookhaven Protein Data Bank entry (1UEA) using a Swiss PDB viewer [55].

stromelysin [44] (Table 1). Among 14 variants, the Gly mutant was the weakest, with the affinity for MMPs lowered by three to five orders of magnitude. This represents a loss of 33–55% of the free energy of interaction. The crystal structure of the MMP-3–TIMP-1 complex indicated that approx. 1300 Å^2 of the accessible surface of each molecule is buried upon formation of the complex [41], but Thr^2 occupies only 8% (108/1300 Å^2) of the area of the TIMP-1–MMP interaction site, even assuming that it is totally exposed in the free inhibitor and completely buried upon binding to an MMP. This disproportionately large effect on binding indicates that Thr^2 in TIMP-1 is a 'hot spot' in the MMP–TIMP interaction interface. The C-terminal domain of TIMP-1 makes a few contacts with the protease, but contributes less than 10% of the free energy of binding.

Negatively charged side chains at position 2 are unfavourable for all three MMPs as compared with the Asn^2 and Gln^2 mutants. Ser, Val and Met are well tolerated, and the best inhibitors for MMP-1, MMP-2 and MMP-3 were the Val^2, Met^2 and Ser^2 variants respectively. The reasonably high affinity of the Arg^2 mutant for MMP-2 and MMP-3 is surprising, since peptides with Arg in the P1' position are extremely poor substrates for these enzymes [45]. The unfavourable interaction of the Arg^2 mutant with MMP-1 may be due to the presence of Arg^{195} in the S1' pocket of MMP-1, whereas Leu is found in this position in MMP-2 and MMP-3.

Although mutation of residue 2 of N-TIMP-1 significantly influences its affinity for three MMPs tested, it is striking that there is a very poor correlation between the effects of a particular amino acid in the P1' position of a peptide substrate and those of the same residue in position 2 of N-TIMP-1 on MMP inhibition. For example, the best N-TIMP-1 mutant for inhibiting MMP-1 is the Val^2 variant, but an octapeptide with Val^2 in the P1' position is a poor substrate. Similarly, Ser and Arg in the P1' position in the substrate are not well tolerated by MMP-2 or MMP-3 [45]. Figure 2 shows an example of very poor correlation when $-\log K_i$ for position 2 mutants of N-TIMP-1 is plotted against $\log(k_{cat}/K_m)$ for octapeptides with the same substitutions in the P1' position. Such a discrepancy may be due to a greater loss of conformational entropy associated with the peptide-substrate–MMP interaction compared with the TIMP–MMP interaction. Because the reactive ridges of TIMPs are rigid due to two disulphide bonds (Cys^1–Cys^{70} and Cys^3–Cys^{99}), the conserved proline at position 5 and the interaction of Cys^1 with the active-site Zn^{2+}, the orientation of residue 2 of TIMP-1 may be greatly influenced, and thus it dictates specificity.

Mutation of residues 4 and 68 of TIMP-1 and combined mutagenesis

The crystal structure of the TIMP-1–MMP-3 complex indicates that the side chain of Val^4 of TIMP-1 interacts with the S3' subsite of the enzyme, and that Ser^{68} interacts with the S2 subsite. Four mutants were designed for each position to test their influence in specificity. Compared with position 2 mutations, the changes observed are moderate [46] (Table 1). V4S and V4K mutants are more selective for MMP-2 than the wild type. Mutations of Ser^{68} have greater effects on MMP binding. The S68Y mutant inhibits MMP-3 much more strongly than MMP-1 or -2. Double and triple mutants with combinations of some selective single-site mutants show further selectivity.

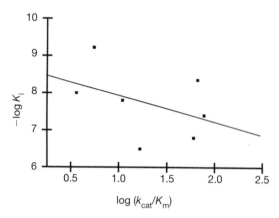

Figure 2 Effects of residues that interact with the S1′ pocket of MMP-I on affinity of N-TIMP-I (-logK_i) and efficiency of substrates [log(k_{cat}/K_m)]. Leu, Phe, Met, Val, Gln, Ser and Ile were substituted in turn for Thr^2 in N-TIMP-I [44] and for Xaa in the substrate Gly-Pro-Gln-Gly↓Xaa-Ala-Gly-Gln (where ↓ is the scissile bond) [45].

The double mutant T2L/V4S maintains excellent affinity for MMP-2, but shows decreased reactivity with MMP-1 and MMP-3 by 350-fold and 30-fold respectively; it does however retain low nanomolar K_i values for MMP-2 and MMP-3. The triple mutant T2S/V4A/S68A has the highest inhibitory activity for MMP-3 among all mutants characterized so far, with a K_i value of 50 pM. It also has an improved affinity for MMP-2 compared with the wild type, but binding to MMP-1 is reduced.

The properties of the multi-site mutants suggest that the effects of some combinations of substitutions are essentially additive, indicating that interactions with the S1′, S3′ and S2 subsites of the MMP contribute independently to the stability of the TIMP–MMP complex. However, some mutants exhibit effects that are not additive. For example, the T2L/V4S/S68Y mutant has a >100-fold higher affinity for MMP-3 than expected, although the K_i values for MMP-1 and MMP-2 are in good agreement with predicted values. The T2S/N4A/S68Y mutant is a 60-fold better inhibitor of MMP-1 and a 68-fold better inhibitor of MMP-2 than predicted values based on additivity. These results indicate interactive effects among the sites in these multisite mutants. This suggests that the local structure of substrate-binding sites differs significantly among the three MMPs, in spite of the fact that the overall polypeptide folds of different MMPs are very similar (see Chapter 1).

Mutational studies of N-TIMP-3

TIMPs are specific inhibitors of matrixins. Exceptions to this are the inhibition of several members of the ADAM and ADAMTS metalloproteinases by TIMP-3, and the inhibition of ADAM-10 by TIMP-1 [23]. TIMP-3 also blocks the shedding of TNFα receptor 1 [47,48], Fas [48], TNF-related apoptosis-inducing ligand receptor 1 [48], L-selectin [49], interleukin-6 receptor [50], syn-decans-1 and -4 [51], and CD163 [52]. Because these processes are not inhibited by other TIMPs, the enzymes responsible for these activities are postulated to be membrane-bound metalloproteinases belonging to the ADAM family. The primary structures of the metalloproteinase domains of ADAMs and ADAMTSs are related, but they have little sequence similarity with MMPs except around the catalytic zinc-binding motif, HEXXHXXGXXH [8]. The apparent K_i values of N-TIMP-3 as an inhibitor of ADAM-17 (TNFα-converting enzyme) [53] and ADAMTS-4 and -5 [22] are subnanomolar. ADAMTS-4 and -5 are also called aggrecanase 1 and 2 respectively, because of their ability to cleave the Glu^{373}–Ala^{374} bond of the core protein of aggrecan. More recently, ADAMTS-1 was also shown to have aggrecanase activity. N-TIMP-3, but not TIMP-1 or TIMP-2, inhibits interleukin-1-induced aggrecan breakdown in articular carti-lage (C. Gendron, M. Kashiwagi and H. Nagase, unpublished work), suggesting that delivery or increased production of TIMP-3 in articular cartilage may pro-tect against cartilage degradation that occurs in various types of arthritis.

To investigate the structural basis of the ability of TIMP-3 to inhibit aggre-canase, we have generated a series of chimaeric N-TIMPs by introducing different portions of TIMP-3 into TIMP-1 from the N-terminus. The chimaera consisting of N-TIMP-3-(1–68) and N-TIMP-1-(71–124) contains all the residues present in the reactive-site ridge of N-TIMP-3, but it does not inhibit ADAMTS-4. Nevertheless, this protein inhibits MMP-1, -2 and -3 with inhibi-tion constants similar to those of wild-type N-TIMP-3, and it is therefore essentially a TIMP-3 rather than a TIMP-1. Replacement of the stretch of the sequence EASESL (residues 62–67) in TIMP-3 by PAMESV (residues 64–69) derived from TIMP-1 did not alter the activity of N-TIMP-3 as an inhibitor of ADAMTS-4 (M. Kashiwagi, O. Polyakova, K. Brew and H. Nagase, unpub-lished work). These results suggest that the MMP reactive ridge alone of TIMP-3 is not sufficient to inhibit ADAMTS-4, but that additional elements are required.

Conclusions and future prospects

The TIMPs are important regulators of ECM catabolism through their ability to control the activities of MMPs and some ADAM and ADAMTS metal-loproteinases. Our mutational studies have provided us with TIMP variants that have selective inhibitory activity for particular MMPs. Although inhibitors that are specific for a single MMP have not yet been designed, the results obtained so far encourage future studies in this direction. Our studies also highlight the unique nature of the reactive-site ridge in TIMPs. The requirement for a hydrophobic side chain in the P1′ position of peptide substrates of MMPs is well established, but this requirement does not correlate with that of the inhibitory TIMP molecule. P1′ side chains that produce a poor substrate are often found to be excellent for inhibitory activity when substituted at the corresponding site (residue 2) in TIMP-1, and in this location significantly influence the inhibitory specificity. This feature of residue 2 and the influence of other residues in the reactive ridge on the affinity of TIMP for different metalloproteinases are worthy of investigation as a route to generating more selective inhibitors.

TIMPs were originally discovered as MMP inhibitors, and are used as reagents to discriminate MMPs from metalloproteinases of other families, such as thermolysin, neprilysin and astacins. However, recent studies have shown that TIMP-3 can inhibit a number of ADAM and ADAMTS metalloproteinases. This finding is of considerable interest to many researchers, as these metalloproteinases play important roles in inflammatory processes, the shedding of cell surface molecules, and the degradation and processing of ECM molecules. The results of our initial mutational studies designed to elucidate possible unique elements for the inhibition of ADAMTS-4 were surprising, as complete replacement of the reactive-ridge region of TIMP-1 by that of TIMP-3 produced a molecule devoid of ADAMTS-4 inhibitory activity, indicating that additional key elements are present in TIMP-3. While the search for such features is under way, the structural basis for this inhibitory specificity requires the determination of the structure of TIMP-3.

Because MMPs are implicated in the progression of many diseases associated with aberrant ECM turnover, numerous synthetic MMP inhibitors have been designed and some have been clinically tested, but with little success [54]. The reasons for the failure of low-molecular-mass inhibitors are not clear, but it could result from the inhibition of non-targeted metalloproteinases or the fact that the disease had progressed to a point where it could not be reversed by the inhibition of MMPs. It is also possible that the inhibitor concentration in the target tissue did not reach an effective level. Animal model studies with selective TIMP variants that exploit tissue-specific gene transfer technology may be useful to investigate which MMPs, ADAMs and ADAMTSs are involved in disease progression. and to further develop therapeutic interventions for diseases linked with enhanced ECM degradation. To this end, and in order to design highly selective inhibitors, we need detailed structural information about the modes of interaction between various target metalloproteinases and TIMP variants.

We thank Dr Robert Visse for preparing the structural figures. The work was supported by the Wellcome Trust grant no. 057508 and NIH grant AR40994.

References

1.   Nagase, H. and Woessner, J.F. (1999) J. Biol. Chem. **274**, 21491–21494
2.   Sternlicht, M.D. and Werb, Z. (2001) Annu. Rev. Cell Dev. Biol. **17**, 463–516
3.   Woessner, J.F. and Nagase, H. (2000) Matrix Metalloproteinases and TIMPs: Protein Profile, Oxford University Press, Oxford
4.   Overall, C.M. (2002) Mol. Biotechnol. **22**, 51–86
5.   Leontovich, A.A., Zhang, J.S., Shimokawa, K., Nagase, H. and Sarras, M.P. (2000) Development **127**, 907–920
6.   Lepage, T. and Gache, C. (1990) EMBO J. **9**, 3003–3012
7.   Maidment, J.M., Moore, D., Murphy, G.P., Murphy, G. and Clark, I.M. (1999) J. Biol. Chem. **274**, 34706–34710
8.   Nagase, H. and Kashiwagi, M. (2003) Arthritis Res. Ther. **5**, 94–103
9.   Woessner, Jr, J.F. (1998) in Matrix Metalloproteinases (Parks, W.C. and Mecham, R.P., eds), pp. 1–14, Academic Press, San Diego
10.   Barrett, A.J. and Starkey, P.M. (1973) Biochem. J. **133**, 709–724

11. Cawston, T.E. and Mercer, E. (1986) FEBS Lett. **209**, 9–12

12. Brew, K., Dinakarpandian, D. and Nagase, H. (2000) Biochim. Biophys. Acta Protein Struct. Mol. Enzymol. **1477**, 267–283

13. Das, S.K., Yano, S., Wang, J., Edwards, D.R., Nagase, H. and Dey, S.K. (1997) Dev. Genet. **21**, 44–54

14. Gomez, D.E., Alonso, D.F., Yoshiji, H. and Thorgeirsson, U.P. (1997) Eur. J. Cell Biol. **74**, 111–122

15. Murphy, G. and Willenbrock, F. (1995) Methods Enzymol. **248**, 496–510

16. Will, H., Atkinson, S.J., Butler, G.S., Smith, B. and Murphy, G. (1996) J. Biol. Chem. **271**, 17119–17123

17. Stracke, J.O., Hutton, M., Stewart, M., Pendas, A.M., Smith, B., López-Otín, C., Murphy, G. and Knäuper, V. (2000) J. Biol. Chem. **275**, 14809–14816

18. Hutton, M., Willenbrock, F., Brocklehurst, K. and Murphy, G. (1998) Biochemistry **37**, 10094–10098

19. Bigg, H.F., Morrison, C.J., Butler, G.S., Bogoyevitch, M.A., Wang, Z.P., Soloway, P.D. and Overall, C.M. (2001) Cancer Res. **61**, 3610–3618

20. Troeberg, L., Tanaka, M., Wait, R., Shi, Y.E., Brew, K. and Nagase, H. (2002) Biochemistry **41**, 15025–15035

21. Willenbrock, F., Crabbe, T., Slocombe, P.M., Sutton, C.W., Docherty, A.J.P., Cockett, M.I., O'Shea, M., Brocklehurst, K., Phillips, I.R. and Murphy, G. (1993) Biochemistry **32**, 4330–4337

22. Kashiwagi, M., Tortorella, M., Nagase, H. and Brew, K. (2001) J. Biol. Chem. **276**, 12501–12504

23. Amour, A., Knight, C.G., Webster, A., Slocombe, P.M., Stephens, P.E., Knäuper, V., Docherty, A.J.P. and Murphy, G. (2000) FEBS Lett. **473**, 275–279

24. Loechel, F., Fox, J.W., Murphy, G., Albrechtsen, R. and Wewer, U.M. (2000) Biochem. Biophys. Res. Commun. **278**, 511–515

25. Amour, A., Slocombe, P.M., Webster, A., Butler, M., Knight, C.G., Smith, B.J., Stephens, P.E., Shelley, C., Hutton, M., Knäuper, V. et al. (1998) FEBS Lett. **435**, 39–44

26. Rodriguez-Manzaneque, J.C., Westling, J. Thai, S.N., Luque, A., Knäuper, V., Murphy, G., Sandy, J.D. and Iruela-Arispe, M.L. (2002) Biochem. Biophys. Res. Commun. **293**, 501–508

27. Hashimoto, G., Aoki, T., Nakamura, H., Tanzawa, K. and Okada, Y. (2001) FEBS Lett. **494**, 192–195

28. Amour, A., Knight, C.G., English, W.R., Webster, A., Slocombe, P.M., Knäuper, V., Docherty, A.J.P., Becherer, J.D., Blobel, C.P. and Murphy, G. (2002) FEBS Lett. **524**, 154–158

29. Morgunova, E., Tuuttila, A., Bergmann, U. and Tryggvason, K. (2002) Proc. Natl. Acad. Sci. U.S.A. **99**, 7414–7419

30. Bigg, H.F., Shi, Y.E., Liu, Y.L.E., Steffensen, B. and Overall, C.M. (1997) J. Biol. Chem. **272**, 15496–15500

31. Butler, G.S., Apte, S.S., Willenbrock, F. and Murphy, G. (1999) J. Biol. Chem. **274**, 10846–10851

32. Goldberg, G.I., Strongin, A., Collier, I.E., Genrich, L.T. and Marmer, B.L. (1992) J. Biol. Chem. **267**, 4583–4591

33. Vater, C.A., Mainardi, C.L. and Harris, Jr, E.D. (1979) J. Biol. Chem. **254**, 3045–3053

34. Welgus, H.G., Stricklin, G.P., Eisen, A.Z., Bauer, E.A., Cooney, R.V. and Jeffrey, J.J. (1979) J. Biol. Chem. **254**, 1938–1943

35. Cawston, T.E., Galloway, W.A., Mercer, E., Murphy, G. and Reynolds, J.J. (1981) Biochem. J. **195**, 159–165

36. Williamson, R.A., Martorell, G., Carr, M.D., Murphy, G., Docherty, A.J.P., Freedman, R.B. and Feeney, J. (1994) Biochemistry **33**, 11745–11759

37. Murzin, A.G. (1993) EMBO J. **12**, 861–867

38. Nagase, H., Suzuki, K., Cawston, T.E. and Brew, K. (1997) Biochem. J. **325**, 163–167
39. Huang, W., Meng, Q., Suzuki, K., Nagase, H. and Brew, K. (1997) J. Biol. Chem. **272**, 22086–22091
40. Caterina, N.C.M., Windsor, L.J., Yermovsky, A.E., Bodden, M.K., Taylor, K.B., Birkedal-Hansen, H. and Engler, J.A. (1997) J. Biol. Chem. **272**, 32141–32149
41. Gomis-Rüth, F.X., Maskos, K., Betz, M., Bergner, A., Huber, R., Suzuki, K., Yoshida, N., Nagase, H., Brew, K., Bourenkov, G.P. et al. (1997) Nature (London) **389**, 77–81
42. Fernandez-Catalan, C., Bode, W., Huber, R., Turk, D., Calvete, J.J., Lichte, A., Tschesche, H. and Maskos, K. (1998) EMBO J. **17**, 5238–5248
43. Nagase, H. (2001) in Matrix Metalloproteinase Inhibitors in Cancer Therapy (Clendeninn, N.J. and Appelt, K., eds), pp. 39–66, Humana Press, Totowa, NJ
44. Meng, Q., Malinovskii, V., Huang, W., Hu, Y.J., Chung, L., Nagase, H., Bode, W., Maskos, K. and Brew, K. (1999) J. Biol. Chem. **274**, 10184–10189
45. Nagase, H. and Fields, G.B. (1996) Biopolymers **40**, 399–416
46. Wei, S., Chen, Y., Chung, L., Nagase, H. and Brew, K. (2003) J. Biol. Chem. **278**, 9831–9834
47. Smith, M.R., Kung, H., Durum, S.K., Colburn, N.H. and Sun, Y. (1997) Cytokine **9**, 770–780
48. Ahonen, M., Poukkula, M., Baker, A.H., Kashiwagi, M., Nagase, H., Eriksson, J.E. and Kähäri, V.M. (2003) Oncogene **22**, 2121–2134
49. Borland, G., Murphy, G. and Ager, A. (1999) J. Biol. Chem. **274**, 2810–2815
50. Hargreaves, P.G., Wang, F.F., Antcliff, J., Murphy, G., Lawry, J., Russell, R.G.G. and Croucher, P.I. (1998) Br. J. Haematol. **101**, 694–702
51. Fitzgerald, M.L., Wang, Z., Park, P.W., Murphy, G. and Bernfield, M. (2000) J. Cell Biol. **148**, 811–824
52. Matsushita, N., Kashiwagi, M., Wait, R., Nagayoshi, R., Nakamura, M., Matsuda, T., Hogger, P., Guyre, P.M., Nagase, H. and Matsuyama, T. (2002) Clin. Exp. Immunol. **130**, 156–161
53. Lee, M.H., Knäuper, V., Becherer, J.D. and Murphy, G. (2001) Biochem. Biophys. Res. Commun. **280**, 945–950
54. Overall, C.M. and López-Otín, C. (2002) Nat. Rev. Cancer **2**, 657–672
55. Guex, N. and Peitsch, M.C. (1997) Electrophoresis **18**, 2714–2723

Biochem. Soc. Symp. **70**, 213–220
(Printed in Great Britain)
© 2003 Biochemical Society

17

Memapsin 2, a drug target for Alzheimer's disease

Gerald Koelsch*, Robert T. Turner III*, Lin Hong*, Arun K. Ghosh† and Jordan Tang*‡[1]

*Protein Studies Program, Oklahoma Medical Research Foundation, University of Oklahoma Medical Center, Oklahoma City, OK 73104, U.S.A., †Department of Chemistry, University of Illinois at Chicago, Chicago, IL 60607, U.S.A., and ‡Department of Biochemistry and Molecular Biology, University of Oklahoma Medical Center, Oklahoma City, OK 73104, U.S.A.

Abstract

Memapsin 2, a β-secretase, is the membrane-anchored aspartic protease that initiates the cleavage of amyloid precursor protein leading to the production of β-amyloid and the onset of Alzheimer's disease. Thus memapsin 2 is a major therapeutic target for the development of inhibitor drugs for the disease. Many biochemical tools, such as the specificity and crystal structure, have been established and have led to the design of potent and relatively small transition-state inhibitors. Although developing a clinically viable memapsin 2 inhibitor remains challenging, progress to date renders hope that memapsin 2 inhibitors may ultimately be useful for therapeutic reduction of β-amyloid.

Introduction

Memapsin 2 [1], also called BACE or ASP-2 [2–5], is the protease known as β-secretase that initiates the pathogenesis of Alzheimer's disease (AD) [6]. In this process, a membrane protein, amyloid precursor protein (APP), is cleaved first by memapsin 2, a membrane-associated aspartic protease, and then by another protease known as γ-secretase to generate a 40/42-residue fragment, β-amyloid (Aβ). Aβ is neurotoxic, and its elevated level in the brain leads ultimately to neural degeneration and the onset of AD. Currently, there is no drug available for treatment of AD that can slow or stop the progression of this disease. One of the most promising therapeutic approaches, which has been shown in animal models of AD to improve cognitive functions [7,8], is to

---

[1]To whom correspondence should be addressed (e-mail Jordan-tang@omrf.ouhsc.edu).

decrease the level of soluble Aβ in the brain. Memapsin 2 is an excellent therapeutic target for Aβ reduction strategy. Since it functions at the first step in the pathogenesis process, the inhibition of its activity would effectively eliminate the production of Aβ and all the subsequent harmful steps. Also, the removal of the memapsin 2 gene in mice produced no apparent deleterious response [9–11], suggesting that the inhibition of memapsin 2 activity can be tolerated physiologically. Finally, the principle of inhibitor design for aspartic proteases is well known, and the successful development of inhibitor drugs against HIV protease provides an encouraging precedent. Therefore the enzymic properties of memapsin 2 are of great interest, and it has been actively studied since its identification 3 years ago.

Target properties of a memapsin 2 inhibitor drug

For the effective development of memapsin 2 inhibitor drugs, it would be useful to consider some target properties in addition to the usual low toxicity and good pharmacological properties. First, this inhibitor drug should be a transition-state analogue of memapsin 2 catalysis. There are very few chemical reactions specific for the active sites of aspartic proteases. The esterification of catalytic aspartyls has been accomplished using EPNP [1,2-epoxy-3-(p-nitrophenoxy)-propane]-like [12] or DAN (diazoacetyl-D,L-norleucine methyl ester)-like [13] compounds, although they are not clinically viable drug candidates. On the other hand, the transition-state isostere hydroxyethylene, found originally in pepstatin A [14], has been successfully applied in all HIV protease inhibitor drugs, suggesting that memapsin 2 inhibitor drugs should also be transition-state analogues. Secondly, to inhibit Aβ production in the brain, the inhibitor needs to penetrate the blood–brain barrier, which usually permits the passage of compounds around 500 Da or less. Within this size range, it would be difficult to incorporate into the inhibitors all desirable properties, such as potency, selectivity and pharmacological characteristics. However, some larger compounds do penetrate the blood–brain barrier. For example, the HIV protease inhibitor drug indinavir, with a molecular mass of 625 Da, can enter the brain [15]. Thus a size around 600 Da seems to be a reasonable target for the design of clinically viable memapsin 2 inhibitors. Thirdly, the potency of the inhibitor may be in the K_i range of 1–10 nM. Memapsin 2 inhibitor drugs are aimed at delaying permanently the onset of the disease, so the strategy of its use in the treatment of AD will be different from that for HIV infection. For the latter, very high inhibitor potency and high dosage are necessary to completely suppress viral replication, to avoid viral resistance. Memapsin 2 inhibitor drugs will probably be used clinically to lower brain Aβ to a non-pathological level. It is known that cells expressing a Swedish mutant of APP, which manifests an early-onset form of AD, produce six to eight times the amount of Aβ as compared with cells expressing native APP [16], thus suggesting that the inhibition of approx. 85% of memapsin 2 activity would delay the onset of AD indefinitely. The inhibition potency mentioned above will provide a near-linear dose–response within the desired inhibition range.

Biochemical tools: keys to inhibitor design

An important approach in the development of a memapsin 2 inhibitor is to utilize specificity- and structure-based design. Although this approach would require the development of many biochemical tools to elucidate memapsin 2 specificity and structure, it seems to have advantages in some respects over inhibitor lead discovery by the screening of a large chemical library. Like other aspartic proteases, memapsin 2 has a large substrate-binding site that accommodates eight residues [17]. Screening is likely to identify many compounds that interact with part of the binding cleft which typically have K_i values in the micromolar range, similar to the K_m of the enzyme. These 'hits' may be too numerous to be followed up. Also, without specifically screening for binding to a transition-state template, the leads may also have a long way to go to attain the desired potency. Such was the experience of the screening approach during the early development of HIV protease inhibitor drugs.

Many of the biochemical tools required for the design of memapsin 2 inhibitors have now been developed; therefore only a brief account will be given here. Production of the recombinant memapsin 2 catalytic domain by expression in *Escherichia coli* [1] was an important first step in supplying a sufficient amount of enzyme for subsequent studies. The kinetic assay of memapsin 2 is most conveniently performed using a fluorogenic assay [18]. Although the conversion of pro-memapsin 2 into the mature protease is effected *in vivo* by furin [19–22], the pro- segment on the recombinant zymogen can be removed *in vitro* by proteases to form a stable memapsin 2 for kinetic/inhibition assays and crystallization [18]. Crystal structures have also been determined for the catalytic unit of memapsin 2 complexed with inhibitors [23,24], which provide a template for inhibitor design.

Specificity of memapsin 2 and its closest homologue, memapsin 1

Like other aspartic proteases, memapsin 2 binds eight substrate residues in an extended binding cleft. To determine the complete residue preference of all eight subsites using the usual kinetic methods is very laborious, which accounted for the fact that no complete specificity had been determined prior to the memapsins. To facilitate such determinations, we devised a new approach, in which the hydrolysis of substrates in mixtures was quantified with MALDI-TOF (matrix-assisted laser-desorption ionization–time-of-flight) MS [17]. The relative rates determined are proportional to the relative k_{cat}/K_m values. The complete subsite specificity of memapsin 2 [17] is shown in Figure 1. These data are in good agreement with those produced from the affinity binding of memapsin 2 to a partial random sequence inhibitor library [17]. The broad specificity of this protease in all eight subsites is reminiscent of the specificity of HIV-1 protease. This may be advantageous for inhibitor drug design, since it predicts the binding of many different inhibitor structures in the active site. The closest homologue of memapsin 2 is memapsin 1 [1] (BACE 2, ASP-1), which also hydrolyses APP at the β-secretase site [25]. We therefore also determined

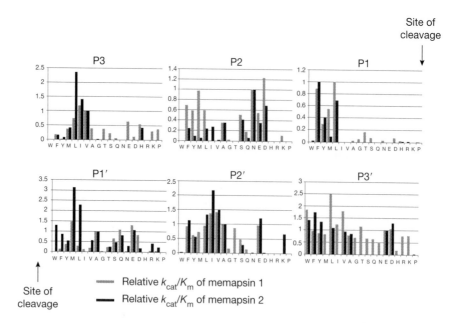

Figure 1 Specificity of memapsin 1 and memapsin 2 for amino acid side chains at positions P3–P3′. The scissile bond is situated between positions P1 and P1′, with the specificity for amino acids at positions in the N- and C-terminal products represented in the upper and lower panels respectively. Specificity constants (k_{cat}/K_m) are expressed relative to the template substrate amino acid.

the complete subsite specificity of memapsin 1 [26]. Figure 1 illustrates that the two proteases share many preferred residues in subsites important for inhibitor design. For example, both proteases prefer Phe and Leu in P1, Asn and Asp in P2 and Ile, Leu and Val in P3. The subsites P1′ to P4′ are very non-stringent. As will be discussed below, the closeness in these two specificities posts a severe challenge for the design of selectivity in memapsin 2 inhibitors.

Binding of inhibitors in the active site of memapsin 2

Based on partial specificity information, we designed first-generation transition-state inhibitors of memapsin 2 [27]. OM99-2, which contains eight residues, EVNL*AAEF, with the scissile bond replaced by a transition-state isostere hydroxyethylene (represented as *), had high potency (K_i 1.6 nM). The 1.9 Å structure of OM99-2 complexed to the catalytic unit of memapsin 2 revealed that the inhibitor is located to the substrate-binding cleft, where the inhibitor backbone interacted extensively with the active site, mostly through hydrogen bonds. Six of the eight inhibitor side chains, from P4 to P2′, clearly interacted with the protease [23]. The side chains of P3′ and P4′, however, had high mobility and poor interaction with memapsin 2. An improved inhibitor, OM00-3 [17], with the sequence ELDL*AVEF (K_i 0.3 nM), was designed based on the specificity information and the optimized side chains. The crystal structure of the complex between memapsin 2 and OM00-3 shows that the inhibitor is essentially in an extended conformation, and all eight side chains are interacting with subsites of the protease (Figure 2). The improved binding of subsites P3' and P4' was due mainly to improved transition-state binding of these two subsites when a valine is present in the S2' pocket [17]. These crystal structures provide a clear template for use in the design of a new generation of inhibitors.

Structure-based design of a memapsin 2 inhibitor

Although the first-generation inhibitors, such as OM99-2, were quite potent, they were considerably larger than the target size. Structure-based design was carried out with aims to (a) substitute the P3' and P4' residues with a small group, since these subsites bind poorly with low stringency, (b) substitute and optimize P4 to P2 with smaller side chains, and (c) optimize the side chains of P1' and P2'. The resulting inhibitor, GT1017 [28,29], is considerably smaller (722 Da), yet still quite potent (K_i 2.5 nM). Modelling of GT1017 in the active site of memapsin 2 revealed that the new inhibitor utilizes the main interactions between the inhibitor backbone and the protease active site. Another important consideration in the design of new inhibitors is selectivity. Among the human aspartic proteases, selectivity against memapsin 1 appears most challenging. This is illustrated by the fact that the early inhibitors designed for memapsin 2, such as OM99-2 and OM00-3, actually inhibited memapsin 1 as

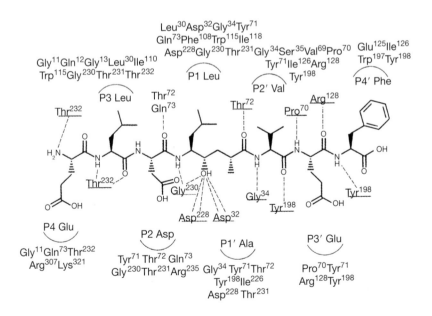

Figure 2 Interactions between inhibitor OM00-3 and memapsin 2.
Residues of memapsin 2 within 4.5 Å of the inhibitor are listed adjacent to the side chains of the OM00-3 structure. Underlined residues indicate those that interact with the backbone atoms of the inhibitor.

well [26]. Since the two memapsins are very similar with regard to their preferred residues in the major subsites, attempts were made to incorporate into the inhibitors less preferred residues that show the greatest difference in affinities between the two enzymes. This approach, however, did not provide the desired selectivity [26]. Improved selectivity was seen with the use of nonnative side chains, which produced inhibitors with a clear preference to inhibit memapsin 2 rather than memapsin 1. These results suggest that the structure-based design of memapsin 2 inhibitors is a powerful tool towards attaining the target inhibitor properties.

Cellular functions and transport of memapsin 2

Several lines of evidence argue that APP is not a physiological substrate of memapsin 2. First, the ultimate product of this activity, Aβ, is neurotoxic, and AD is not a positive selection in evolution. Secondly, the kinetic parameters for the hydrolysis of native APP by memapsin 2 are very poor [1]. For a physiological activity, the catalytic efficiency would be likely to have improved through evolution. Thirdly, the sequence around the β-secretase site of APP is not well conserved, suggesting a lack of functional preservation. Thus it seems probable that cleavage of APP by memapsin 2 is a tolerable, controlled physiological error, and that the pathological process occurs only when Aβ is overproduced, as in the case of the APP$_{Swedish}$ mutation, or when Aβ overaccumulates in the brain by other mechanisms, as in the case of lipoprotein E4 polymorphism. These points argue for the presence of as yet unidentified physiological functions of memapsin 2. Although APP may not be a physiological substrate of memapsin 2, the localization and mechanism of APP hydrolysis may serve as a model for the physiological substrates.

The cellular functions of memapsin 2 may be intimately linked to its route of transport in the cells. Pro-memapsin 2 is biosynthesized, transported and activated to memapsin 2 through the secretory pathway via the endoplasmic reticulum and Golgi to reach the cell surface. Memapsin 2 is then endocytosed from the cell surface to endosomes, where the acidic environment, about pH 4, is optimal for its activity. Supporting the notion that the endosome is the primary site of the memapsin 2-mediated hydrolysis of APP are cellular localization studies and the inhibition of Aβ formation by bafilomycin, an inhibitor of endosomal acidification. However, β-secretase-mediated cleavage of APP and the formation of Aβ in the endoplasmic reticulum and Golgi have been reported by many laboratories. These compartments have pH values of between 6 and 7, a condition in which memapsin 2 has very little activity [18]. Since these studies were performed with cells overexpressing both memapsin 2 and APP$_{Swedish}$, it is possible that some cleavage takes place in these compartments in the presence of high concentrations of both enzyme and substrate in spite of an unfavourable pH.

The cytosolic domain of memapsin 2 is required for its endocytosis [22,30]. Recently, the cytosolic domain of memapsin 2, including a Leu-Leu motif and an Asp, was demonstrated to bind to the VHS ('VPS-27, Hrs and STAM') domain of GGA ('Golgi-localized, γ-ear containing, ADP-ribosylation factor-binding') proteins [31], which are implicated in the mechanism of memapsin 2 endocytosis. Since GGA proteins also mediate the transport of mannose-6-phosphate receptors from the *trans*-Golgi to endosomes, this opens up an interesting possibility that memapsin 2 may be also directly transported by this route.

Future perspectives

The biochemical tools developed for the design of structure-based inhibitors of memapsin 2 have shown considerable promise. The new generations of inhibitors are approaching the target size and potency. Although to attain clinically viable inhibitors remains challenging, the results so far provide encouragement that useful drugs against this target will eventually be possible. Little progress has been made so far on understanding of the physiological functions of memapsin 2. One may argue that the lack of phenotype caused by gene deletion would suggest that memapsin 2 inhibitor drugs can be developed without the functional knowledge. However, AD is a chronic disease, and treatment using memapsin 2 inhibitor drugs will require sustainable long-term treatment. Thus an understanding of the physiological functions of memapsin 2 and the pathological consequences of its inhibition is likely to be important in the clinical management of the disease.

This research was supported in part by NIH grant AG-18933 and an American Alzheimer's Association Pioneer Award. R.T.T. is the recipient of a Predoctoral Fellowship from the Oklahoma Medical Research Foundation and is a Glenn/American Foundation of Aging Research Scholar. G.K. is a Scientist Development Awardee of the American Heart Association. J.T. is the holder of J.G. Puterbaugh Chair in Biomedical Research at the Oklahoma Medical Research Foundation.

References

1. Lin, X., Koelsch, G., Wu, S., Downs, D., Dashti, A. and Tang, J. (2000) Proc. Natl. Acad. Sci. U.S.A. **97**, 1456–1460
2. Vassar, R., Bennett, B.D., Babu-Khan, S., Kahn, S., Mendiaz, E.A., Denis, P., Teplow, D.B., Ross, S., Amarante, P., Loeloff, R. et al. (1999) Science **286**, 735–741
3. Yan, R., Bienkowski, M.J., Shuck, M.E., Miao, H., Tory, M.C., Pauley, A.M., Brashier, J.R., Stratman, N.C., Mathews, W.R., Buhl, A.E. et al. (1999) Nature (London) **402**, 533–537
4. Sinha, S., Anderson, J. P., Barbour, R., Basi, G.S., Caccavello, R., Davis, D., Doan, M., Dovey, H.F., Frigon, N., Hong, J. et al. (1999) Nature (London) **402**, 537–540
5. Hussain, J., Powell, D., Howlett, D.R., Tew, D.G., Meek, T.D., Chapman, C., Gloger, I.S., Murphy, K.E., Southan, C.D., Ryan, D.M. et al. (1999) Mol. Cell. Neurosci. **14**, 419–427
6. Selkoe, D.J. (1999) Nature (London) **399A**, 23–31
7. Morgan, D., Diamond, D.M., Gottschall, P.E., Ugen, K.E., Dickey, C., Hardy, J., Duff, K., Jantzen, P., DiCarlo, G., Wilcock, D. et al. (2000) Nature (London) **408**, 982–985
8. Dodart, J.C., Bales, K.R., Gannon, K.S., Greene, S.J., DeMattos, R.B., Mathis, C., DeLong, C.A., Wu, S., Wu, X., Holtzman, D.M. and Paul, S.M. (2002) Nat. Neurosci. **5**, 452–457
9. Cai, H., Wang, Y., McCarthy, D., Wen, H., Borchelt, D.R., Price, D.L. and Wong, P.C. (2001) Nat. Neurosci. **4**, 233–234

220

G. Koelsch et al.

10. Luo, Y., Bolon, B., Kahn, S., Bennett, B.D., Babu-Khan, S., Denis, P., Fan, W., Kha, H., Zhang, J., Gong, Y. et al. (2001) Nat. Neurosci. **4**, 231–232

11. Roberds, S.L., Anderson, J., Basi, G., Bienkowski, M.J., Bienkowski, M.J., Branstetter, D.G., Chen, K.S., Freedman, S.B., Frigon, N.L., Games, D. et al. (2001) Hum. Mol. Genet. **10**, 1317–1324

12. Chen, K.C.S. and Tang, J. (1972) J. Biol. Chem. **247**, 2566–2571

13. Ragagopolan, T.G., Stein, W.H. and Moore, S. (1966) J. Biol. Chem. **241**, 4295–4297

14. Marciniszyn, Jr, J., Hartsuck, J.A. and Tang, J. (1976) J. Biol. Chem. **251**, 7088–7092

15. Martin, C., Sonnerborg, A., Svensson, J.O. and Stahle, L. (1999) AIDS **13**, 1227–1232

16. Citron, M., Oltersdorf, T., Haass, C., McConlogue, L., Hung, A.Y., Seubert, P., Vigo-Pelfrey, C., Lieberburg, I. and Selkoe, D.J. (1992) Nature (London) **360**, 672–674

17. Turner, R.T., Koelsch, G., Hong, L., Castenheira, P., Ghosh, A. and Tang, J. (2001) Biochemistry **40**, 10001–10006

18. Ermolieff, J., Loy, J. A., Koelsch, G. and Tang, J. (2000) Biochemistry **39**, 12450–12456

19. Bennett, B.D., Denis, P., Haniu, R., Teplow, D.B., Kahn, S., Louis, J.C., Citron, M. and Vassar, R. (2000) J. Biol. Chem. **275**, 37712–37717

20. Creemers, J.W., Dominguez, D.I., Plets, E., Serneels, L., Taylor, N.A., Multharp, G., Craessaerts, K., Annaert, W. and De Strooper, B. (2000) J. Biol. Chem. **276**, 4211–4217

21. Capell, A., Steiner, H., Willem, M., Kaiser, H., Meyer, C., Walter, J., Lammich, S., Multhaup, G. and Haass, C. (2000) J. Biol. Chem. **275**, 30849–30854

22. Huse, J.T., Pijak, D.S., Leslie, G.J., Lee, V.M.-Y. and Doms, R.W. (2000) J. Biol. Chem. **275**, 33729–33737

23. Hong, L., Koelsch, G., Lin, X., Wu, S., Terzyan, S., Ghosh, A.K., Zhang, X.C. and Tang, J. (2000) Science **290**, 150–153

24. Hong, L., Turner, R.T., Koelsch, G., Shin, D., Ghosh, A.K. and Tang, J. (2002) Biochemistry **41**, 10963–10967

25. Farzan, M., Schnitzler, C.E., Vasilieva, N., Leung, D. and Choe, H. (2000) Proc. Natl. Acad. Sci. U.S.A. **97**, 9712–9717

26. Turner, R.T., Loy, J.A., Nguyen, C., Devasamudram, T., Ghosh, A., Koelsch, G. and Tang, J. (2002) Biochemistry **41**, 8742–8746

27. Ghosh, A.K., Shin, D., Downs, D., Koelsch, G., Lin, X., Ermolieff, J. and Tang, J. (2000) J. Am. Chem. Soc. **122**, 3522–3523

28. Hong, L., Turner, R.T., Koelsch, G., Ghosh, A.K. and Tang, J. (2002) Biochem. Soc. Trans. **30**, 530–534

29. Ghosh, A.K., Bilcer, G., Harwood, C., Kawahama, R., Shin, D., Hussain, K.A., Hong, L., Loy, J.A., Nguyen, C., Koelsch, G. et al. (2001) J. Med. Chem. **44**, 2865–2868

30. Pastorino, L., Ikin, A.F., Nairn, A.C., Pursnani, A. and Buxbaum, J.D. (2002) Mol. Cell. Neurosci. **19**, 175–185

31. He, X., Chang, W., Koelsch, J. and Tang, J. (2002) FEBS Lett. **524**, 183–187

Biochem. Soc. Symp. **70**, 221–231
(Printed in Great Britain)
© 2003 Biochemical Society

18

Control of lipid metabolism by regulated intramembrane proteolysis of sterol regulatory element binding proteins (SREBPs)

Robert B. Rawson[1]

Department of Molecular Genetics, University of Texas Southwestern Medical
Center, 5323 Harry Hines Blvd, Dallas, TX 75390-79046, U.S.A.

Abstract

In mammalian cells, the supply of lipids is co-ordinated with demand through the transcriptional control of genes encoding proteins required for synthesis or uptake. The sterol regulatory element binding proteins (SREBPs) are responsible for increased transcription of these genes when lipid levels fall. Mammals have three SREBPs (-1a, -1c and -2), which are the products of two distinct genes. Synthesized as ~120 kDa precursors, they are inserted into membranes of the endoplasmic reticulum (ER) in a hairpin fashion. Both the N-terminal transcription factor domain and the C-terminal regulatory domain face the cytoplasm. These are connected by two transmembrane helices separated by a short loop projecting into the ER lumen. The C-terminal domain of SREBP interacts with the C-terminal domain of SREBP-cleavage-activating protein (SCAP). The N-terminal half of SCAP contains eight transmembrane helices, five of which (helices 2–6) form the sterol-sensing domain. In response to cellular demand for lipid, this complex exits the ER and transits to the Golgi apparatus, where two distinct proteases cleave the SREBP precursor to release the transcriptionally active N-terminus. This process was the first example of regulated intramembrane proteolysis for which the proteases were identified. Recent work has additionally uncovered integral membrane proteins, insig-1 and insig-2, that are required to retain the SREBP–SCAP complex in the ER in the presence of sterols, thus providing a more complete understanding of the control of proteolysis in this complex regulatory pathway.

[1]e-mail Rawson@UTSW.SWMed.Edu

Introduction

Until recently, few people would have considered proteolysis to be a vital mechanism for cellular signal transduction. Yet recent results, coming from a wide range of biological research, point to just such a conclusion. The most surprising examples concern proteases that cleave their substrates within transmembrane helices in a process termed regulated intramembrane proteolysis (Rip) [1]. Here I will consider in detail the role of Rip in the sterol regulatory element binding protein (SREBP) pathway controlling lipid metabolism in mammalian cells, with special consideration of recent discoveries. To place this example in context, I will briefly mention several other examples of Rip that have been discovered in the past few years.

Rip

In the best understood examples of Rip, two distinct proteases cleave their substrate sequentially to release protein fragments that activate the transcription of target genes. The first cleavage occurs in the extracytoplasmic domain of the substrate, 20–30 amino acids away from the membrane. The second cleavage, which happens only after the first cleavage has taken place, occurs within a transmembrane helix, releasing a cytoplasmic domain of the substrate.

Class 1 Rip involves substrates that are type 1 membrane proteins (i.e. N-terminus is extracytoplasmic and C-terminus is cytoplasmic). Cleavage within the transmembrane helices of these proteins requires γ-secretase. This enzyme is a multi-component complex of integral membrane proteins that includes presenilin-1 or -2 and nicastrin, as well as other proteins (reviewed in [2]). Interaction of γ-secretase inhibitors with conserved aspartate residues in transmembrane helices of presenilin led to the suggestion that presenilins are aspartyl proteases [3]. Examples of class 1 Rip include signalling via the Notch receptor, processing of the receptor tyrosine kinase ErbB-4 [4] and, the most notorious example, the generation of β-amyloid (Aβ) from amyloid precursor protein (APP). Accumulation of Aβ leads to the formation of the amyloid plaques that characterize Alzheimer's disease [5]. While Aβ, the extracytoplasmic fragment of APP, has received most attention, the cytoplasmic fragment of APP appears to play a role in transcriptional activation, as observed in most other examples of Rip [6–8].

Substrates for class 2 Rip are type 2 membrane proteins (i.e. N-terminus cytoplasmic, C-terminus extracytoplasmic), and include the SREBPs that regulate the genes of lipid synthesis [9], as well as activating transcription factor 6α (ATF6α) and ATF6β that mediate the transcriptional programme of the unfolded protein response [10,11]. The former will be discussed in detail below.

Examples of both class 1 and class 2 Rip are known from invertebrates as well as mammals [12–16]. Rip also occurs in prokaryotes. SpoIV FB, a distant homologue of mammalian site-2 protease (S2P), is required to release the pro-σ^k transcription factor from the membrane during sporulation in *Bacillus subtilis* [17]. In *Escherichia coli*, YaeL protein, another S2P homologue, cleaves the anti-σ factor RseA to activate a stress response dependent on σ^E [18,19]. In *Enterococcus*, Rip is required for extracellular signalling. In this organism, the S2P homologue *eep* cleaves its substrate within a transmembrane domain to produce active aggregation pheromone [1].

Signal peptide peptidase (SPP)

Processing of signal peptides also employs Rip. A unique protein, SPP, cleaves the intramembrane helices that are a product of signal peptidase cleavage of secreted proteins [20]. Based on studies with protease inhibitors, SPP was predicted to be an aspartyl protease. Intriguingly, SPP shares with presenilins two conserved motifs within its transmembrane helices: a Tyr-Asp motif in helix 4 and a Leu-Gly-Leu-Gly-Asp motif in helix 5 [20]. Both of the conserved aspartates are necessary for activity in presenilin [3,21,22]. The two proteins are not otherwise similar; in fact, the predicted membrane orientation of the active sites in the two proteins is opposite [20], as is the orientation of their substrates. The presence of two conserved aspartates in conserved context, each in its own transmembrane helix, strongly supports the conclusion that both SPP and the presenilins are intramembrane aspartyl proteases [3,20]. SPP-mediated cleavage is thought to generate signalling molecules [23,24] and may serve a general 'housekeeping' function as well.

Rhomboid

Drosophila rhomboid is an intramembrane serine protease necessary for signalling through epidermal growth factor (EGF) receptor homologues [25,26]. This example of intramembrane proteolysis differs from the preceding examples in that intramembrane cleavage by rhomboid does not require a primary cleavage by a distinct enzyme to produce the intramembrane substrate. Rather, proteolysis by rhomboid is controlled by substrate localization [25,26]. Rhomboid homologues are found in prokaryotes as well as in other eukaryotes, and many of these can be functionally substituted for rhomboid in mutant flies [27,28].

Rip and regulation of lipid metabolism

Rip has rapidly emerged as a general and ancient approach to generating signals that control diverse cellular processes [1,29–31]. The best-studied role for Rip is in the regulation of lipid metabolism in animal cells by cleavage of SREBPs. Results reported in the past year have significantly advanced our understanding of the complex regulation of cellular lipid metabolism and the mechanisms by which proteolysis is controlled in this signalling pathway.

SREBPs and lipid metabolism

Animal cells control lipid levels primarily by regulating the transcription of the genes involved in lipid biosynthesis and uptake. Transcriptional up-regulation of these genes depends on SREBPs, a family of membrane-bound transcription factors. In mammals, there are three isoforms of SREBP, which are the products of two separate genes. The isoforms have overlapping but distinct domains of transcriptional activation. While SREBP-1a and -1c preferentially activate the genes of fatty acid synthesis, SREBP-2 activates the genes of cholesterol synthesis [9].

The SREBPs are synthesized as ~120 kDa precursors that are located in the membrane of the endoplasmic reticulum (ER). They span the membrane twice, with both the N-terminal transcription factor domain and the C-terminal regulatory domain facing the cytosol. These domains are connected by two transmembrane helices that are themselves separated by a loop of about 30 amino acids projecting into the lumen of the ER (Figure 1). In the ER, SREBP forms a complex with SREBP-cleavage-activating protein (SCAP) via their respective C-terminal domains.

The pathway regulating SREBP cleavage and, therefore, transcriptional activation of the genes of lipid biosynthesis is best understood for cholesterol. Cholesterol is an essential structural component of mammalian membranes. When cellular supplies of cholesterol are adequate, the SREBP–SCAP complex remains in the ER; however, when demand for sterols rises, the SREBP–SCAP complex exits the ER by way of incorporation into COPII (coatomer complex) vesicles and moves to the Golgi apparatus [32–34]. Once there, SREBP undergoes two sequential proteolytic cleavages, due to the action of two distinct enzymes. As a result, the N-terminal transcription factor domain is released from the membrane and is free to translocate to the nucleus, where it binds to sterol regulatory elements in the promoter regions of target genes and mediates their increased transcription.

The crucial regulatory event in this process is the sensing of sterol levels by SCAP and the subsequent inclusion of SREBP–SCAP complexes into COPII vesicles. Movement of SREBP from the ER to the Golgi brings it into contact with the active proteases needed to complete processing. Therefore, regulation of proteolysis occurs at the level of accessibility of the substrate to the protease, rather than activation of or changes in the location of the proteases themselves. By placing the active proteases in a compartment separate from that in which the substrate resides in the absence of sterols, this system makes use of a distinctive feature of eukaryotic cells: subcellular compartmentalization by endomembrane systems. Recent work has detailed additional components of the regulatory machinery needed to complete this programme. Here I will discuss each of the components of the SREBP signalling machinery, with emphasis on those identified recently.

Components of the SREBP pathway

Site-1 protease (S1P)

S1P is a membrane-anchored subtilisin-like serine protease of 1052 amino acids. The bulk of the enzyme, including the active site, is disposed in the lumen of the Golgi apparatus, with only a short, basic tail in the cytoplasm [35]. Figure 1 gives a schematic illustration of the membrane topology of components of the SREBP pathway. In common with other proteases of the secretory

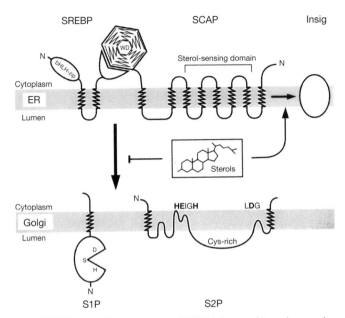

Figure 1 SREBP signalling pathway. SREBP is inserted into the membrane of the ER in a hairpin fashion. It forms a complex with SCAP by interaction between their C-terminal domains. The C-terminal domain of SCAP consists of multiple copies of a protein–protein interaction motif (termed WD repeats due to the presence of Trp and Asp residues) that form propeller-like structures (WD). The N-terminal domain of SCAP consists of eight transmembrane helices, of which helices 2–6 comprise the sterol-sensing domain (indicated). In the presence of adequate cellular supplies of sterols, the SCAP sterol-sensing domain interacts with insig proteins. This interaction serves to retain the SREBP–SCAP complex within the ER, preventing its movement to the Golgi apparatus. When cellular demand for cholesterol increases, the interaction between SCAP and insig is disrupted and the SREBP–SCAP complex is free to be packaged into COPII vesicles and move to the Golgi apparatus. Once there, SREBP becomes the substrate for two different proteases that act in sequence. First, S1P cleaves SREBP at site-1, in the middle of the luminal loop. S1P is a subtilisin-like protein and has residues of the classic serine protease catalytic triad (Ser, His, Asp). Cleavage by S1P separates the two halves of SREBP. The N-terminal half is then cleaved at site-2, within the transmembrane helix. This cleavage requires S2P, a hydrophobic, membrane-embedded protein with an HEXXH motif (indicated) characteristic of many metalloproteases. Also indicated is a motif, LDG, that is common to all members of the S2P family and that probably provides a additional co-ordinating ligand for the active-site metal atom. The putative metal ligand residues are shown in bold. The metal ligand motifs are separated in the linear sequence of S2P by a large, relatively hydrophilic domain containing multiple, conserved Cys residues (Cys-rich). The function of this domain is unknown. As a result of these sequential cleavages, the N-terminal basic helix–loop–helix leucine zipper transcription factor domain (bHLH-zip) of SREBP is released into the cytoplasm and translocates to the nucleus to direct the increased transcription of genes of lipid metabolism. For references, see the text. The Figure is based on membrane topology data derived from the following protease protection and glycosylation studies: SREBP [71], SCAP [72], S1P [35,40] and S2P [49]. No topological data have been reported for insig-1 or -2.

pathway, S1P is synthesized in the rough ER as an inactive proenzyme. Two autocatalytic cleavages serve to render the enzyme active in the Golgi apparatus [35,36]. In contrast with other serine proteases of the secretory pathway, S1P cleaves its substrate within a luminal domain after a hydrophobic residue rather than a basic residue, i.e. RXXL↓L rather than RX(R/K)R↓X [37,38], where ↓ indicates the scissile bond.

S1P was isolated by an expression/complementation strategy in mutant Chinese hamster ovary cells that are auxotrophic for cholesterol (SRD-12A and -12B cells) [39,40]. In these cells,, SREBPs are not cleaved at site 1. As a result, no nuclear SREBP is formed, and the transcription of the genes of cholesterol biosynthesis cannot be up-regulated. This result explained the auxotrophic phenotype of SRD-12 cells and confirmed the essential role of S1P in the regulation of lipid metabolism [39].

In addition to the SREBPs, S1P also cleaves ATF6α at an RHLL↓G sequence within its luminal domain [41]. S1P may also have roles in the secretory pathway other than simply processing transmembrane transcription factors. At about the same time that we described the identification of S1P, the analogous enzyme from mice and rats, termed SKI-1, was isolated by virtue of its ability to cleave brain-derived neurotrophic factor [42]. Seidah and co-workers [43] found that, in some circumstances, S1P may be shed into the medium. Subsequently, Lenz et al. [44] showed that S1P is necessary for the production of infective Lassa virus particles. S1P may also cleave other viral proteins, such as Crimean–Congo haemorrhagic fever virus [45]. In mice, S1P is an essential gene; mice lacking the enzyme die as embryos [46]. Mice with a liver-specific conditional knockout of the S1P gene are viable and express only 10–30% of normal levels of S1P. In these animals, lipid synthesis and plasma cholesterol levels are reduced [46], demonstrating that S1P is involved in the normal physiological regulation of lipid metabolism in the liver, just as observed in our studies with mutant fibroblasts [39,40].

S2P

S2P is a highly unusual protein of 519 amino acids that contains an HEXXH (His-Glu-Xaa-Xaa-His) motif. This sequence is characteristic of many families of metalloproteases, where the two histidines provide co-ordinating ligands for the active-site metal atom and the glutamate provides a nucleophile for hydrolysis of the scissile bond [47]. S2P is unique among metalloproteases in having the HEXXH motif within an otherwise hydrophobic stretch of amino acids [48,49]. Analysis of mutant Chinese hamster ovary cells auxotrophic for cholesterol led to the identification and complementation cloning of S2P. In these cells, SREBP cleavage does not occur. No SREBP gets to the nucleus and transcription of cholesterol biosynthetic genes cannot be up-regulated. Instead, an intermediate form of SREBP, the product of cleavage at site-1, accumulates in the membranes in the absence of sterols [50].

Subsequent studies with site-directed mutant versions of the S2P cDNA in the S2P-null cells demonstrated that both histidines and the glutamate of the HEXXH motif are essential for proteolysis of SREBPs. An additional aspartate residue residing within the sequence LDG (Leu-Asp-Gly) is also essential for

S2P function. All S2P homologues in species from prokaryotes to humans also contain both the HEXXH and ΦDG motifs (where Φ is a hydrophobic amino acid, typically leucine) [1,51]. Together, these observations support the notion that this aspartate residue provides an additional co-ordinating ligand for the active-site metal atom [49].

Biochemical studies on the membrane topology of S2P, using protease protection and glycosylation, support a model in which both the HEXXH and LDG motifs are within the bilayer or, perhaps, just at its cytoplasmic surface. All of the hydrophilic regions of the protein are luminally disposed, whereas both the C- and N-termini are facing the cytoplasm [49]. These results are in contrast with predictions based on computational considerations [51].

Cleavage of SREBPs by S2P rapidly follows cleavage by S1P; the intermediate product is not normally observed in wild-type cells. The same result is observed for ATF6α [41]. Since active S1P is located in the Golgi apparatus, active S2P is thought to reside there as well. Immunofluorescence studies with overexpressed, epitope-tagged S2P indicate a localization within the Golgi for the tagged protein [52].

SCAP

SCAP is a 1276-amino-acid protein that comprises two domains. The N-terminal domain consists of eight transmembrane helices. Helices 2–6 comprise a sterol-sensing domain. 3-Hydroxy-3-methylglutaryl-CoA reductase, the rate-limiting enzyme of cholesterol biosynthesis, contains a similar domain, which is necessary for the sterol-modulated stability of the enzyme [53]. Similar domains are also present in two other proteins associated with cholesterol: the Niemann–Pick Type C gene product (NPC1) [54,55] and Patched, the receptor for Hedgehog. Hedgehog is the only protein known to contain a covalently attached cholesterol moiety [56].

Additional evidence for the important role of these transmembrane helices in transducing the signal from cellular cholesterol to SREBP cleavage comes from analysis of mutant cells resistant to the killing effects of 25-hydroxycholesterol. 25-Hydroxycholesterol is a potent suppressor of SREBP cleavage and, therefore, of the transcriptional activation of the sterol biosynthetic genes that are targets of nuclear SREBP [57]. Owing to the presence of the additional hydroxy group, 25-hydroxycholesterol cannot substitute functionally for cholesterol within the membrane bilayer. As a result, wild-type cells grown in the continual presence of 25-hydroxycholesterol starve to death, as they can neither make nor take up cholesterol, and they cannot use the sterol that is present [58]. One class of mutations that confer resistance to the killing effects of 25-hydroxycholesterol are point mutations within the sterol-sensing domain of one copy of the two SCAP genes normally expressed. Such mutations confer dominant resistance to 25-hydroxycholesterol. In fact, SCAP was cloned owing to one such dominant point mutation within the sterol-sensing domain, D443N [59].

Subsequently, two additional sterol-insensitive mutations were mapped to the sterol-sensing domain of SCAP [60,61]. We now know that one effect of these point mutations is to interfere with the interaction between the SCAP sterol-sensing domain and other proteins. This interaction is normally required to retain SCAP in the ER in the presence of sterols (see below). Brown et al. [62] have recently shown that the membrane-spanning helices of SCAP undergo a conformational change upon addition of cholesterol to membranes. Mutant SCAP harbouring the Y298C mutation that renders it insensitive to sterols is somewhat more resistant to this change than is wild-type SCAP. This suggests that sterol sensing by SCAP involves a conformational change within the sterol-sensing domain, and that this change may serve to block interaction between SCAP and its ER retention partner, insig (see below).

The C-terminal domain of SCAP is composed of multiple WD (Trp-Asp) repeats, which are normally found in protein-interaction domains. SCAP and SREBP form a complex via their respective C-terminal domains. Formation of this complex is constitutive, occurring in both the absence and the presence of cholesterol. In the absence of complex formation, both SREBP and SCAP are much less stable than in its presence [63].

Mutant Chinese hamster ovary cells that completely lack SCAP protein (SRD-13A cells) fail to process SREBPs and are cholesterol auxotrophs as well [63]. Similarly, mice harbouring a liver-specific knockout of the SCAP gene show profound deficits in lipid synthesis [64]. Transgenic mice carrying a SCAP allele harbouring the activating point mutation D443N further confirm the crucial role of SCAP in lipid homoeostasis in the liver. Just as observed in mutant cells, SREBP cleavage and, therefore, lipid synthesis is strongly up-regulated in the livers of these animals [65].

Insig

In August 2002, Yang et al. [66] reported the identification of an important new component of the SREBP cleavage regulatory machinery, insig-1. Insig-1 mRNA was originally identified by Taub and co-workers [67,68] as an mRNA that increased when rat hepatoma cells were cultured in the presence of insulin. The search for this protein was stimulated by previous work that indicated that SREBP–SCAP complexes had to interact (via the SCAP sterol-sensor domain) with another protein in the ER in order to be retained in the presence of sterols [69]. Insig-1 was identified by co-precipitating proteins that bound to membrane-spanning helices 1–6 of SCAP and identifying the co-precipitated proteins by tandem MS. Yang et al. [66] demonstrated that insig-1 is located in the ER in both the presence and the absence of sterols. They also showed that binding of SCAP and insig-1 is induced by sterols, and that over-expression of insig-1 sensitizes cells to the sterol-mediated suppression of SREBP proteolysis. These results strongly support the identification of insig as the SCAP retention factor.

Yabe et al. [70] then identified another ER retention protein for the SREBP–SCAP complex, insig-2, on the basis of its sequence identity with insig-1. Two differences between the insig proteins are that (1) while the insig-1 gene is a transcriptional target of SREBP-2, the insig-2 gene is not, and (2) insig-1 will, when expressed at sufficiently high levels, bind to SCAP even in the absence of sterols, whereas insig-2 will bind to SCAP only in the presence of sterols [66,70]. These characteristics mean that the insig proteins add additional levels of regulatory feedback complexity to the proteolytic cleavage of SREBPs.

Interestingly, as touched on above, the three point mutations known to render SCAP insensitive to sterols, and therefore cause the constitutive proteolysis of SREBPs, also block interaction between SCAP and insig proteins [61]. An antibody against insig protein efficiently precipitates SREBP in the presence of wild-type SCAP, but in the presence of mutant SCAP (Y298C, L315F or D443N), SREBP is not precipitated with insig [61]. These results further indicate the crucial role that the SCAP–insig interaction plays in the sterol-regulated proteolytic processing of SREBPs.

The insigs are hydrophobic, integral membrane proteins [66,70]. However, they lack any obvious ER retention signal, raising the question of how the insig proteins are themselves retained within the ER. It seems likely that other, as-yet-undiscovered, proteins may also be involved in the complex protein machinery that controls lipid metabolism in animal cells.

Conclusions

The SREBP pathway exemplifies a classic biochemical feedback mechanism. It also provides a detailed example of the way a newly appreciated strategy, Rip, can play a crucial role in vital signal transduction pathways. By controlling the movement of a substrate (SREBP) to active proteases (S1P and S2P), sterols regulate their own biosynthesis and uptake via transcriptional activation of the relevant genes. As the details of SREBP proteolysis have been worked out, similar proteolytic signalling processes have been identified in other systems, ranging from development (e.g. the Notch/γ-secretase and star/spitz/rhomboid systems) to neurodegeneration (e.g. Alzheimer's disease.) Additional examples are now known from bacterial systems (eep, SpoIV FB and YaeL) and from signal peptide processing (SPP). This list of examples is likely to continue to grow as additional examples of Rip are identified.

We now know of examples of intramembrane cleaving proteases that appear to be metalloproteases (S2P), aspartyl proteases (γ-secretase/presenilin, SPP) and serine proteases (rhomboid). In closing, an intriguing question comes to mind: are there any intramembrane cysteine proteases?

R.B.R. is supported by grants from the National Institutes of Health (HL20948) and the American Heart Association (0130010N).

References

1. Brown, M.S., Ye, J., Rawson, R.B. and Goldstein, J.L. (2000) Cell **100**, 391–398
2. Kopan, R. and Goate, A. (2002) Neuron **33**, 321–324
3. Wolfe, M.S., Xia, W., Ostaszewski, B.L., Diehl, T.S., Kimberly, W.T. and Selkoe, D.J. (1999) Nature (London) **398**, 513–517
4. Ni, C.Y., Murphy, M.P., Golde, T.E. and Carpenter, G. (2001) Science **294**, 2179–2181
5. Selkoe, D.J. (1997) Science **275**, 630–631
6. Cao, X. and Sudhof, T.C. (2001) Science **293**, 115–120
7. Biederer, T., Cao, X., Sudhof, T.C. and Liu, X. (2002) J. Neurosci. **22**, 7340–7351

8. Cupers, P., Orlans, I., Craessaerts, K., Annaert, W. and De Strooper, B. (2001) J. Neurochem. **78**, 1168–1178

9. Horton, J.D., Goldstein, J.L. and Brown, M.S. (2002) J. Clin. Invest. **109**, 1125–1131

10. Haze, K., Yoshida, H., Yanagi, H., Yura, T. and Mori, K. (1999) Mol. Biol. Cell **10**, 3787–3799

11. Haze, K., Okada, T., Yoshida, H., Yanagi, H., Yura, T., Negishi, M. and Mori, K. (2001) Biochem. J. **355**, 19–28

12. Seegmiller, A.C., Dobrosotskaya, I., Goldstein, J.L., Ho, Y.K., Brown, M.S. and Rawson, R.B. (2002) Dev. Cell **2**, 229–238

13. Hu, Y., Ye, Y. and Fortini, M.E. (2002) Dev. Cell **2**, 69–78

14. Struhl, G. and Greenwald, I. (2001) Proc. Natl. Acad. Sci. U.S.A. **98**, 229–234

15. Levitan, D. and Greenwald, I. (1995) Nature (London) **377**, 351–354

16. Greenwald, I. (1998) Genes Dev. **12**, 1751–1762

17. Rudner, D.Z., Fawcett, P. and Losick, R. (1999) Proc. Natl. Acad. Sci. U.S.A. **96**, 14765–14770

18. Alba, B.M., Leeds, J.A., Onufryk, C., Lu, C.Z. and Gross, C.A. (2002) Genes Dev. **16**, 2156–2168

19. Kanehara, K., Ito, K. and Akiyama, Y. (2002) Genes Dev. **16**, 2147–2155

20. Weihofen, A., Binns, K., Lemberg, M.K., Ashman, K. and Martoglio, B. (2002) Science **296**, 2215–2218

21. Kimberly, W.T., Xia, W., Rahmati, T., Wolfe, M.S. and Selkoe, D.J. (2000) J. Biol. Chem. **275**, 3173–3178

22. Berezovska, O., Jack, C., McLean, P., Aster, J.C., Hicks, C., Xia, W., Wolfe, M.S., Kimberly, W.T., Weinmaster, G., Selkoe, D.J. and Hyman, B.T. (2000) J. Neurochem. **75**, 583–593

23. Lemberg, M.K., Bland, F.A., Weihofen, A., Braud, V.M. and Martoglio, B. (2001) J Immunol. **167**, 6441–6446

24. Weihofen, A., Lemberg, M.K., Ploegh, H.L., Bogyo, M. and Martoglio, B. (2000) J. Biol. Chem. **275**, 30951–30956

25. Lee, J.R., Urban, S., Garvey, C.F. and Freeman, M. (2001) Cell **107**, 161–171

26. Urban, S., Lee, J.R. and Freeman, M. (2001) Cell **107**, 173–182

27. Urban, S., Schlieper, D. and Freeman, M. (2002) Curr. Biol. **12**, 1507–1512

28. Gallio, M., Sturgill, G., Rather, P. and Kylsten, P. (2002) Proc. Natl. Acad. Sci. U.S.A. **99**, 12208–12213

29. Mumm, J.S. and Kopan, R. (2000) Dev. Biol. **228**, 151–165

30. Esler, W.P. and Wolfe, M.S. (2001) Science **293**, 1449–1454

31. Urban, S. and Freeman, M. (2002) Curr. Opin. Genet. Dev. **12**, 512–518

32. DeBose-Boyd, R.A., Brown, M.S., Li, W.P., Nohturfft, A., Goldstein, J.L. and Espenshade, P.J. (1999) Cell **99**, 703–712

33. Nohturfft, A., Yabe, D., Goldstein, J.L., Brown, M.S. and Espenshade, P.J. (2000) Cell **102**, 315–323

34. Espenshade, P.J., Li, W.P. and Yabe, D. (2002) Proc. Natl. Acad. Sci. U.S.A. **99**, 11694–11699

35. Espenshade, P.J., Cheng, D., Goldstein, J.L. and Brown, M.S. (1999) J. Biol. Chem. **274**, 22795–22804

36. Toure, B.B., Munzer, J.S., Basak, A., Benjannet, S., Rochemont, J., Lazure, C., Chretien, M. and Seidah, N.G. (2000) J. Biol. Chem. **275**, 2349–2358

37. Duncan, E.A., Brown, M.S., Goldstein, J.L. and Sakai, J. (1997) J. Biol. Chem. **272**, 12778–12785

38. Cheng, D., Espenshade, P.J., Slaughter, C.A., Jaen, J.C., Brown, M.S. and Goldstein, J.L. (1999) J. Biol. Chem. **274**, 22805–22812

39. Rawson, R.B., Cheng, D., Brown, M.S. and Goldstein, J.L. (1998) J. Biol. Chem. **273**, 28261–28269

40. Sakai, J., Rawson, R.B., Espenshade, P.J., Cheng, D., Seegmiller, A.C., Goldstein, J.L. and Brown, M.S. (1998) Mol. Cell **2**, 505–514

41. Ye, J., Rawson, R.B., Komuro, R., Chen, X., Dave, U.P., Prywes, R., Brown, M.S. and Goldstein, J.L. (2000) Mol. Cell **6**, 1355–1364

42. Seidah, N.G., Mowla, S.J., Hamelin, J., Mamarbachi, A.M., Benjannet, S., Toure, B.B., Basak, A., Munzer, J.S., Marcinkiewicz, J., Zhong, M. et al. (1999) Proc. Natl. Acad. Sci. U.S.A. **96**, 1321–1326

43. Elagoz, A., Benjannet, S., Mammarbassi, A., Wickham, L. and Seidah, N.G. (2002) J. Biol. Chem. **277**, 11265–11275

44. Lenz, O., ter Meulen, J., Klenk, H.D., Seidah, N.G. and Garten, W. (2001) Proc. Natl. Acad. Sci. U.S.A. **98**, 12701–12705

45. Sanchez, A.J., Vincent, M.J. and Nichol, S.T. (2002) J. Virol. **76**, 7263–7275

46. Yang, J., Goldstein, J.L., Hammer, R.E., Moon, Y.A., Brown, M.S. and Horton, J.D. (2001) Proc. Natl. Acad. Sci. U.S.A. **98**, 13607–13612

47. Rawlings, N.D. and Barrett, A.J. (1995) Methods Enzymol. **248**, 183–228

48. Rawson, R.B., Zelenski, N.G., Nijhawan, D., Ye, J., Sakai, J., Hasan, M.T., Chang, T.Y., Brown, M.S. and Goldstein, J.L. (1997) Mol. Cell **1**, 47–57

49. Zelenski, N.G., Rawson, R.B., Brown, M.S. and Goldstein, J.L. (1999) J. Biol. Chem. **274**, 21973–21980

50. Sakai, J., Duncan, E.A., Rawson, R.B., Hua, X., Brown, M.S. and Goldstein, J.L. (1996) Cell **85**, 1037–1046

51. Lewis, A.P. and Thomas, P.J. (1999) Protein Sci. **8**, 439–442

52. Shen, J., Chen, X., Hendershot, L. and Prywes, R. (2002) Dev. Cell **3**, 99–111

53. Gil, G., Faust, J.R., Chin, D.J., Goldstein, J.L. and Brown, M.S. (1985) Cell **41**, 249–258

54. Carstea, E.D., Morris, J.A., Coleman, K.G., Loftus, S.K., Zhang, D., Cummings, C., Gu, J., Rosenfeld, M.A., Pavan, W.J., Krizman, D.B. et al. (1997) Science **277**, 228–231

55. Loftus, S.K., Morris, J.A., Carstea, E.D., Gu, J.Z., Cummings, C., Brown, A., Ellison, J., Ohno, K., Rosenfeld, M.A., Tagle, D.A. et al. (1997) Science **277**, 232–235

56. Porter, J.A., Young, K.E. and Beachy, P.A. (1996) Science **274**, 255–259

57. Goldstein, J.L., Rawson, R.B. and Brown, M.S. (2002) Arch. Biochem. Biophys. **397**, 139–148

58. Kandutsch, A.A. and Chen, H.W. (1975) J. Cell. Physiol. **85**, 415–424

59. Hua, X., Nohturfft, A., Goldstein, J.L. and Brown, M.S. (1996) Cell **87**, 415–426

60. Nohturfft, A., Brown, M.S. and Goldstein, J.L. (1998) Proc. Natl. Acad. Sci. U.S.A. **95**, 12848–12853

61. Yabe, D., Xia, X.-P., Adames, C. M. and Rawson, R.B. (2003) Proc. Natl. Acad. Sci. U.S.A., in the press

62. Brown, A., Sun, L., Feramisco, J., Brown, M. and Goldstein, J. (2002) Mol. Cell **10**, 237–245

63. Rawson, R.B., DeBose-Boyd, R., Goldstein, J.L. and Brown, M.S. (1999) J. Biol. Chem. **274**, 28549–28556

64. Matsuda, M., Korn, B.S., Hammer, R.E., Moon, Y.A., Komuro, R., Horton, J.D., Goldstein, J.L., Brown, M.S. and Shimomura, I. (2001) Genes Dev. **15**, 1206–1216

65. Korn, B.S., Shimomura, I., Bashmakov, Y., Hammer, R.E., Horton, J.D., Goldstein, J.L. and Brown, M.S. (1998) J. Clin. Invest. **102**, 2050–2060

66. Yang, T., Espenshade, P., Wright, M., Yabe, D., Gong, Y., Aebersold, R., Goldstein, J. and Brown, M. (2002) Cell **110**, 489–500

67. Mohn, K.L., Laz, T.M., Hsu, J.C., Melby, A.E., Bravo, R. and Taub, R. (1991) Mol. Cell. Biol. **11**, 381–390

68. Diamond, R.H., Du, K., Lee, V.M., Mohn, K.L., Haber, B.A., Tewari, D.S. and Taub, R. (1993) J. Biol. Chem. **268**, 15185–15192

69. Yang, T., Goldstein, J.L. and Brown, M.S. (2000) J. Biol. Chem. **275**, 29881–29886

70. Yabe, D., Brown, M.S. and Goldstein, J.L. (2002) Proc. Natl. Acad. Sci. U.S.A. **99**, 12753–12758

71. Hua, X., Sakai, J., Ho, Y.K., Goldstein, J.L. and Brown, M.S. (1995) J. Biol. Chem. **270**, 29422–29427

72. Nohturfft, A., Brown, M.S. and Goldstein, J.L. (1998) J. Biol. Chem. **273**, 17243–17250

Biochem. Soc. Symp. **70**, 233–242
(Printed in Great Britain)
© 2003 Biochemical Society

19

Caspase activation

Kelly M. Boatright*† and Guy S. Salvesen*†[1]

*The Program in Apoptosis and Cell Death, The Burnham Institute, 10901 North Torrey Pines Road, La Jolla, CA 92037, U.S.A., and †Department of Molecular Pathology, University of California San Diego, La Jolla, CA 92037, U.S.A.

Abstract

Caspase activation is the 'point of no return' commitment to cell death. Synthesized as inactive zymogens, it is essential that the caspases remain inactive until the death signal is received. It is known for the downstream executioner caspases-3 and -7 that the activation event is proteolytic cleavage, and this had been assumed to apply to the initiator caspases as well. However, recent studies conducted on caspases-2, -8 and -9 have challenged this tenet of caspase activation. In this review we focus on the molecular details of caspase activation, with emphasis on recent work that provides a pleasing explanation for the differential requirements for the activation of executioner and initiator caspases.

Introduction

Apoptosis, the major form of programmed cell death, is responsible for the elimination of cells during development, during immune system education and in response to cellular damage. This is an orderly form of cell death, resulting in the packaging of cellular contents that can be disposed of neatly by the organism. This death pathway is of great advantage to the organism, as it prevents inflammatory and autoimmune responses to the leakage of cellular contents. Its misregulation is linked to multiple pathologies, including neurodegenerative diseases and cancer, and it has been estimated that loss of the cellular control of apoptosis is a factor in as many as 70% of known diseases [1]. Apoptotic cell death is carried out through the action of specific members of the caspase family, which are <u>c</u>ysteine-dependent <u>asp</u>artate-specific pro<u>teases</u> [2]. Humans have 11 caspases: caspases-2, -3, -6, -7, -8, -9 and -10 are involved in apoptosis, caspases-1 and -5 (and possibly caspase-4) are involved in the inflammatory response, and caspase-14 is possibly involved in skin differentiation (at least in mice; reviewed in [3]). The apoptotic caspases are subdivided into two groups on the basis of their point of entry into the death

[1]To whom correspondence should be addressed (e-mail gsalvesen@burnham.org).

pathway. The 'apical' (or 'initiator') caspases are the first to be activated by a particular death pathway, and include caspases-8, -9 and -10. Recent work provides a strong argument that caspase-2 belongs to this category as well [4]. Once activated, the apical caspases then activate the downstream 'executioner' caspases, which are responsible for the cleavage of the cellular proteins, resulting in cell death. This group includes caspases-3, -6 and -7. The execution phase of apoptosis is under the control of members of the IAP (inhibitor of apoptosis protein) family, which act as reversible inhibitors of caspase activity [5].

The caspases are synthesized as inactive zymogens, known as pro-caspases, and it is essential for the survival of the cell that they remain inactive until the proper signal is received. Pro-caspases contain N-terminal extensions that play various roles in activation, and in some instances are removed during maturation. They also possess a conserved aspartic acid at a site termed the 'linker region'. Cleavage at this site is mediated either autocatalytically or by another caspase during the maturation process of most caspases. This cleavage releases subdomains, termed the large subunit and the small subunit, which remain associated within the active protease. In all cases known so far, active caspases are dimers of catalytic domains, and since each domain is composed of a large and a small subunit, they are sometimes called tetramers. However, in this review, for simplicity we shall refer to a monomer as a single catalytic domain, and a dimer as the characteristic form seen in three-dimensional structures of the active protein (Figure 1).

This review will focus on the mechanisms responsible for the activation of the caspases. It is well established that proteolytic processing within their linker region by one of the apical caspases activates the executioner caspases. However, in the death cascade there are no proteases that act on the apical caspases, so Nature has evolved an alternative mechanism by which to activate these proteases: zymogen clustering mediated by multiprotein complexes.

Keys in the ignition: activation of the apical caspases

The nature of the death signal determines which of the apical caspases are activated. On the one hand, the extrinsic death pathway is triggered by ligation of a death receptor on the cell surface. The initiators of this pathway are caspase-8 and its close paralogue caspase-10 in humans. Note that mice do not contain caspase-10. On the other hand, the intrinsic death pathway is key in the orderly elimination of damaged cells, and leads to the activation of apical caspase-9. Additionally, there is now evidence for a role for caspase-2 as an apical caspase in certain forms of stress-induced apoptosis [4,6]. Regardless of the origin of the apoptotic signals, the intrinsic and the extrinsic pathways converge with the activation of the executioner caspases-3, -6 and -7 (Figure 2).

Intrinsic pathway (caspases-2 and -9)

The intrinsic pathway is activated in response to cellular stress initiated by internal death signals, such as DNA damage, treatment with cytotoxic drugs, UV irradiation and growth factor withdrawal. The initiator of this pathway is caspase-9, although there is emerging evidence that caspase-2 may be the

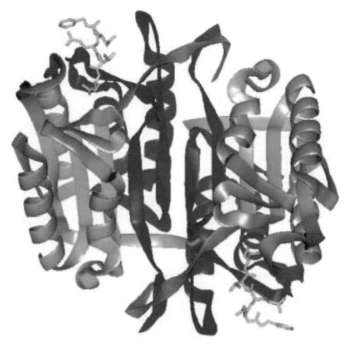

Figure 1 Three-dimensional structure of a caspase dimer. This ribbon model of inhibitor-bound caspase-8 (PDB ID code 1QDU) demonstrates the organization of an active caspase. Note that the caspase is a dimer composed of two catalytic units, with each unit consisting of a large subunit (grey) and a small subunit (blue). A tripeptide ketone inhibitor (orange) is shown bound to the two independent active sites.

apical caspase for certain forms of stress-induced apoptosis. Caspase-9 exerts its function as an apical caspase by activating the executioner caspases via proteolytic cleavage within their linker region. Caspase-2 is not thought to be capable of activating executioner caspases directly. Rather, it seems to activate the intrinsic death pathway by somehow inducing mitochondrial permeabilization, leading to caspase-9 activation [7]. Therefore it may be classified as an initiator, if not as a classical apical caspase.

Caspases-2 and -9 possess an N-terminal caspase recruitment domain (CARD). This domain is responsible for their recruitment to their respective activating complexes during activation of the intrinsic pathway. The activator complex for caspase-2 is unknown, but that for caspase-9 has been well described at the biochemical and low-resolution structural level. This complex is known as the 'apoptosome' [8], and has at its heart the protein Apaf-1 (apoptotic protease activating factor-1). In the presence of cytochrome *c* and ATP, Apaf-1 oligomerizes to a higher structure, of the order of seven protomers [9], capable of recruiting and activating pro-caspase-9. The requirement for cytochrome *c* implicates mitochondrial protein flux as a triggering event, and this may be mediated by members of the Bcl-2 family acting as integrators of the intrinsic pathway [10,11]. Although it is clear that the CARD of Apaf-1

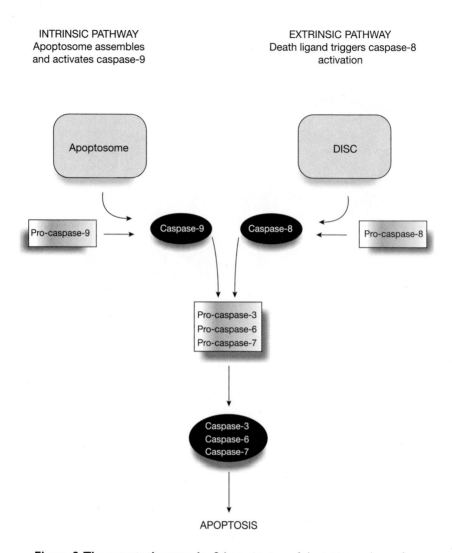

Figure 2 The apoptotic cascade. Schematic view of the initiator phase of caspase activation induced through either the intrinsic (caspase-9) or the extrinsic (caspase-8) death pathway, and culminating in the activation of the executioner caspases-3, -6 and -7. Caspase-10, a close orthologue of caspase-8 in humans, may also contribute to initiation of the extrinsic pathway.

recruits caspase-9 via its own CARD, the pathway to proteolytic activation within the apoptosome is still unknown, although recent structural and bio-chemical evidence has now shed light on the event.

Pro-caspase-9 exists as a monomer, and the simple act of dimerization can trigger its proteolytic activity [12]. The structure of the monomeric, zymogen, form of caspase-9 has not yet been determined, but the crystal structure of inhibitor-bound active caspase-9 lacking the CARD (ΔCARD) reveals a homodimeric enzyme in which only one of the catalytic units has a competent active site. The second unit of the dimer is in a zymogen-like conformation in which the all-important active-site cysteine residue is distorted from its catalytic position, the substrate-binding sites are disrupted, and the oxyanion hole is annihilated. Interestingly, the transition from this distortion of the catalytic apparatus to the active form can occur by relatively simple loop rearrangements without any perturbation of secondary structure.

During activation of the intrinsic pathway, caspase-9 is cleaved in its linker region between the large and small subunits. Significantly, this is neither sufficient nor required for caspase-9 activation [9,13,14]. Although cleavage of caspase-9 in the apoptosome may enhance its activity [15], it is not the fundamental activating event. Indeed, cleavage of pro-caspase-9 only increases its activity by 10-fold, whereas binding to cytosolic factors in apoptotic extracts increases its activity by 2000-fold [14]. Rather, cleavage of caspase-9 is thought to influence its ability to be inhibited by XIAP (X-linked inhibitor of apoptosis) [16], as discussed towards the end of this review.

Several interesting possibilities for caspase-9 activation await further investigation. One is that monomeric caspase-9 zymogens are activated due to dimerization when binding at the apoptosome increases their local concentration. This could be due simply to the conversion of a bimolecular reaction (weak monomer/dimer equilibrium of pro-caspase-9 monomers) into a unimolecular reaction within the confines of the apoptosome, with a tremendous entropic advantage of forming the active, dimeric, conformation. Also plausible is the idea that, somehow, binding at the apoptosome activates caspase-9 by altering its conformation, in a manner independent of dimerization. Hopefully, future studies of zymogen caspase-9, as well as caspase-9 complexed within the apoptosome, will provide answers to these questions.

Recent work by Read and colleagues [17] suggests that monomeric zymogen latency also plays a role in the mechanism of activation of caspase-2. They showed that pro-caspase-2 exists within the cell as a monomer, and is activated by the formation of a high-molecular-mass complex in a manner reminiscent of the activation of caspase-9 in the apoptosome. Reports that caspase-3 cleaves caspase-2 in cytochrome c/dATP-treated cytosolic extracts had placed caspase-2 as a downstream executioner caspase [18,19]. However, probing for the active form of caspase-2 in apoptotic cytosolic extracts revealed that it consists of both cleaved and uncleaved material [17]. Therefore it appears as though monomeric zymogen clustering at an activation complex, and not cleavage, is responsible for activation of caspase-2. Future work in this area will surely yield exciting new insight into caspase-2 activation within this complex.

Extrinsic pathway (caspases-8 and -10)

The extrinsic death pathway is used for the deletion of unwanted cells during immune system development, and is implicated in a number of leukaemic pathologies. Once activated, caspases-8 and -10 in turn activate the executioner caspases by proteolytic cleavage within their linker region. Caspase-8 may also engage the intrinsic pathway by cleaving the pro-apoptotic Bcl-2 family member Bid, which then permeabilizes the mitochondria [20]. It is generally accepted that caspases-8 and -10 work co-operatively in the initiation of the extrinsic death pathway. However, caspase-10 is unable to initiate cell death in some cell lines in which caspase-8 is mutated, suggesting that it cannot functionally substitute for caspase-8 [21].

Caspases-8 and -10 each possess two tandem death effector domains (DEDs) at their N-termini. These domains are important for recruitment of the caspases to the death-inducing signalling complex (DISC) that is formed following death receptor ligation in an assembly bridged by the adaptor molecule FADD (Fas-associated protein with death domain). DISC formation is triggered by ligation of a member of the tumour necrosis factor receptor family of type I transmembrane proteins. Upon ligation, the death receptor oligomerizes, signalling for the adaptor molecule FADD to bind to the cytosolic tail of the receptor via a homophilic interaction with the receptor's death domain. The DED of FADD then recruits pro-caspases-8 and -10 to the DISC via homophilic interactions with their N-terminal DEDs. Caspase-8 (and caspase-10) is then activated by zymogen clustering at the cytosolic face of the DISC. Thus the death receptors act as a conduit for the transmission of external signals to the cell's death machinery through activation of the apical caspases-8 and -10.

The mode of activation of caspase-8 within the DISC has been termed 'induced proximity', to signify that close apposition of pro-caspase-8 molecules is required to drive the zymogen→active transition [22,23]. During activation of caspase-8 at the DISC, a cleavage event occurs in the linker region of the caspase between the large and small subunits [21,24]. At least for caspase-8, this cleavage event appears to occur *in trans*. The second cleavage removes the DEDs, which remain associated with the DISC, releasing the activated caspase into the cytosol. Initially it was thought that the cleavage was essential for the generation of catalytic activity, but recent reports have demonstrated that this is not the case, and that the mode of activation is essentially identical to that of caspase-9. Like pro-caspase-9, pro-caspase-8 is a monomer, and the active form is a dimer [25,26]. The transition to the active forms of both caspases-8 and -9 can be induced by kosmotropic salts, which have the ability to order protein structures [25]. In the case of these two caspases, activation probably occurs by driving the weak monomer/dimer equilibrium towards the active dimeric species. The cleavage event that occurs between the large and small subunits of the catalytic domain during activation at best increases the stability of the dimer, and the fundamental activating event is dimerization.

An interesting facet of caspase-8 activation is the role of c-FLIP$_L$ (flice-like inhibitory protein). c-FLIP$_L$ is a caspase-8 homologue, but it lacks the key catalytic cysteine, rendering it proteolytically inactive. Initial reports using overexpression of c-FLIP$_L$ in mammalian cells demonstrated that c-FLIP$_L$ possessed both pro- and anti-apoptotic activities (reviewed in [27]). Work by Chang and colleagues [28] reconciled these contradictory findings by carefully analysing the effects of FLIP transfection levels. They found that physiologically relevant levels of expression of c-FLIP$_L$ lead to enhanced caspase-8 processing at the DISC, while at high expression levels c-FLIP$_L$ acts as an inhibitor of apoptosis, presumably by blocking caspase-8-binding sites at the DISC. Recent studies have demonstrated that c-FLIP$_L$ and caspase-8 are capable of forming heterodimers that possess catalytic activity [29]. An attractive hypothesis, therefore, is that c-FLIP$_L$ activates caspase-8 by providing a dimerization interface, rendering the caspase-8 unit of the heterodimer catalytically competent. Future studies in this area will no doubt provide exciting mechanistic details.

Foot on the accelerator: the executioner caspases

Much of our current understanding of executioner caspase zymogen activation comes from studies on caspase-7, which has been crystallized as a zymogen, an active protease, and an inhibitor-bound protease. Each of these structures fortuitously shares the same space grouping (P3$_2$21), allowing for comparison between them [30–32]. Together, the structures provide a molecular explanation for executioner caspase zymogen latency.

In stark contrast with the apical caspases, the zymogens of executioner caspases-3, -6 and -7 are homodimers of two single chains [25,33,34]. Crystal structures of active caspases reveal a common fold, with each catalytic unit containing six β-sheets and five α-helices. From these core structural elements emanates a four-loop bundle that comprises the active site and the substrate-binding cleft. The core structural elements of pro-caspase-7 are nearly identical to those of inhibitor-bound caspase-7. The main difference between these two conformations encompasses the active site and substrate-binding loops, which in pro-caspase-7 are in a conformation almost identical to that seen in the catalytically incompetent domain in the structure of caspase-9 [12].

The crystal structure of active, unbound caspase-7 unveils an additional aspect of executioner caspase activation [30]. In this configuration, two loops have moved to a conformation similar to that seen in the structure of inhibitor-bound caspase-7. However, a third loop containing aspects of the catalytic machinery is misaligned. Thus it appears that, although cleavage in the linker region readies the executioner caspase for substrate binding, it is the actual act of substrate binding that induces adoption of the active conformation. This is an example of substrate-induced activation that is not seen in caspase-9 [12] or in the distant caspase homologue Arg-gingipain [35].

That cleavage in the linker region activates executioner pro-caspases is well accepted, and is supported by a deluge of experimental evidence. For example, the activity of pro-caspase-3 is increased >10000-fold by cleavage [36]. However, one aspect of executioner caspase activation is not so clear: the role of the N-terminal extension, or propeptide. No electron density was found for the propeptide in the crystal structure of pro-caspase-7, suggesting that this domain is flexible. Studies performed using recombinant material demonstrated that the propeptides of caspases-3 and -7 had no influence on either their intrinsic activity or their ability to be activated *in vitro* [36]. However, for endogenous caspases-3 and -7, it has been shown that removal of their prodomains is a requirement for maximum activity [37,38]. Therefore the roles of these prodomains in the activation of executioner caspases is still under investigation.

Summary

Although the mechanisms of activation of the apical caspases (dimerization) and the executioner caspases (cleavage) at first glance seem dissimilar, there are conserved elements underlying the acquisition of catalytic activity. Both types of caspases require the translocation of an activation loop, carrying the substrate-binding machinery, from one dimer into an acceptor pocket on the partner dimer [12,30,31]. It is simply a matter of how this transition from the latent to the active form is orchestrated. In the executioner caspases, the preformed dimer contains a blocking segment that must be removed proteolytically, but in the apical caspases the dimer must form first. The issue of why proteolysis is not required for activation of the apical caspases may simply be a matter of the length and composition of the inter-chain connector, which is minimal in executioner caspases-3 and -7, but longer in the apical caspases – possibly long enough to avoid blocking translocation of the activation loop.

To achieve dimerization of apical caspases, the cell has evolved activation complexes where zymogen clustering can occur. For caspase-9 this complex is the apoptosome, and for caspases-8 and -10 it is the DISC. The activating complex for caspase-2 has yet to be purified and dissected, although future work in this area will hopefully identify its components.

Finally, there are numerous examples in the literature of investigators assuming that cleavage of an apical caspase is tantamount to its activation. However, for apical caspases-8 and -9 this has been shown not to be the case [13,14,25,26], and recent work on caspase-2 suggests that cleavage is not required for activation of this protease either. This has led to convoluted models of death pathways implicating the involvement of apical caspases when it has not been established that this cleavage event is anything other than the simple processing of cellular proteins during the death of the cell. Therefore it is important that efforts be made to distinguish 'activation' from 'cleavage', especially when dealing with caspases-2, -8 and -9, for which it is now known that cleavage is not responsible for activation. This can be accomplished by using commercially available affinity probes that will bind only to activated caspases [17,25].

Finally, the mechanisms employed by caspases to retain their inactive forms, and the pathways to their activation, are now becoming clear. However, of equal importance is the regulation of their active forms by endogenous inhibitors – but that is another story [39,40].

We thank the members of the Salvesen laboratory who have worked on the caspase activation project. This work was supported by NIH grants CA69381 and HL51399, and the California Breast Cancer Research Program Fellowship 8GB-0137.

References

1.   Reed, J.C. (1999) Curr. Opin. Oncol. 11, 68–75
2.   Alnemri, E.S., Livingston, D.J., Nicholson, D.W., Salvesen, G., Thornberry, N.A., Wong, W.W. and Yuan, J. (1996) Cell 87, 171
3.   Lamkanfi, M., Declercq, W., Kalai, M., Saelens, X. and Vandenabeele, P. (2002) Cell Death Differ. 9, 358–361

4. Lassus, P., Opitz-Araya, X. and Lazebnik, Y. (2002) Science **297**, 1352–1354
5. Salvesen, G.S. and Duckett, C.S. (2002) Nat. Rev. Mol. Cell Biol. **3**, 401–410
6. Kumar, S. and Vaux, D.L. (2002) Science **297**, 1290–1291
7. Guo, Y., Srinivasula, S.M., Druilhe, A., Fernandes-Alnemri, T. and Alnemri, E.S. (2002) J. Biol. Chem. **277**, 13430–13437
8. Zou, H., Li, Y., Liu, X. and Wang, X. (1999) J. Biol. Chem. **274**, 11549–11556
9. Acehan, D., Jiang, X., Morgan, D.G., Heuser, J.E., Wang, X. and Akey, C.W. (2002) Mol. Cell **9**, 423–432
10. Green, D.R. and Reed, J.C. (1998) Science **281**, 1309–1312
11. Korsmeyer, S.J., Wei, M.C., Saito, M., Weiler, S., Oh, K.J. and Schlesinger, P.H. (2000) Cell Death Differ. **7**, 1166–1173
12. Renatus, M., Stennicke, H.R., Scott, F.L., Liddington, R.C. and Salvesen, G.S. (2001) Proc. Natl. Acad. Sci. U.S.A. **98**, 14250–14255
13. Rodriguez, J. and Lazebnik, Y. (1999) Genes Dev. **13**, 3179–3184
14. Stennicke, H.R., Deveraux, Q.L., Humke, E.W., Reed, J.C., Dixit, V.M. and Salvesen, G.S. (1999) J. Biol. Chem. **274**, 8359–8362
15. Zou, H., Yang, R., Hao, J., Wang, J., Sun, C., Fesik, S.W., Wu, J.C., Tomaselli, K.J. and Armstrong, R.C. (2003) J. Biol. Chem. **278**, 8091–8098
16. Srinivasula, S.M., Hegde, R., Saleh, A., Datta, P., Shiozaki, E., Chai, J., Lee, R.A., Robbins, P.D., Fernandes-Alnemri, T., Shi, Y. and Alnemri, E.S. (2001) Nature (London) **410**, 112–116
17. Read, S.H., Baliga, B.C., Ekert, P.G., Vaux, D.L. and Kumar, S. (2002) J. Cell Biol. **159**, 739–745
18. Slee, E.A., Harte, M.T., Kluck, R.M., Wolf, B.B., Casiano, C.A., Newmeyer, D.D., Wang, H.G., Reed, J.C., Nicholson, D.W., Alnemri, E.S. et al. (1999) J. Cell Biol. **144**, 281–292
19. O'Reilly, L.A., Ekert, P., Harvey, N., Marsden, V., Cullen, L., Vaux, D.L., Hacker, G., Magnusson, C., Pakusch, M., Cecconi, F. et al. (2002) Cell Death Differ. **9**, 832–841
20. Luo, X., Budihardjo, I., Zou, H., Slaughter, C. and Wang, X. (1998) Cell **94**, 481–490
21. Sprick, M.R., Rieser, E., Stahl, H., Grosse-Wilde, A., Weigand, M.A. and Walczak, H. (2002) EMBO J. **21**, 4520–4530
22. Muzio, M., Stockwell, B.R., Stennicke, H.R., Salvesen, G.S. and Dixit, V.M. (1998) J. Biol. Chem. **273**, 2926–2930
23. Salvesen, G.S. and Dixit, V.M. (1999) Proc. Natl. Acad. Sci. U.S.A. **96**, 10964–10967
24. Scaffidi, C., Medema, J.P., Krammer, P.H. and Peter, M.E. (1997) J. Biol. Chem. **272**, 26953–26958
25. Boatright, K.M., Renatus, M., Scott, F.L., Sperandio, S., Shin, H., Pedersen, I.M., Ricci, J.E., Edris, W.A., Sutherlin, D.P., Green, D.R. and Salvesen, G.S. (2003) Mol. Cell **11**, 529–541
26. Donepudi, M., Sweeney, A.M., Briand, C. and Grutter, M.G. (2003) Mol. Cell **11**, 543–549
27. Krueger, A., Baumann, S., Krammer, P.H. and Kirchhoff, S. (2001) Mol. Cell. Biol. **21**, 8247–8254
28. Chang, D.W., Xing, Z., Pan, Y., Algeciras-Schimnich, A., Barnhart, B.C., Yaish-Ohad, S., Peter, M.E. and Yang, X. (2002) EMBO J. **21**, 3704–3714
29. Micheau, O., Thome, M., Schneider, P., Holler, N., Tschopp, J., Nicholson, D.W., Briand, C. and Grutter, M.G. (2002) J. Biol. Chem. **277**, 45162–45171
30. Chai, J., Wu, Q., Shiozaki, E., Srinivasula, S.M., Alnemri, E.S. and Shi, Y. (2001) Cell **107**, 399–407
31. Riedl, S.J., Fuentes-Prior, P., Renatus, M., Kairies, N., Krapp, R., Huber, R., Salvesen, G.S. and Bode, W. (2001) Proc. Natl. Acad. Sci. U.S.A. **98**, 14790–14795
32. Wei, Y., Fox, T., Chambers, S.P., Sintchak, J., Coll, J.T., Golec, J.M., Swenson, L., Wilson, K.P. and Charifson, P.S. (2000) Chem. Biol. **7**, 423–432
33. Bose, K. and Clark, A.C. (2001) Biochemistry **40**, 14236–14242
34. Kang, B.H., Ko, E., Kwon, O.K. and Choi, K.Y. (2002) Biochem. J. **364**, 629–634

35. Eichinger, A., Beisel, H.G., Jacob, U., Huber, R., Medrano, F.J., Banbula, A., Potempa, J., Travis, J. and Bode, W. (1999) EMBO J. **18**, 5453–5462
36. Stennicke, H.R., Jurgensmeier, J.M., Shin, H., Deveraux, Q., Wolf, B.B., Yang, X., Zhou, Q., Ellerby, H.M., Ellerby, L.M., Bredesen, D. et al. (1998) J. Biol. Chem. **273**, 27084–27090
37. Yang, X., Stennicke, H.R., Wang, B., Green, D.R., Janicke, R.U., Srinivasan, A., Seth, P., Salvesen, G.S. and Froelich, C.J. (1998) J. Biol. Chem. **273**, 34278–34283
38. Meergans, T., Hildebrandt, A.K., Horak, D., Haenisch, C. and Wendel, A. (2000) Biochem. J. **349**, 135–140
39. Stennicke, H.R., Ryan, C.A. and Salvesen, G.S. (2002) Trends Biochem. Sci. **27**, 94–101
40. Fesik, S.W. and Shi, Y. (2001) Science **294**, 1477–1478

Biochem. Soc. Symp. **70**, 243–251
(Printed in Great Britain)
© 2003 Biochemical Society

20

Separase regulation during mitosis

Frank Uhlmann[1]

Lincoln's Inn Fields Laboratories, Cancer Research UK, 44 Lincoln's Inn Fields, London WC2A 3PX, U.K.

Abstract

The final, irreversible step in the duplication and distribution of genomes to daughter cells takes place when chromosomes split at the metaphase-to-anaphase transition. A protease of the CD clan, separase (C50 family), is the key regulator of this transition. During metaphase, cohesion between sister chromatids is maintained by a chromosomal protein complex, cohesin. Anaphase is triggered when separase cleaves the Scc1 subunit of cohesin at two specific recognition sequences. As a result of this cleavage, the cohesin complex is destroyed, allowing the spindle to pull sister chromatids into opposite halves of the cell. Because of the final and irreversible nature of Scc1 cleavage, this reaction is tightly controlled. Several independent mechanisms impose regulation on separase activity, as well as on the susceptibility of the cleavage target Scc1 to cleavage by separase. This chapter provides an overview of these multiple levels of regulation.

Introduction

The DNA that comprises eukaryotic genomes is packaged into chromosomes. These must be replicated accurately to produce exact copies during S-phase, and then distributed correctly during mitosis. Errors in distribution lead to cells with supernumerary or missing chromosomes. The resulting aneuploidy is associated with many cancers, and is a leading cause of human birth defects. It is crucial that the products of DNA replication, the sister chromatids, remain physically linked by sister chromatid cohesion after their synthesis. This facilitates the repair of DNA lesions by recombination using the sister chromatid as a template [1]. Sister cohesion is also fundamental to the bipolar alignment of chromosomes on the metaphase spindle, as it counteracts the pulling force of microtubules toward the spindle poles (reviewed in [2]). At

[1]e-mail frank.uhlmann@cancer.org.uk

the start of anaphase, cohesion is abolished abruptly by a tightly regulated proteolytic cascade that activates the CD clan protease, separase. Separase cleaves one of the subunits of the cohesin complex, the Scc1 subunit, thereby destroying the complex. This review focuses on the regulation of separase activity, and on how cells ensure that cohesin cleavage occurs at the right time and place, making possible the complete and accurate distribution of chromosomes.

Separase, the protease that cleaves cohesin

Separase, the protease responsible for cleaving cohesin at anaphase onset [3,4], is a protein that was genetically identified and implicated in the regulation of chromosome segregation some time ago. Separase homologues probably exist in all eukaryotes, and mutations in the separases in *Schizosaccharomyces pombe* (Cut1), *Saccharomyces cerevisiae* (Esp1) and *Aspergillus nidulans* (BimB) have been characterized [5–7]. All prevent chromosome segregation at anaphase without halting the continuation of the cell cycle. This leads to cells with re-replicated chromosomes and excess spindle pole bodies, explaining the original phenotypic description of *Extra Spindle Poles* (*esp1*) [8].

The primary defect in *esp1* mutant cells only became apparent following the discovery of cohesin [9]. During anaphase, two of cohesin's subunits, Scc1 and Scc3, suddenly disappear from the chromosomes of wild-type cells (Figure 1) [10,11]. In *esp1* mutant cells, however, these subunits fail to dissociate from chromosomes, and sister chromatids remain paired even after they should have separated [12]. This observation led to the hypothesis that separases are cohesin removal factors (Figure 1). Meanwhile, the separases in a number of other organisms had been identified by genome sequencing projects. Separases are generally large proteins of close to 200 kDa, and only a C-terminal domain seems to be conserved among them. This conserved 'separase domain' contains the signature motif for cysteine proteases of the CD clan, the superfamily of proteases that also includes the caspases. Separases have been assigned to family C50. Indeed, separases purified from both budding yeast and human cells possess proteolytic activity against Scc1 [4,13]. There are two specific cleavage sites within budding yeast Scc1 that display a characteristic consensus motif [3]. This motif has also aided the identification of the cleavage sites within Scc1 homologues in other species [9,14], the most important determinants of which are an arginine in the P1 position and a negatively charged amino acid in the P3 position. In addition, a specific Scc1-derived peptide inhibitor has been developed against yeast separase, based on caspase inhibitors containing an activated chloromethyl ketone or acyloxymethyl ketone that covalently binds to and inhibits the separase active-site cysteine [4].

Securins: cellular separase inhibitors

The best known regulators of separases are the securins. First discovered in both budding and fission yeast [15,16], securins have since also been characterized in metazoans ([17,18]; reviewed in [19]). They are functionally

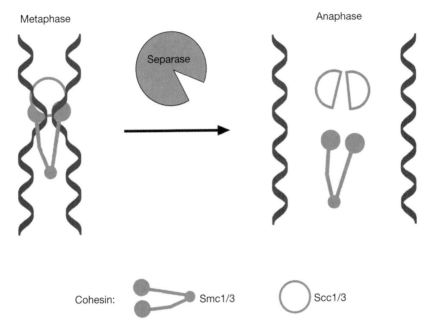

Metaphase

Anaphase

Separase

Cohesin: Smc1/3   Scc1/3

Figure 1 Model for how cleavage of cohesin by separase destroys sister chromatid cohesion at anaphase onset. Separase recognizes and cleaves two distinct sites in the Scc1 subunit. This destroys the interactions within the cohesin complex, leading to dissociation of the Scc1 and Scc3 subunits from chromosomes. Cohesion is lost, and the pulling force of the mitotic spindle segregates the sister chromatids towards opposite poles.

conserved proteins, although there is little conservation of their primary amino acid sequence. Securins bind to and inhibit separase for most of the cell cycle [3,12], but are degraded at the onset of anaphase, thus releasing separase (Figure 2). Their degradation is triggered via ubiquitylation by the anaphase-promoting complex (APC) [15,20]. Although securins are potent separase inhibitors, in budding yeast securin is not essential for cell cycle regulation of Scc1 cleavage, indicating that other control mechanisms exist (see below).

Yeast and human securins have recently been characterized as *bona fide* protease inhibitors for separase [21,22]. Securin prevents an intramolecular interaction between the N- and C-termini within separase. The protease active site resides in the separase C-terminus, and the interaction with the N-terminus may be required to induce an activating conformational change. Thus securin may prevent separase's own activation [21]. This also suggests a function for the large N-terminal extensions of separases in the regulated activation of the C-terminal protease domains. In addition, securin prevents binding of separase to its substrate Scc1 [21].

Securins are not simply inhibitors of separase. In fission yeast and *Drosophila*, the absence of securin does not lead to a prematurely active separase as one might predict, but, rather paradoxically, to an apparent lack of separase activity. This suggests a dual role for securins: the priming of separase

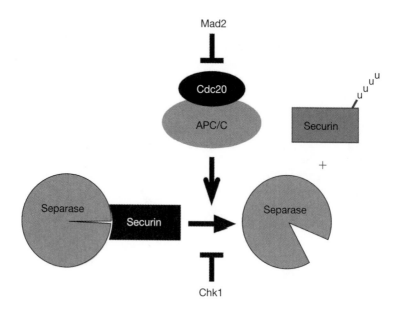

Figure 2 Separase regulation by securin. From late G1 until metaphase, securin binds to and inhibits separase. At anaphase, securin is targeted for destruction via ubiquitylation by the APC. The APC is activated by the regulatory subunit Cdc20/Fizzy. As long as chromosomes are not aligned properly on the mitotic spindle, Mad2 prevents activation of the APC. After DNA damage, the kinase Chk1 phosphorylates securin and prevents its destruction. Reproduced from [50], with permission.

activity via binding, and the inhibition of separase until securin's degradation by the APC [15,23]. Even in budding yeast where securin is not essential, separase function is impaired in its absence [12]. Securin acts to concentrate separase in the nucleus where Scc1 must be cleaved during anaphase [21,24]. Securin also enhances the specific proteolytic activity of separase after its own destruction [21,25], indicating that it may act as a molecular chaperone, helping separase to acquire its active conformation.

When human securin was identified, it was found to be the product of the *pituitary tumour-transforming* gene, which is overexpressed in certain tumours and exhibits transforming activity in NIH 3T3 cells [17]. Chromosome mis-segregation is thought to be a cause of the genetic instabilities found associated with many cancers [26], and overexpressed securin might lead to incomplete chromosome segregation due to inhibition of separase activity during anaphase.

Securin regulation via the APC

The APC is a protein ubiquitin ligase that is essential for mitotic progression. It controls the degradation of numerous proteins in addition to securin at this stage (reviewed in [27]). Whereas in yeast the APC exerts its effect on sister chromatid separation solely by targeting securin [12,28] (Figure 2), the activation of *Xenopus* and human separases may also require the APC-dependent degradation of mitotic cyclins [29]. The APC is activated at anaphase by the Cdc20/Fizzy protein, whose expression in turn is cell cycle regulated (see [27]). Cell cycle-dependent phosphorylation of APC subunits is also required for the activation of the APC complex [30,31].

APC activation is also the entry point for the Mad2-dependent checkpoint pathway that monitors the bipolar attachment of chromosomes to the mitotic spindle (Figure 2) (reviewed in [32]). Unattached kinetochores send a signal via Mad2 that keeps the APC inactive, potentially through the binding of Mad2 to the APC activator Cdc20/Fizzy. Indeed, the budding yeast securin, Pds1, had initially been identified as a protein required to prevent sister chromatid separation when the Mad2-dependent checkpoint pathway is activated [28]. In yeast, this pathway for regulating securin destruction only becomes essential once actual damage to spindle kinetochore attachment occurs. In contrast, in higher eukaryotes it acts during each cell cycle to ensure timely sister chromatid separation [33–35].

Controlling the onset of anaphase based on the state of chromosome attachment to the mitotic spindle seems to be of greatest importance. But it is not the only control. If DNA is damaged, anaphase onset is delayed to allow repair before sister sequences are separated from each other. In budding yeast, DNA damage elicits a response pathway that uses two routes that act together to prevent anaphase [36–38]. One route again acts via securin that is stabilized in response to DNA damage through the action of the kinase Chk1. Chk1 directly phosphorylates the budding yeast securin, Pds1 [38]. In higher eukaryotic cells, the majority of sister DNA sequences become separated during chromosome condensation in prophase. Accordingly, the DNA damage response mainly down-regulates Cdk activity, which blocks cells from entering prophase (see [32]).

Phosphorylation of the cleavage target

As described above, budding yeast securin is essential for the prevention of anaphase onset in response to spindle or DNA damage. During undisturbed cell cycle progression, however, yeast securin is entirely dispensable. Cleavage of cohesin still occurs in a regulated fashion with unchanged kinetics [39]. Is there a second regulator besides securin that can inhibit premature activation of separase? Probably not, since the overall separase activity in yeast cells lacking securin no longer undergoes detectable changes during the cell cycle. Instead, regulation occurs at the level of the separase cleavage target. Scc1 is a phosphoprotein whose phosphorylation is crucial for its cleavage by separase (Figure 3) [4,40,41]. The polo-like kinase Cdc5 in budding yeast is responsible for phosphorylation of Scc1 during metaphase. Preventing Scc1 phosphorylation decreases the rate of Scc1 cleavage *in vivo*. This effect is especially pronounced in the absence of securin, possibly due to the impairment of separase activity [39]. Of several sites phosphorylated in Scc1 by Cdc5, two phosphorylated serines lie adjacent to the

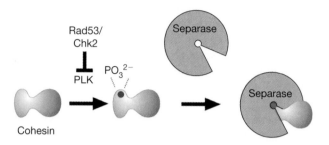

Figure 3 Regulation of Scc1 cleavage at the level of the cleavage substrate. Phosphorylation of the Scc1 subunit of cohesin by polo-like kinase (PLK) is required for its efficient cleavage. Phosphorylation might be prevented by Rad53/Chk2 after DNA damage. Reproduced from [50], with permission.

separase cleavage sites, and the affinity of separase for the cleavage sites is increased dramatically by phosphorylation of these residues [39].

Separase regulation in higher eukaryotes

While total levels of separase in budding or fission yeast do not undergo obvious changes during the cell cycle [12,42], the abundance of human separase fluctuates. In human cells, separase levels are high during metaphase and decline during anaphase [13]. Once activated, separase cleaves itself [22,29]. The two cleavage products stay associated with each other, and cleavage initially does not alter the activity of separase in cleaving Scc1 [43]. However, the C-terminal cleavage product, containing the separase active site, becomes unstable after cleavage and is degraded (Figure 4). This leads to down-regulation of separase after anaphase, which might be necessary to allow the rapid rebinding of cohesin to chromosomes during telophase that is observed in vertebrate cells [13,44,45]. In budding yeast, separase remains active throughout G1-phase [3], presumably until it is inactivated by the resynthesis of securin shortly before the next S-phase.

A remarkable situation is found in *Drosophila*. Here, separase appears to be split into two polypeptides. The protein containing the separase protease domain is short compared with separases in other organisms. However, it forms a tight complex with another protein, Three rows, that may correspond to the long N-terminal extensions found in most separases [46]. It is the Three rows protein that is cleaved by separase in anaphase, and the C-terminal Three rows fragment is rendered unstable and disappears from cells. When a site-specific mutation was introduced into Three rows that made it uncleavable by separase, this led to mitotic defects and other phenotypes that were probably caused by a failure to down-regulate separase activity after anaphase [47].

Another level of regulation of *Xenopus* and human separase that is not found in budding yeast is the inhibition of separase by mitotic phosphorylation. High cyclin-dependent kinase activity in *Xenopus* egg extracts inhibits separase, and human separase in metaphase is inhibited by phosphorylation on

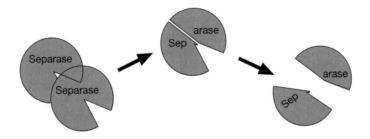

Figure 4 Human and *Xenopus* separases undergo self-cleavage during anaphase. The cleavage products stay associated with each other, and cleavage initially does not change the protease activity of separase. The C-terminal separase fragment, however, is rendered unstable by the cleavage, leading to down-regulation of separase after anaphase.

a specific serine residue [29]. The phosphate has to be removed in anaphase, when cyclin-dependent kinase activity decreases, before separase can become active. This mechanism might, at least in vertebrates, help to regulate separase in the absence of securin.

Conclusions

We now understand the molecular principle of how chromosome segregation is triggered at anaphase onset. A specific protease, separase, cleaves a protein that is required to hold the chromosomes together. We are also beginning to understand how this reaction is regulated on a number of levels to ensure that chromosomes are not separated prematurely. It may be that additional controls become important under certain conditions. For example, calcium waves are thought to play a role in triggering anaphase, and separase has a potential calcium-binding site [48], suggesting that separase activity might be regulated by calcium levels. In addition, different-sized complexes of separase with securin have been detected in fission yeast cells, but their possible roles have not yet been explored [42]. Finally, separase cleaves not only cohesin and itself at anaphase onset, but also the kinetochore- and microtubule-associated protein Slk19 [49]. Thus separase may have multiple roles to ensure a smooth transition from metaphase into anaphase. It would not be surprising if more tasks are discovered for this formidable protease.

I thank all the members of my laboratory for discussions and for the contributions that they have made to our understanding of separase regulation. Thanks are also due to *EMBO Reports* for permission to reproduce some of the figures in modified form from [50].

References
1. Sjögren, C. and Nasmyth, K. (2001) Curr. Biol. 11, 991–995
2. Nasmyth, K., Peters, J.-M. and Uhlmann, F. (2000) Science 288, 1379–1384

3.   Uhlmann, F., Lottspeich, F. and Nasmyth, K. (1999) Nature (London) 400, 37–42

4.   Uhlmann, F., Wernic, D., Poupart, M.-A., Koonin, E.V. and Nasmyth, K. (2000) Cell 103, 375–386

5.   Uzawa, S., Samejima, I., Hirano, T., Tanaka, K. and Yanagida, M. (1990) Cell 62, 913–925

6.   McGrew, J.T., Goetsch, L., Byers, B. and Baum, P. (1992) Mol. Biol. Cell 3, 1443–1454

7.   May, G.S., McGoldrick, C.A., Holt, C.L. and Denison, S.H. (1992) J. Biol. Chem. 267, 15737–15743

8.   Baum, P., Yip, C., Goetsch, L. and Byers, B. (1988) Mol. Cell. Biol. 8, 5386–5397

9.   Nasmyth, K. (2001) Annu. Rev. Genet. 35, 673–745

10.   Michaelis, C., Ciosk, R. and Nasmyth, K. (1997) Cell 91, 35–45

11.   Tóth, A., Ciosk, R., Uhlmann, F., Galova, M., Schleiffer, A. and Nasmyth, K. (1999) Genes Dev. 13, 320–333

12.   Ciosk, R., Zachariae, W., Michaelis, C., Shevchenko, A., Mann, M. and Nasmyth, K. (1998) Cell 93, 1067–1076

13.   Waizenegger, I.C., Hauf, S., Meinke, A. and Peters, J.-M. (2000) Cell 103, 399–410

14.   Hauf, S., Waizenegger, I.C. and Peters, J.-M. (2001) Science 293, 1320–1323

15.   Funabiki, H., Yamano, H., Kumada, K., Nagao, K., Hunt, T. and Yanagida, M. (1996) Nature (London) 381, 438–441

16.   Yamamoto, A., Guacci, V. and Koshland, D. (1996) J. Cell Biol. 133, 85–97

17.   Zou, H., McGarry, T.J., Bernal, T. and Kirschner, M.W. (1999) Science 285, 418–422

18.   Leismann, O., Herzig, A., Heidmann, S. and Lehner, C.F. (2000) Genes Dev. 14, 2192–2205

19.   Yanagida, M. (2000) Genes Cells 5, 1–8

20.   Cohen-Fix, O., Peters, J.-M., Kirschner, M.W. and Koshland, D. (1996) Genes Dev. 10, 3081–3093

21.   Hornig, N.C.D., Knowles, P.P., McDonald, N.Q. and Uhlmann, F. (2002) Curr. Biol. 12, 973–982

22.   Waizenegger, I.C., Gimenez-Abian, J.F., Wernic, D. and Peters, J.-M. (2002) Curr. Biol. 12, 1368–1378

23.   Stratmann, R. and Lehner, C.F. (1996) Cell 84, 25–35

24.   Jensen, S., Segal, M., Clarke, D.J. and Reed, S.I. (2001) J. Cell Biol. 152, 27–40

25.   Jallepalli, P.V., Waizenegger, I.C., Bunz, F., Langer, S., Speicher, M.R., Peters, J.-M., Kinzler, K.W., Vogelstein, B. and Lengauer, C. (2001) Cell 105, 445–457

26.   Lengauer, C., Kinzler, K.W. and Vogelstein, B. (1997) Nature (London) 386, 623–627

27.   Zachariae, W. and Nasmyth, K. (1999) Genes Dev. 13, 2039–2058

28.   Yamamoto, A., Guacci, V. and Koshland, D. (1996) J. Cell Biol. 133, 99–110

29.   Stemmann, O., Zou, H., Gerber, S.A., Gygi, S.P. and Kirschner, M.W. (2001) Cell 107, 715–726

30.   Shteinberg, M., Protopopov, Y., Listovsky, T., Brandeis, M. and Hershko, A. (1999) Biochem. Biophys. Res. Commun. 260, 193–198

31.   Kotani, S., Tanaka, H., Yasuda, H. and Todokoro, K. (1999) J. Cell Biol. 146, 791–800

32.   Clarke, D.J. and Gimenez-Abian, J.F. (2000) Bioessays 22, 351–363

33.   Basu, J., Bousbaa, H., Logarinho, E., Li, Z., Williams, B.C., Lopes, C., Sunkel, C.E. and Goldberg, M.L. (1999) J. Cell Biol. 146, 13–28

34.   Kitagawa, R. and Rose, A.M. (1999) Nat. Cell Biol. 1, 514–521

35.   Dobles, M., Liberal, V., Scott, M.L., Benezra, R. and Sorger, P.K. (2000) Cell 101, 635–645

36.   Cohen-Fix, O. and Koshland, D. (1997) Proc. Natl. Acad. Sci. U.S.A. 94, 14361–14366

37.   Gardner, R., Putnam, C.W. and Weinert, T. (1999) EMBO J. 18, 3173–3185

38.   Sanchez, Y., Bachant, J., Wang, H., Hu, F., Liu, D., Tetzlaff, M. and Elledge, S.J. (1999) Science 286, 1166–1171

39.   Alexandru, G., Uhlmann, F., Poupart, M.-A., Mechtler, K. and Nasmyth, K. (2001) Cell 105, 459–472

40.   Birkenbihl, R.P. and Subramani, S. (1995) J. Biol. Chem. 270, 7703–7711

41. Tomonaga, T., Nagao, K., Kawasaki, Y., Furuya, K., Murakami, A., Morishita, J., Yuasa, T., Sutani, T., Kearsey, S.E., Uhlmann, F. et al. (2000) Genes Dev. **14**, 2757–2770

42. Funabiki, H., Kumada, K. and Yanagida, M. (1996) EMBO J. **15**, 6617–6628

43. Zou, H., Stemmann, O., Anderson, J.S., Mann, M. and Kirschner, M.W. (2002) FEBS Lett. **528**, 246–250

44. Darwiche, N., Freeman, L.A. and Strunnikov, A. (1999) Gene **233**, 39–47

45. Losada, A., Yokochi, T., Kobayashi, R. and Hirano, T. (2000) J. Cell Biol. **150**, 405–416

46. Jäger, H., Herzig, A., Lehner, C.F. and Heidmann, S. (2001) Genes Dev. **15**, 2572–2584

47. Herzig, A., Lehner, C.F. and Heidmann, S. (2002) Genes Dev. **16**, 2443–2454

48. Kumada, K., Nakamura, T., Nagao, K., Funabiki, H., Nakagawa, T. and Yanagida, M. (1998) Curr. Biol. **8**, 633–641

49. Sullivan, M., Lehane, C. and Uhlmann, F. (2001) Nat. Cell Biol. **3**, 771–777

50. Uhlmann, F. (2001) EMBO Rep. **2**, 487–492

Biochem. Soc. Symp. **70**, 253–262
(Printed in Great Britain)
© 2003 Biochemical Society

21

Membrane-type 1 matrix metalloproteinase and cell migration

Motoharu Seiki[1], Hidetoshi Mori, Masahiro Kajita, Takamasa Uekita and Yoshifumi Itoh[2]

Division of Cancer Cell Research, Institute of Medical Science, University of Tokyo, 4-6-1 Shirokane-dai, Minato-ku, Tokyo 108-8639, Japan

Abstract

Membrane-type 1 matrix metalloproteinase (MT1-MMP) is an integral membrane proteinase that performs processing of cell surface proteins and degradation of extracellular matrix (ECM) components. Through these proteolytic events, MT1-MMP regulates various cellular functions, including ECM turnover, promotion of cell migration and invasion, and morphogenic responses to extracellular stimuli. MT1-MMP has to be regulated strictly to accomplish its function appropriately at various steps, including at the transcriptional and post-translational levels. MT1-MMP was originally identified as an invasion-promoting enzyme expressed in malignant tumour cells, and also as a specific activator of proMMP-2, which is believed to play a role in invasion of the basement membrane. Since then, it has attracted attention as a membrane-associated MMP that promotes cancer cell invasion and angiogenesis by endothelial cells. Although MT1-MMP has now become one of the best characterized enzymes in the MMP family, there remain numerous unanswered questions. In this chapter, we summarize our recent findings on how MT1-MMP is regulated during cell migration, and how cell migration is regulated by MT1-MMP.

Introduction

Membrane-type 1 matrix metalloproteinase (MT1-MMP; also known as MMP-14) was the first MT-MMP to be identified as an activator of proMMP-2 on the surface of cancer cells [1]. Another five MT-MMPs have since been iden-

[1]To whom correspondence should be addressed (e-mail mseiki@ims.u-tokyo.ac.jp).
[2]Present address: Department of Matrix Biology, Kennedy Institute of Rheumatology Division, Faculty of Medicine, Imperial College London, London W6 8LH, U.K.

tified: MT2-MMP (MMP-15) [2], MT3-MMP (MMP-16) [3], MT4-MMP (MMP-17) [4,5], MT5-MMP (MMP-24) [6,7] and MT6-MMP (MMP-25) [8,9]. Among the six MT-MMPs, MT1-MMP is most frequently expressed in human tumours, and has the ability to promote invasion and metastasis when expressed in cancer cells [10].

To promote cancer invasion, MT1-MMP has to degrade the extracellular matrix (ECM) barrier. MT1-MMP can digest fibronectin, vitronectin, laminin-1, laminin-5, fibrin and dermatan sulphate proteoglycans [11–15]. The enzyme also degrades gelatin, casein and elastin [15–17], and shows activity against collagen types I, II and III [14]. A deficiency of MT1-MMP in mice has emphasized the importance of the degradation of the ECM by MT1-MMP during development [18,19]. The animals showed inadequate collagen turnover, resulting in dwarfism, osteopenia, arthritis and connective tissue disease.

MT1-MMP also co-operates with other MMPs to degrade complex ECM components. Most importantly, MT1-MMP activates proMMP-2 on the cell surface. As a first step, proMMP-2 binds to cells expressing MT1-MMP. However, MT1-MMP cannot bind proMMP-2 directly, but uses TIMP-2 (tissue inhibitor of metalloproteinases-2) as an adaptor molecule. In addition to an N-terminal inhibitory domain that binds to and inhibits the catalytic domains of MMPs, TIMP-2 has a C-terminal domain that has a specific ability to bind to the haemopexin-like (PEX) domain of MMP-2. Thus, after MT1-MMP is inhibited by TIMP-2, the complex provides a binding site for proMMP-2 on the cell surface [16,20]. The proMMP-2 in the complex can then be activated by another MT1-MMP molecule adjacent to the complex [21,22]. Thus the levels of TIMP-2 on the cell surface play a critical regulatory role in proMMP-2 activation. Without TIMP-2, MT1-MMP cannot mediate the activation, but an excess of TIMP-2 inhibits all of the activity of MT1-MMP, including the activation of proMMP-2 [21]. Although there appears to be a contradiction between the model in which TIMP-2 is required for MT1-MMP to activate proMMP-2 and the role of TIMP-2 as a general MMP inhibitor, there is genetic evidence that proMMP-2 is not activated efficiently in TIMP-2-deficient mice [23,24]. Activation of proMMP-2 by cancer cells is presumably important for invasion of the basement membrane because of its type IV collagenase activity. In the type I collagen-rich environment in the stroma, MT1-MMP/MMP-2 can act as a potent type I collagen degradation system through a combination of the collagenase activity of MT1-MMP and the gelatinase activity of MMP-2. MT1-MMP can also activate proMMP-13 in a cell-mediated manner, and proMMP-9 indirectly [25].

CD44 regulates localization of MT1-MMP at the migration front

When cancer cells migrate or invade tissue, MT1-MMP is localized at the leading edge, and this appears to be necessary for degradation of the ECM barrier. How is such localization of MT1-MMP regulated? A relatively stable association of MT1-MMP with the actin cytoskeleton was found when cells were treated with cytochalasin D, which disrupts polymerized actin and causes actin to aggregate [26]. Localization of MT1-MMP on the cell surface coincided exactly with the actin aggregates within the cells treated with the drug. Thus the localization of MT1-MMP seems to be regulated by dynamic remodelling of the actin cytoskeleton during cell locomotion. MT1-MMP has a short cytoplasmic tail, but this is not the site for actin association; unexpectedly, the PEX domain was found to be responsible for this association. Thus it was speculated that the association is mediated by another cell surface protein. CD44, a major hyaluro-nan receptor that associates with actin within cells, also localizes at the leading edge of migrating cells, and was identified as a linker that mediates the association of MT1-MMP with actin [26] (Figure 1). CD44 was co-immunoprecipitated with MT1-MMP, and this could be prevented by overexpression of the MT1-MMP PEX fragment, but not the catalytic fragment. Finally, the recombinant PEX fragment binds directly to CD44 polypeptide produced in *Escherichia coli*. The complex-forming ability of the two molecules correlated well with the localiza-tion of MT1-MMP to the ruffling edge. Deletion of the PEX domain of MT1-MMP abolished its localization at the leading edge, and overexpression of a mutant CD44 lacking the cytoplasmic portion also blocked the distribution of MT1-MMP to the edge. Thus CD44 plays a critical role in the regulation of the polarized distribution of MT1-MMP during cell migration and invasion, and of its association with the actin cytoskeleton.

Regulation of MT1-MMP activity at the leading edge

The activation of proMMP-2 requires at least two MT1-MMP molecules, and this is accomplished by the formation of a homophilic oligomer through the PEX domain [27,28] (Figure 2). To detect the MT1-MMP oligomer *in vivo*, we prepared MT1-MMP tagged either with a FLAG epitope or with Myc peptide [27]. When the two molecules were co-expressed and immunoprecipitated using antibodies against one of the tags, the other MT1-MMP was associated with the precipitate, indicating that a homophilic complex had been formed. The PEX domain was again responsible for oligomer formation. As expected, formation of the oligomer was critical for the efficient activation of proMMP-2, because replacement of the PEX domain of MT1-MMP with that of MT4-MMP, which does not form a homo-oligomer, abolished the ability of MT1-MMP to activate proMMP-2. Interestingly, the chimaera containing the MT4-MMP PEX domain retained the ability to form a trimolecular complex of MT1-MMP, TIMP-2 and proMMP-2 on the cell surface.

How is the formation of such an oligomer regulated in relation to cell func-tion? To visualize oligomer formation *in situ*, a chimaeric protein composed of the extracellular portion of MT1-MMP and the transmembrane and cytoplasmic portions of the nerve growth factor receptor (NGF-R) was constructed [27]. The binding of ligand to NGF-R is known to induce dimerization of the receptor and to cause autophosphorylation of the cytoplasmic tyrosine residues. Thus, if the MT1-MMP portion of the chimaera formed homo-oligomers, this would cause autophosphorylation of the NGF-R cytoplasmic portion. Co-expression of constitutively active Rac1 and the MT1-MMP/NGF-R chimaera in COS-1 cells induced the formation of lamellipodia, and the chimaera was found to localize at

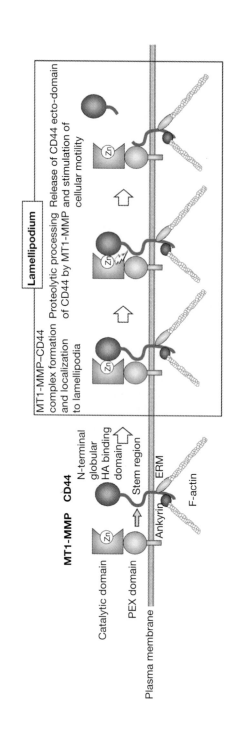

Figure 1 Interaction of MT1-MMP and CD44 on the cell surface. MT1-MMP interacts with CD44 through its PEX domain and the stem region of CD44. CD44 interacts with F-actin through protein that belongs to ezrin/radixin/moesin (ERM) family and/or ankyrin. As a result, MT1-MMP is indirectly associated with F-actin. CD44 then pulls MT1-MMP to the lamellipodia of the plasma membrane, and MT1-MMP proteolytically processes CD44, stimulating cellular motility. HA, hyaluronic acid.

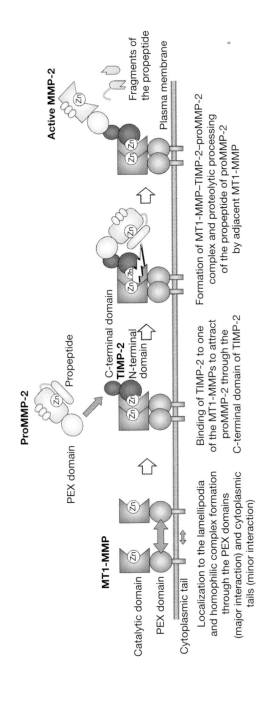

Figure 2 Mechanism of activation of proMMP-2 by MT1-MMP. See the text for details.

these sites with intense autophosphorylation signals. At the same time, the expression of active Rac1 enhanced the activation of proMMP-2 [27,29]. Thus the ruffled edge that forms the migration front is likely to be the place for oligomer formation of MT1-MMP and activation of proMMP-2.

The cytoplasmic portion of MT1-MMP may play some role in the formation of the oligomer [28,30], although our chimaera that formed homo-oligomers lacked the cytoplasmic portion. It may modulate the localization of MT1-MMP in certain situations and eventually affect oligomer formation. Collagen binding to the PEX domain also modulates the efficiency of proMMP-2 activation [31].

Internalization forms part of the mechanism of MT1-MMP turnover on the cell surface

After being expressed as an active enzyme on the cell surface and performing its function, MT1-MMP is inactivated either by binding to TIMPs or by proteolytic degradation. Cells expressing MT1-MMP at high levels frequently generate degradation fragments, of which major forms can be detected as bands in the range 43–45 kDa on SDS/PAGE [32–34]. This degradation is mainly caused by autocatalytic processing, and therefore such a pattern of degradation accurately reflects the active state of MT1-MMP on the cell surface, including its ability to activate proMMP-2. How are these molecules cleared from the surface and replaced by newly synthesized forms? MT1-MMP has a short cytoplasmic tail of 20 amino acids, and this portion was found to mediate internalization of the enzyme [35,36]. The cytoplasmic tail has a binding site for the $\mu 2$ subunit of adaptor protein 2 (AP2), which mediates incorporation of the target protein into a clathrin-coated pit for internalization [35]. Although internalization is not selective for inactivated MT1-MMP molecules, it is presumably an important part of the mechanism involved in MT1-MMP turnover. TIMP-2 has been reported to be internalized and degraded during activation of proMMP-2 [37], and this process is also likely to be mediated by MT1-MMP internalization [35].

Regulation of cell migration

Cellular invasion requires ECM degradation coupled with cell locomotion. MT1-MMP acts not only to eliminate the ECM barrier, but also to modulate the migratory behaviour of cells in multiple ways. We found that MT1-MMP binds CD44 and forms a complex at the leading edge [26]. CD44 is also processed by MT1-MMP accompanying promotion of cell migration [38] (Figure 1). Expression of a mutant CD44 that cannot be cleaved by MT1-MMP prevented the promotion of cell migration by MT1-MMP. Although the mechanism by which cell migration is promoted by MT1-MMP is not clear, processing of CD44 may regulate cell adhesion properties to the level appropriate for migration. Another scenario is that the processing generates signals to promote cell migration via CD44 itself or associated molecules. Indeed, activation of extracellular-signal-regulated kinase (ERK) is observed when cell migration is induced by MT1-MMP [39]. The cytoplasmic portion of the processed CD44 is reported to translocate into the nucleus after additional processing, and acts as a transcription factor by binding to the regulatory regions of genes [40]. Targets of transcriptional activation by the cytoplasmic tail of CD44 may include genes encoding migration-related proteins.

Laminin-5, a major component of the basement membrane, is known to support the migration of various types of cells, including tumour cells. MMP-2 was shown to cleave the γ2 chain of laminin-5 and to promote the migration of breast epithelial cells [41]. Subsequently, cells constitutively motile on laminin-5 were found to express MT1-MMP rather than MMP-2 [12]. Antisense oligonucleotides against the MT1-MMP gene inhibited both processing of the γ2 chain of laminin-5 and cell migration. Thus cleavage of the γ2 chain by MT1-MMP appears to convert it from an inactive into an active form as regards the promotion of migration, presumably exposing a new functional domain that is cryptic before processing. This system might support the sustained locomotion of tumour cells that express both MT1-MMP and the laminin-5 γ2 chain.

Integrins also form a well known adhesion system for cell migration [42]. Integrin αvβ3, which binds to vitronectin, is expressed in endothelial cells and invasive tumour cells. The αv chain, which is translated as a single polypeptide and converted into a two-chain form by protein convertases, is processed alternatively by MT1-MMP into a functional form [43]. The expression of integrin αvβ3 and MT1-MMP in human breast carcinoma MCF-7 cells did not alter adhesion to vitronectin, but stimulated migration on the matrix accompanying phosphorylation of focal adhesion kinase [44]. Additionally, the cell surface tissue transglutaminase that associates with the integrin β1 or β3 chain is also cleaved at multiple sites by MT1-MMP [45]. Tissue transglutaminase binds to fibronectin as a co-receptor of integrins, and cleavage of the transglutaminase by MT1-MMP suppressed cell adhesion to fibronectin and migration. On the other hand, MT1-MMP promoted migration of the same cells on type I collagen [45].

Invasion-promoting activity of MT1-MMP

Expression of MT1-MMP in cancer cells enhances their invasive and metastatic potential [1,46]. Cancer cell lines that express MT1-MMP constitutively are in general more invasive than non-MT1-MMP-producing cells [10,47].

In type I collagen gels, MDCK cells form a tubular structure in response to hepatocyte growth factor. Induced expression of MT1-MMP in the collagen gel plays a critical role in the tubule formation that requires invasion of the gel, because an antisense oligonucleotide to MT1-MMP abolished the response almost completely [48]. However, the expression of excess amounts of MT1-MMP in MDCK cells strongly enhanced their invasive properties and disturbed tubule formation [49]. In contrast, secreted MMPs, including the collagen-degrading MMP-1 and a soluble form of a MT1-MMP mutant, failed to promote invasion [49]. Thus membrane anchoring appears to be crucial in order for MT1-MMP to promote invasion, and excessive invasion hampers the morphogenic response to hepatocyte growth factor.

In addition its catalytic activity, other domains of MT1-MMP also regulate invasion-promoting activity. One such domain is the PEX domain, which is responsible for the formation of a homo-oligomer and binding to CD44. Overexpression of the PEX domain, which anchors to the plasma membrane, inhibited activation of proMMP-2 and the invasion of HT1080 cells into Matrigel [27]. The PEX domain also has collagen binding activity, and overexpression of a PEX domain fragment together with the linker sequence strongly inhibited collagenolysis and collagen-induced proMMP-2 activation by MT1-MMP [31]. The cytoplasmic tail of MT1-MMP is also indispensable for its invasion-promoting activity in Matrigel [35,50], but not in collagen gels [49]. Interestingly, mutations in the cytoplasmic tail that abolished this activity of MT1-MMP coincided with those that impaired internalization [35]. These mutant enzymes retain proteolytic activity on the cell surface at a level comparable with that of wild-type MT1-MMP. Excessive accumulation of inactivated MT1-MMP at the leading edge of invading cells may hamper the normal function of MT1-MMP there. It is also possible that an important functional region in the regulation of invasion overlaps with the motif for internalization, because there are possible sites for phosphorylation and lipid modification close to this motif.

Conclusions

MT1-MMP is a potent ECM-degrading enzyme that forms part of the invasion machinery of cancer cells. In this regard, MT1-MMP has to be closely regulated in order to co-operate with other components of the invasion-related machinery so it can accomplish its roles appropriately. For example, MT1-MMP is delivered to the leading edge of migratory cells to open up the route of invasion, and once there it assembles other MMPs with different substrate specificities to degrade complex ECM components in the basement membrane and in the stroma. MT1-MMP also modulates the migratory properties of cells by carrying out the functional conversion of target molecules such as CD44, integrin αv chain, tissue transglutaminase and laminin-5 γ2chain. Thus the complex roles of MT1-MMP and its regulation during cancer invasion are now becoming clear. These regulatory steps represent targets in developing strategies for cancer therapy.

References

1. Sato, H., Takino, T., Okada, Y., Cao, J., Shinagawa, A., Yamamoto, E. and Seiki, M. (1994) Nature (London) 370, 61–65
2. Will, H. and Hinzmann, B. (1995) Eur. J. Biochem. 231, 602–608
3. Takino, T., Sato, H., Shinagawa, A. and Seiki, M. (1995) J. Biol. Chem. 270, 23013–23020
4. Puente, X.S., Pendas, A.M., Llano, E., Velasco, G. and Lopez-Otin, C. (1996) Cancer Res. 56, 944–949
5. Kajita, M., Kinoh, H., Ito, N., Takamura, A., Itoh, Y., Okada, A., Sato, H. and Seiki, M. (1999) FEBS Lett. 457, 353–356

6. Llano, E., Pendas, A.M., Freije, J.P., Nakano, A., Knauper, V., Murphy, G. and Lopez-Otin, C. (1999) Cancer Res. **59**, 2570–2576

7. Pei, D. (1999) J. Biol. Chem. **274**, 8925–8932

8. Pei, D. (1999) Cell Res. **9**, 291–303

9. Velasco, G., Cal, S., Merlos-Suarez, A., Ferrando, A.A., Alvarez, S., Nakano, A., Arribas, J. and Lopez-Otin, C. (2000) Cancer Res. **60**, 877–882

10. Seiki, M. (1999) APMIS **107**, 137–143

11. Hiraoka, N., Allen, E., Apel, I.J., Gyetko, M.R. and Weiss, S.J. (1998) Cell **95**, 365–377

12. Koshikawa, N., Giannelli, G., Cirulli, V., Miyazaki, K. and Quaranta, V. (2000) J. Cell Biol. **148**, 615–624

13. Shofuda, K., Yasumitsu, H., Nishihashi, A., Miki, K. and Miyazaki, K. (1997) J. Biol. Chem. **272**, 9749–9754

14. Ohuchi, E., Imai, K., Fujii, Y., Sato, H., Seiki, M. and Okada, Y. (1997) J. Biol. Chem. **272**, 2446–2451

15. Pei, D. and Weiss, S.J. (1996) J. Biol. Chem. **271**, 9135–9140

16. Imai, K., Ohuchi, E., Aoki, T., Nomura, H., Fujii, Y., Sato, H., Seiki, M. and Okada, Y. (1996) Cancer Res. **56**, 2707–2710

17. Will, H., Atkinson, S.J., Butler, G.S., Smith, B. and Murphy, G. (1996) J. Biol. Chem. **271**, 17119–17123

18. Holmbeck, K., Bianco, P., Caterina, J., Yamada, S., Kromer, M., Kuznetsov, S.A., Mankani, M., Robey, P.G., Poole, A.R., Pidoux, I. et al. (1999) Cell **99**, 81–92

19. Zhou, Z., Apte, S.S., Soininen, R., Cao, R., Baaklini, G.Y., Rauser, R.W., Wang, J., Cao, Y. and Tryggvason, K. (2000) Proc. Natl. Acad. Sci. U.S.A. **97**, 4052–4057

20. Strongin, A.Y., Collier, I., Bannikov, G., Marmer, B.L., Grant, G.A. and Goldberg, G.I. (1995) J. Biol. Chem. **270**, 5331–5338

21. Kinoshita, T., Sato, H., Okada, A., Ohuchi, E., Imai, K., Okada, Y. and Seiki, M. (1998) J. Biol. Chem. **273**, 16098–16103

22. Seiki, M. (2002) Curr. Opin. Cell Biol. **14**, 624–632

23. Wang, Z., Juttermann, R. and Soloway, P.D. (2000) J. Biol. Chem. **275**, 26411–26415

24. Caterina, J.J., Yamada, S., Caterina, N.C., Longenecker, G., Holmback, K., Shi, J., Yermovsky, A.E., Engler, J.A. and Birkedal-Hansen, H. (2000) J. Biol. Chem. **275**, 26416–26422

25. Knauper, V., Will, H., Lopez-Otin, C., Smith, B., Atkinson, S.J., Stanton, H., Hembry, R.M. and Murphy, G. (1996) J. Biol. Chem. **271**, 17124–17131

26. Mori, H., Tomari, T., Koshikawa, N., Kajita, M., Itoh, Y., Sato, H., Tojo, H., Yana, I. and Seiki, M. (2002) EMBO J. **21**, 3949–3959

27. Itoh, Y., Takamura, A., Ito, N., Maru, Y., Sato, H., Suenaga, N., Aoki, T. and Seiki, M. (2001) EMBO J. **20**, 4782–4793

28. Lehti, K., Lohi, J., Juntunen, M.M., Pei, D. and Keski-Oja, J. (2002) J. Biol. Chem. **277**, 8440–8448

29. Zhuge, Y. and Xu, J. (2001) J. Biol. Chem. **276**, 16248–16256

30. Rozanov, D.V., Deryugina, E.I., Ratnikov, B.I., Monosov, E.Z., Marchenko, G.N., Quigley, J.P. and Strongin, A.Y. (2001) J. Biol. Chem. **276**, 25705–25714

31. Tam, E.M., Wu, Y.I., Butler, G.S., Stack, M.S. and Overall, C.M. (2002) J. Biol. Chem. **277**, 39005–39014

32. Stanton, H., Gavrilovic, J., Atkinson, S.J., d'Ortho, M.P., Yamada, K.M., Zardi, L. and Murphy, G. (1998) J. Cell Sci. **111**, 2789–2798

33. Lehti, K., Lohi, J., Valtanen, H. and Keski-Oja, J. (1998) Biochem. J. **334**, 345–353

34. Toth, M., Hernandez-Barrantes, S., Osenkowski, P., Bernardo, M.M., Gervasi, D.C., Shimura, Y., Meroueh, O., Kotra, L.P., Galvez, B.G., Arroyo, A.G. et al. (2002) J. Biol. Chem. **277**, 26340–26350

35. Uekita, T., Itoh, Y., Yana, I., Ohno, H. and Seiki, M. (2001) J. Cell Biol. **155**, 1345–1356

36. Jiang, A., Lehti, K., Wang, X., Weiss, S.J., Keski-Oja, J. and Pei, D. (2001) Proc. Natl. Acad. Sci. U.S.A. **98**, 13693–13698

37. Maquoi, E., Frankenne, F., Baramova, E., Munaut, C., Sounni, N.E., Remacle, A., Noel, A., Murphy, G. and Foidart, J.M. (2000) J. Biol. Chem. **275**, 11368–11378

38. Kajita, M., Itoh, Y., Chiba, T., Mori, H., Okada, A., Kinoh, H. and Seiki, M. (2001) J. Cell Biol. **153**, 893–904

39. Gingras, D., Bousquet-Gagnon, N., Langlois, S., Lachambre, M.P., Annabi, B. and Beliveau, R. (2001) FEBS Lett. **507**, 231–236

40. Okamoto, I., Kawano, Y., Murakami, D., Sasayama, T., Araki, N., Miki, T., Wong, A.J. and Saya, H. (2001) J. Cell Biol. **155**, 755–762

41. Giannelli, G., Falk-Marzillier, J., Schiraldi, O., Stetler-Stevenson, W.G. and Quaranta, V. (1997) Science **277**, 225–228

42. Webb, D.J., Parsons, T. and Horwitz, A.F. (2002) Nat. Cell Biol. **4**, 97–100

43. Ratnikov, B.I., Rozanov, D.V., Postnova, T.I., Baciu, P.G., Zhang, H., DiScipio, R.G., Chestukhina, G.G., Smith, J.W., Deryugina, E.I. and Strongin, A.Y. (2002) J. Biol. Chem. **277**, 7377–7385

44. Deryugina, E.I., Ratnikov, B.I., Postnova, T.I., Rozanov, D.V. and Strongin, A.Y. (2002) J. Biol. Chem. **277**, 9749–9756

45. Belkin, A.M., Akimov, S.S., Zaritskaya, L.S., Ratnikov, B.I., Deryugina, E.I. and Strongin, A.Y. (2001) J. Biol. Chem. **276**, 18415–18422

46. Tsunezuka, Y., Kinoh, H., Takino, T., Watanabe, Y., Okada, Y., Shinagawa, A., Sato, H. and Seiki, M. (1996) Cancer Res. **56**, 5678–5683

47. Pulyaeva, H., Bueno, J., Polette, M., Birembaut, P., Sato, H., Seiki, M. and Thompson, E.W. (1997) Clin. Exp. Metastasis **15**, 111–120

48. Kadono, Y., Shibahara, K., Namiki, M., Watanabe, Y., Seiki, M. and Sato, H. (1998) Biochem. Biophys. Res. Commun. **251**, 681–687

49. Hotary, K., Allen, E., Punturieri, A., Yana, I. and Weiss, S.J. (2000) J. Cell Biol. **149**, 1309–1323

50. Lehti, K., Valtanen, H., Wickstrom, S., Lohi, J. and Keski-Oja, J. (2000) J. Biol. Chem. **275**, 15006–15013

Biochem. Soc. Symp. **70**, 263–276
(Printed in Great Britain)
© 2003 Biochemical Society

22

Cathepsin B and its role(s) in cancer progression

Izabela Podgorski* and Bonnie F. Sloane*†[1]

*Department of Pharmacology, Wayne State University School of Medicine, Detroit, MI 48201, U.S.A. and †Barbara Ann Karmanos Cancer Institute, Wayne State University School of Medicine, Detroit, MI 48201, U.S.A.

Abstract

Experimental and clinical evidence has linked cathepsin B with tumour invasion and metastasis. Cathepsin B expression is increased in many human cancers at the mRNA, protein and activity levels. In addition, cathepsin B is frequently overexpressed in premalignant lesions, an observation that associates this protease with local invasive stages of cancer. Increased expression of cathepsin B in primary cancers, and especially in preneoplastic lesions, suggests that this enzyme might have pro-apoptotic features. Expression of cathepsin B is regulated at many different levels, from gene amplification, use of alternative promoters, increased transcription and alternative splicing, to increased stability and translatability of transcripts. During the transition to malignancy, a change in the localization of cathepsin B occurs, as demonstrated by the presence of cathepsin B-containing vesicles at the cell periphery and at the basal pole of polarized cells. Due to increased expression of cathepsin B and changes in intracellular trafficking, increased secretion of procathepsin B from tumours is observed. Active cathepsin B is also secreted from tumours, a mechanism likely to be facilitated by lysosomal exocytosis or extracellular processing by surface activators. Cathepsin B is localized to caveolae on the tumour surface, where binding to the annexin II heterotetramer occurs. Activation of cathepsin B on the cell surface leads to the regulation of downstream proteolytic cascade(s).

Introduction

Tumour progression is a multi-step process that is accompanied by invasive proteolytic activity. During this process, tumour cells attach to and invade through the basement membrane and stromal extracellular matrix and migrate to distant sites. Various classes of proteolytic enzymes, including matrix metal-

[1]To whom correspondence should be addressed (e-mail bsloane@med.wayne.edu).

loproteases (e.g. gelatinases), serine proteases [urokinase-type plasminogen activator (uPA), plasmin], aspartic proteases (cathepsin D) and cysteine proteases (cathepsins B and L), have been implicated in tumour progression and invasion. The invasive proteolytic processes are complex, and depend on the sequential and simultaneous actions of many proteases. Tumour–stromal interactions add even more complexity to this system. Fibroblasts and inflammatory cells may also secrete proteolytic enzymes, and therefore aid in matrix-degrading activities. In addition, these specialized cells may also be capable of sending signals to the tumour cells to induce them to produce and secrete digestive enzymes.

An association of cathepsin B with malignant progression was reported more than two decades ago. In these early reports, cathepsin B was linked to the progression of human breast cancer [1]. Since then, increased expression, activity and secretion and changes in the localization of cathepsin B have been observed in many different tumours, including colorectal, gastric, lung and prostate tumours, melanoma, chondrosarcoma and many others [1]. In this review, we will focus on the relationship between cathepsin B expression in human cancers and malignant progression. We will discuss the molecular mechanisms responsible for the increased expression of this protease in tumours. We will characterize changes in the intracellular trafficking of cathepsin B during the transition to malignancy, and discuss the association of this enzyme with the plasma membrane. Finally, we will provide our view on the significance of cathepsin B in activating proteolytic cascade(s).

Expression of cathepsin B in advanced cancers and premalignant lesions

Cathepsin B is found in cells under normal physiological conditions, where it plays various roles in the maintenance of normal cellular metabolism. Under these normal conditions, cathepsin B is regulated at multiple levels, including transcription, post-transcriptional processing, translation, post-translational processing and finally trafficking. In malignancy, one or more of these levels becomes misregulated, leading to increased expression and activity, and altered distribution of the enzyme within the cell.

An involvement of cathepsin B in tumour invasion has been widely demonstrated in many human carcinomas [2,3]. Cathepsin B has both endopeptidase and exopeptidase activities [2] and is capable of degrading various components of the extracellular matrix, including type IV collagen, laminin and fibronectin [1,2]. Cathepsin B is synthesized as an inactive proenzyme and requires activation, which may occur autocatalytically or through the action of other proteases, including cathepsins D and G, uPA, tissue-type plasminogen activator (tPA) and elastase [1]. The resulting active cathepsin B can activate other proteases. For example, pro-uPA can be activated by cathepsin B, which then leads to the conversion of plasminogen into plasmin, and subsequent degradation of many components of the stroma [4].

Cathepsin B is not randomly expressed in tumour tissues, a finding consistent with studies suggesting that its expression is regulated by stroma–tumour interactions. In prostate cancer, increased levels of cathepsin B mRNA are found at the invasive edges of the tumours [5]. Interaction of human prostate cancer cells with matrix type I collagen leads to increased levels of the enzyme (mRNA, protein and activity), and induced secretion is observed (I. Podgorski, B. Linebaugh, M.L. Cher and B.F. Sloane, unpublished work). Similar increases in secretion upon interaction with type I collagen were reported for human breast fibroblasts [6]. Up-regulation of mRNA, protein or activity was demonstrated similarly in tumour cells at the invasive edges of colon and bladder cancers and in glioblastoma cells [3]. Increased expression of cathepsin B was also observed in fibroblasts and invading macrophages in colon and breast carcinomas [7,8].

Numerous studies have been dedicated to finding a correlation between cathepsin B expression/activity and stages of malignant progression. Campo et al. [7] have shown that increases in staining for cathepsin B protein are predictive of shortened patient survival. Similar correlations with progression and invasive capability were reported for gliomas and gastric carcinomas [1]. In Clara cell type carcinoma, cathepsin B staining correlated with invasion of the lymph nodes, distant metastasis and poor prognosis [3]. There is a line of evidence suggesting that cathepsin B might have a role in the transition of a preneoplastic to a neoplastic lesion. In the MCF-10 model of human breast cancer progression, diploid MCF-10A cells are spontaneously immortalized human breast epithelial cells and are non-tumorigenic. A daughter cell line, MCF-10AneoT, obtained by transfection with an activated *ras* oncogene, exhibits a transformed phenotype and is capable of forming preneoplastic lesions in mice, which progress to neoplasias in 30% of cases [9]. Fernandez et al. [10] observed increased expression of cathepsins B and S in the pre-invasive stages of transformed prostate epithelium. Further, amplification and overexpression of cathepsin B were reported in stage I oesophageal adenocarcinomas and 5% of Barrett's oesophagus [1]. In addition, elevated cathepsin B protein levels were detected in sera of patients with premalignant diseases of the liver and pancreas [1]. In a study by Murnane et al. [3a], increased cathepsin B mRNA levels were observed in colorectal carcinomas when compared with normal colon tissues. More interestingly, these increases were greater for Duke's A and B tumours, which were in the process of invading through the bowel wall, than for the more advanced Duke's C and D tumours. Collectively, these results suggest that cathepsin B expression in human tumours is correlated with local invasive stages of the cancer.

Cathepsin B as a mediator of apoptotic cell death

Frequent overexpression of cathepsin B in primary tumours, and especially in premalignant stages of the disease, suggests that this enzyme may possess pro-apoptotic characteristics. In tumour development and progression, the regulation of cell survival and cell death, including apoptosis, plays an extremely important role. Mammalian organisms use programmed cell death to remove excessive, infected or dangerous cells, and as a result maintain normal cell metabolism. Molecular pathways of apoptosis are complex and not yet

fully understood. Several proteolytic systems have been suggested to participate in this process, with proteases from the caspase family being the most
critical [11–13]. Caspase-mediated apoptosis can be initiated in several ways. In
the extrinsic pathway, cell death is provoked by tumour necrosis factor (TNF),
which binds to TNF receptors (death receptors) and causes receptor oligomerization, recruitment of several intracellular proteins and subsequent activation
of the cysteine protease caspase-8, also known as the initiator caspase [11,14].
The intrinsic pathway, which is stress-induced, involves release of cytochrome
c from mitochondria, induction of the cytosolic factor APAF-1 (apoptotic protease activating factor-1), apoptosome formation and once again activation of
caspase-8 or -9 [11]. In the granzyme B pathway, the apoptotic signal is an
effect of the delivery of the cytotoxic cell protease granzyme B to the target
cell. Following the induction of initiator caspases, direct activation of executing
caspases -3, -6 and -7 and subsequent killing of the cell occurs [11,13]. This final
step is common to all three pathways described above.

Interestingly, in recent years it has become evident that the triggering of
apoptosis is not exclusive to caspases. Many laboratories have demonstrated
that lysosomes and their associated non-caspase proteases, including cathepsins
B and D, are essential downstream effectors of caspases [11,12,15]. The importance of lysosomes in cell death has been suggested by several sources.
Lysosomal rupture has been readily associated with oxidative stress.
Accordingly, oxidative-stress-induced apoptosis caused by hydrogen peroxide
and serum deprivation have been linked to such ruptures of lysosomes [14,16].
Lysosomal integrity is often compromised in pathological states as well as during the normal aging process, thus resulting in leakage of proteases into the
cytosol [11]. It has been suggested that such leakage may be involved in the
activation of caspases [14]. For instance, cathepsin B has been shown to activate
pro-inflammatory caspase-11 and -1 [17]. Interestingly, some studies have
shown that apoptosis can occur in the absolute absence of caspases [12,18]

One of the first reports indicating the involvement of cathepsins in the
death pathways came from Roberts et al. [19], who suggested a role for cathepsin B in bile salt-induced apoptosis in rat hepatocytes. The same group has
shown in subsequent studies that cathepsin B was not an activator of caspases
in these cells, but that cathepsin B activity was caspase-dependent [20].
Therefore cathepsin B seems to act downstream of caspases in this particular
apoptotic pathway.

The apoptotic role of cathepsin B in hepatocytes was elucidated further
by Guicciardi et al. [17], who investigated the role of cathepsin B in TNF-
mediated killing. In their study, treatment of mouse hepatocytes with
actinomycin D and TNFα resulted in release of cathepsin B into the cytosol
and subsequent induction of enzyme activity. Release of cathepsin B into the
cytosol also was induced by active caspase-2 or -8, and was reduced in hepatocytes from cathepsin B-deficient mice. Interestingly, during treatment of
wild-type hepatocytes with TNFα, release of mitochondrial cytochrome c into
the cytosol was observed and the time course of this release was similar to that
of cathepsin B. Cytochrome c release was reduced in cathepsin B-deficient
cells, which places cathepsin B upstream of mitochondria in this apoptotic

pathway [17]. One molecular mechanism by which cathepsin B may affect mitochondria and the release of cytochrome c is via selective proteolysis of Bid [14]. Bid is a Bcl-2 family member, which upon activation (cleavage) by caspase-8 or granzyme B induces cytochrome c release. In the study of Stoka et al. [14], recombinant Bid was cleaved by a lysosomal extract, which led to induced cytochrome c release from mitochondria. In addition, levels of cytochrome c were similar to those released by the caspase-8-activated Bid.

Additional evidence for an involvement of cathepsin B in TNF-mediated killing was demonstrated by Foghsgaard et al. [12]. This study utilized WEHI-S fibrosarcoma cells and ME180as cervix carcinoma cells, both of which are sensitive to TNF-mediated cell death. In WEHI-S cells, cathepsin B was shown to be the dominant execution protease even in the presence of a pan-caspase inhibitor. In ME180as cells, cathepsin B acted as a downstream effector of receptor-activated caspases. The involvement of cathepsin B in various apoptotic pathways is summarized in Table 1.

The molecular identity of all mediators of programmed cell death remains to be elucidated. It is now clear that a single death receptor can trigger multiple mechanisms leading to cell death. Evidence now clearly suggests that both caspase-dependent and caspase-independent apoptotic pathways exist. It is possible that the caspase-independent pathway is a back-up system in tumour cells, in which caspase-dependent mechanisms have been impaired due to mutations or overexpression of survival proteins [21]. Cathepsin B seems to be involved in both upstream and downstream death signals, suggesting the importance of this protease in all stages of apoptosis. Tumour cells generally have increased levels of cathepsin B. Interestingly, a death receptor-triggered, cathepsin B-dependent apoptosis pathway seems to be more prominent in tumour cells than in primary cells [12]. On the other hand, natural cysteine protease inhibitors (e.g. cystatin A) may inhibit the apoptotic process due to inactivation of the effector species. This possibility is supported by reports of the aggressiveness of tumours overexpressing cystatin A [12]. It is therefore possible that lysosomal proteases, especially cathepsin B, play dual roles in malignant progression, and that the entire process depends on a delicate balance between the protease and its inhibitor, as well as between pro-apoptotic and pro-invasive properties of the protease. It is also becoming evident that a variety of cells participate in malignant transition.

Molecular mechanisms behind cathepsin B expression in tumours

Whether at a certain stage of the disease cathepsin B exhibits pro-apoptotic or invasion-promoting characteristics might be the result of various genetic rearrangements and mutations. Cathepsin B is synthesized as a preproenzyme in the rough endoplasmic reticulum and, after being co-translationally glycosylated, moves through the Golgi complex to lysosomes. Trafficking to lysosomes can occur via a mannose 6-phosphate receptor (MPR)-mediated pathway or an MPR-independent pathway [22]. Multiple levels of regulation exist during the synthesis and processing of the enzyme, including transcription, maturation of

Table 1 Involvement of cathepsin B in apoptotic pathways. $1,25(OH_2)D_3$, 1,25-dihydroxyvitamin D_3; TNF-RI, TNF receptor-I.

System	Inducer	Observation	Reference
Rat hepatocytes	Bile salt	Cathepsin B is a downstream mediator of caspase-mediated death*	[19]
Mouse hepatocytes	TNF/actinomycin D	Release of cathepsin B by TNFα and caspases-2 and -8; cathepsin B acts upstream of mitochondria and releases cytochrome c*	[17]
WEHI-S fibrosarcoma cells	TNF	Cathepsin B as a dominant execution protease required for TNF-mediated killing*	[12]
ME180as cervix carcinoma cells	TNF	Downstream effector of receptor-activated caspases*	[12]
MCF-7 and T47D breast cancer cells	TNF/$1,25(OH_2)D_3$	Increase in cathepsin B and TNF-RI, an execution protease	[18]
PC12 cells	Serum deprivation	Cathepsin B inhibits pro-apoptotic activity of cathepsin D	[16]
Neuroblastoma cells	Protease inhibitors	Inhibition of cathepsins B and D induces caspase-dependent apoptosis	[15]

*Associated with translocation of lysosomes to cytosol and nucleus.

RNA, translation, trafficking and interaction with natural inhibitors. Cathepsin B expression, activity and localization in tumour cells may be modulated at any one or many of these levels.

The cathepsin B gene is a single-copy gene mapping to chromosome 8p22 [3,22]. The protease has a housekeeping-gene-type promoter that is TATA- and CAAT-less and GC-rich [3]. This promoter has a single transcription start site and multiple binding sites that include USF (upstream stimulatory factor), six closely spaced Sp1 and four Ets binding sites. The promoter activity is up-regulated by the binding of transcription factors ([3]; S. Yan and B.F. Sloane, unpublished work). The 5′-flanking region of the cathepsin B gene contains additional transcription start sites [23], which suggests the existence of more than one promoter. Alternative promoters have also been demonstrated in the mouse [24]. Perhaps the different promoter regions of the cathepsin B gene are regulated independently depending on the microenvironment and the presence or relative levels of transcription factors or other regulatory species.

Transcription factors, including Sp1 family members, appear to play an important role in the regulation of cathepsin B gene expression. The activity of Sp1 itself is modulated by various regulatory proteins and nuclear factors, regulation that could be critical for cathepsin B expression and activity [25]. In addition, it has been shown that Sp1-regulated promoters may be regulated by members of various transduction pathways, such as c-Fos tyrosine-protein kinase or cAMP-dependent protein kinase [26]

Another transcription factor implicated in the regulation of proteases important for malignant progression is Ets1, a member of the Ets superfamily. Expression of Ets1 has been demonstrated to correlate positively with invasive phenotypes of lung carcinomas and breast carcinoma cell lines [26]. Ets1 has been implicated in the regulation of uPA, a serine protease known to play a role in motility and tumour invasion [26]. In addition, Ets1 has been reported to regulate the transcription of some matrix metalloproteinases (MMPs), i.e. MMP-1 and MMP-3 [27]. Furthermore, correlations between mRNA levels of Ets1 and another metalloproteinase, MMP-7, have been suggested to be predictive of the metastatic progression of lung cancer [28] and a marker of poor prognosis in ovarian carcinoma [29]. A role for Ets1 in two important malignant processes, apoptosis and angiogenesis, has also been suggested [30,31]. Transcription of cathepsin B is most likely to be mediated by multiple transcription factors, and it is possible that the interplay between Sp1 and Ets1 regulates cathepsin B expression. Such synergistic Sp1/Ets1 activity has already been demonstrated to regulate the promoter activity of megakaryocytic and parathyroid hormone-related genes [26].

The increased expression of cathepsin B mRNA in tumours is often associated with gene amplification. A novel 8p22–8p23 amplicon, corresponding to the cathepsin B gene locus, was recently identified and linked to the overexpression of cathepsin B in oesophageal adenocarcinoma [1]. Amplified and overexpressed cathepsin B was observed in premalignant lesions (Barrett's oesophagus), suggesting a role for cathepsin B in malignant progression. This is likely to be a regulatory mechanism in other tumours, as suggested by amplification of the cathepsin B gene in transformed rat ovarian cells [1].

Cathepsin B is a product of a single gene, yet several transcript species of this enzyme have been reported [32]. Multiple transcripts can be a result of alternative splicing, a regulatory mechanism mediating cathepsin B expression in arthritis and cancer [22]. Six mRNA transcript species are produced due to alternative splicing of the 5'-untranslated region, whereas splicing of the 5'-translated region produces two additional transcripts [22]. A truncated form of procathepsin B is produced from the latter two transcripts. Lacking a signal propeptide, such a protein cannot enter the vesicular pathway. Thus translation from such transcript species may affect the cellular localization of cathepsin B and account for its presence in the cytoplasm [22,32]. Diffuse cytoplasmic staining for cathepsin B was observed in COS cells transfected with the truncated form of the proenzyme [32]. Changes in the post-translational processing of primary transcripts can lead to the production of more stable mRNA species, and ultimately to increased levels of cathepsin B mRNA and protein. In some cases, elevated mRNA levels result from an increased rate of transcription caused by the altered levels of transcription factors [33].

Various studies have demonstrated changes in cathepsin B mRNA expression during differentiation. An increase in mRNA and protein levels has been observed in human U937 promonocytic cells induced to differentiate by granulocyte/macrophage colony-stimulating factor or phorbol ester [34]. Cathepsin B mRNA expression in the HL-60 leukaemia cell line is induced in a dose-dependent manner by inducers of both monocytic and granulocytic differentiation [34]. An increase in the level of cathepsin B transcripts occurs before the achievement of the differentiated phenotype, suggesting that the above inducers regulate gene expression by directly or indirectly triggering certain signal transduction cascades. Cathepsin B mRNA expression can therefore be regulated at both the transcriptional and post-transcriptional levels [34].

The increased levels of cathepsin B protein and activity observed in many cancer cell lines and tumours [3,23] do not always result from elevated levels of mRNA. In human breast epithelial cells transfected with the c-Ha-*ras* oncogene, increased levels of cathepsin B protein and activity along with changes in trafficking and an association with the plasma membrane are observed, while mRNA levels are unchanged [35]. A similar pattern is observed in colon epithelial cells transfected with K-*ras*4B[val12], suggesting post-translational regulation of cathepsin B expression [36].

Localized expression and secretion of cathepsin B in human cancer

Under normal conditions, cathepsin B is associated with lysosomes, where as a housekeeping enzyme it is involved mainly in protein degradation. A small fraction of the proenzyme (5–10%) can be secreted from normal cells [1,22]. For malignant tumours to invade, proteolytic activity is required in the vicinity of infiltrating cells and at the invasive edge of the tumour. Accordingly, localization of cathepsin B in cancer cells changes from perinuclear lysosomes to vesicles in the peripheral cytoplasm and cell processes and to the plasma membrane. The presence of cathepsin B on the surface of tumour cells has been demonstrated for B16a melanoma and glioma cells [37] and Ki-*ras*-expressing colorectal carcinoma cells (D. Cavallo-Medved and B.F. Sloane, unpublished work). A striking change in localization of the enzyme during malignant transition has been demonstrated in the MCF-10A breast cancer model [38]. By means of immunogold staining, the presence of cathepsin B on the surface, in perinuclear vesicles and in cell processes of MCF-10AneoT cells was shown, a phenomenon not observed in the parental MCF-10A cells. Interestingly, the breast carcinoma cell lines MCF-7 and BT-20 showed staining patterns similar to those of MCF-10AneoT cells, and cathepsin B was localized to the interior basal and outer basal surface of the cells [38].

Localized expression of cathepsin B has been extensively demonstrated in urinary bladder carcinomas, glioblastomas, and colon and prostate carcinomas [1–3]. Expression of cathepsin B in tumour cells at the invading edge as well as in the surrounding stroma [7] usually correlates with increased secretion of the enzyme and subsequent degradation of the extracellular matrix [1]. Cathepsin B has been shown to degrade extracellular matrix proteins, including laminin, collagen IV, fibronectin [1] and, recently, tenascin-C [39]. Interestingly, in gastric cancer, increased expression of surface cathepsin B is correlated with decreased levels of laminin and a higher invasive potential [40].

Cancer cells and tumours secrete both pro- and mature forms of cathepsin B, as demonstrated in breast and colon carcinomas, gliomas and murine melanomas [1,22]. Procathepsin B is generally secreted constitutively by an MPR pathway. Saturation of MPRs due to increased expression of cathepsin B in cancers will most probably result in increased secretion of proenzyme. Such a mechanism has already been demonstrated for the secretion of procathepsin D [41]. Increased secretion of procathepsin B can also occur due to defective or down-regulated MPRs, as observed in an invasive murine SCC-VII squamous carcinoma cell line and in pancreatic islets of transgenic mice, both of which are deficient in MPRs [42,43].

The secretion of active forms of cathepsin B occurs via inducible pathways, mechanisms for which are still being investigated. It is possible that the activators on the cell surface process cathepsin B extracellularly. Alternatively, the enzyme may be processed through the classical lysosomal pathway and secreted by a retrograde mechanism. Lysosomal exocytosis and secretion of mature active cathepsin B has been observed in rat exocrine pancreas due to the fusion of lysosomes with secretory granules [32]. It is well known that catalytic activity of cathepsin B is pH-dependent and complex [32]. The distribution and secretion of active enzyme also depends on extracellular pH. In particular, in B16 melanoma cells, slightly acidic conditions induce the migration of cathepsin B-containing vesicles to the cell periphery, resulting in enhanced secretion [32]. Movement of lysosomes to the cell surface upon a reduction in pH has also been demonstrated in macrophages and fibroblasts [44]. In this case, lysosomes move to the cell surface on microtubules, a process linked to the movement of cathepsin B to the surface of MCF-10AneoT cells [9]. Growth of tumour cells on extracellular matrices led to the observation that large acidic vesicles are formed and participate in matrix digestion [32]. A correlation between the invasiveness of various cancer lines and their ability to phagocytose extracellular matrix has been shown by Coopman et al. [45]. Recently, exocytosis of cathepsin B-containing vesicles was observed upon stimulation of MCF-10Aneo cells with phorbol ester [46]. In this case the participation of cathepsin B in a proteolytic cascade involving uPA and plasmin(ogen) was demonstrated, a cascade implicated in metastatic progression.

Association of cathepsin B with the cell surface

In addition to proteases secreted from tumour cells and tumour-associated cells, surface-bound proteases are important players in local proteolysis. Many proteases implicated in extracellular matrix degradation and tumour progression, including tPA, plasmin(ogen), uPA, membrane-type MMPs and membrane-type serine proteases, are integral membrane proteins or are associated with receptors/binding proteins on the tumour cell surface [47,48]. Recently we have demonstrated that p11, one of the two subunits of annexin II heterotetramer (AIIt), is a binding partner for procathepsin B on the tumour cell surface [49]. Interestingly, expression of AIIt has been demonstrated to be highly up-regulated in many tumours and, similarly to cathepsin B, parallels malignancy [49]. Moreover, AIIt binds to, among others, matrix type I collagen, tPA, plasmin(ogen) and tenascin C [49]. Association of AIIt with tenascin C is particularly attractive, since we have shown recently that cathepsin B is capable of degrading this extracellular matrix protein [39]. In addition, both proteins are overexpressed in malignant anaplastic astrocytomas and glioblastomas, and can also be detected in neovessels, suggesting their participation in angiogenesis. Since tenascin C has been shown to be highly overexpressed at the stroma–tumour interface of various tumours [39], interaction of this protein with cathepsin B *in vivo* may be critical to metastatic progression.

An interesting aspect of the association of cathepsin B with the cell surface is the recent finding that this protease is localized to caveolae (lipid raft) fractions isolated from colorectal carcinoma cells (D. Cavallo-Medved and B.F. Sloane, unpublished work). Caveolae are believed to be involved in cell surface proteolysis. Enhanced plasminogen activation has been demonstrated as a result of clustering of uPA and its receptor in caveolae [50]. In addition, AIIt has also been shown to localize to caveolae (D. Cavallo-Medved and B.F. Sloane, unpublished work). Caveolin-1, the major structural component of caveolae, regulates many signalling molecules involved in cell adhesion, migration and invasion [51], and has been reported to possess tumour suppressor characteristics [52]. Caveolin-1 is overexpressed in various tumours of the prostate, colon, breast and bladder and, interestingly, an involvement in the metastatic progression of prostate cancer has been attributed to its anti-apoptotic activities [53]. In addition, Ki-*ras*-expressing HCT-116 colorectal carcinoma cells overexpress caveolin-1, as compared with HKh-2 cells in which the Ki-*ras* allele was disrupted by homologous recombination (D. Cavallo-Medved and B.F. Sloane, unpublished work). In parallel with caveolin expression, increased localization of cathepsin B to caveolae has been observed in HCT-116 cells, changes that were correlated with increased invasiveness. Collectively, these data suggest the possibility of functional interplay between caveolin-1 and cathepsin B on the

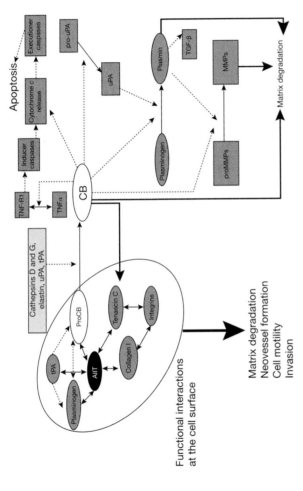

Figure 1 Involvement of cathepsin B in various proteolytic cascades implicated in metastatic progression. Procathepsin B (ProCB) is associated on the cell surface with AIIt, a protein that interacts with collagen I, tenascin C, tPA and plasminogen. Procathepsin B and plasminogen compete for activation by tPA and, as a result of various functional interactions occurring at the cell surface, processes such as matrix degradation, neovessel formation and invasion occur. Procathepsin B is activated by various proteases, whereupon active cathepsin B (CB) cleaves pro-uPA to uPA, and uPA subsequently cleaves plasminogen to plasmin. A series of metalloproteinases are then activated by plasmin, a process that results in degradation of the extracellular matrix. Components of the extracellular matrix can also be degraded directly by cathepsin B and plasmin. Cathepsin B is also involved in pro-apoptotic pathways, acting both upstream and downstream of initiator caspases. Double arrows indicate binding interactions; double dotted arrows denote competitive interactions; solid thin arrows show conversion into active form; dotted arrows denote participation in activation; and solid bold arrows indicate matrix-degrading activity. TNF-R1, TNF receptor 1; TGF-β, transforming growth factor-β.

tumour cell surface. Perhaps caveolae serve as a centre of proteolytic activity, where various components of proteolytic cascades co-cluster, interact, and ultimately execute their functions in metastatic progression.

Conclusions, and significance of cathepsin B in malignant progression

Cancer is a complex disease that results from multiple genetic alterations, and involves a functional interplay between various cell types and signalling molecules expressed by them. An important factor in metastasis is the ability of tumour cells to degrade and move through extracellular matrices. Matrix degradation is co-operatively influenced by various members of proteolytic cascade(s), and each member may have an important, but distinct, role(s) in tumour growth and invasion. Our current view of the interactions of cathepsin B with the various systems implicated in malignant progression is depicted in Figure 1.

The tumour environment consists of not only tumour cells, but also stromal and inflammatory cells and components of the extracellular matrix, and it is now well recognized that all of these contribute to malignancy. There is no doubt that the roles that cathepsin B plays in tumour progression need to be elucidated further, especially with regard to interactions with other proteolytic enzymes and regulation of cathepsin B expression. Evidence gathered so far, however, leaves no doubt about the association of this enzyme with tumour progression, especially with the transition from premalignant lesions to invasive ones.

This work was supported by National Institutes of Health (NIH) grants CA 36481 and CA 56586, and DOD grants PC991261 and BC013005. Due to a limitation on the number of references, we have cited mainly review articles, and therefore apologize to the authors of many important primary papers, which have not been cited.

References

1. Koblinski, J.E., Ahram, M. and Sloane, B.F. (2000) Clin. Chim. Acta **291**, 113–135
2. Ahram, M. and Sloane, B.F. (1997) in Proteolysis in Cell Functions (Hopsu-Havu, V.K. Jarvinen, M. and Kirschke, H., eds), pp 455–462, IOS Press, Amsterdam
3. Yan, S., Sameni, M. and Sloane, B.F. (1998) Biol. Chem. **379**, 113–123
3a. Murnane, M.J., Sheahan, K., Ozdermirli, M. and Shuja, S. (1991) Cancer Res. **51**, 1137–1142
4. Kobayashi, H., Schmitt, M., Goretzki, L., Chucholowski, N., Calvete, J., Kramer, M., Gunzler, W.A., Janicke, F. and Graeff, H. (1991) J. Biol. Chem. **266**, 5147–5152
5. Sinha, A.A., Gleason, D.F., Deleon, O.F., Wilson, M.J. and Sloane, B.F. (1993) Anat. Rec. **235**, 233–240
6. Koblinski, J.E., Dosescu, J., Sameni, M., Moin, K., Clark, K. and Sloane, B.F. (2002) J. Biol. Chem. **277**, 32220–32227
7. Campo, E., Munoz, J., Miquel, R., Palacin, A., Cardesa, A., Sloane, B.F. and Emmert-Buck, M.R. (1994) Am. J. Pathol. **145**, 301–309
8. Castiglioni, T., Merino, M.J., Elsner, B., Lah, T.T., Sloane, B.F. and Emmert-Buck, M.R. (1994) Hum. Pathol. **25**, 857–862

9. Sloane, B.F., Moin, K., Sameni, M., Tait, L.R., Rozhin, J. and Ziegler, G., (1994) J. Cell Sci. 107, 373–384

10. Fernandez, P.L., Farre, X., Nadal, A., Fernandez, E., Peiro, N., Sloane, B.F., Shi, G.P., Chapman, H.A., Campo, E. and Cardesa, A. (2001) Int. J. Cancer 95, 51–55

11. Turk, B., Stoka, V., Rozman-Pungercar, J., Cirman, T., Droga-Mazovec, G., Oreslic, K. and Turk, V. (2002) Biol. Chem. 383, 1035–1044

12. Foghsgaard, L., Wissing, D., Mauch, D., Lademann, U., Bastholm, L., Boes, M., Elling, F., Leist, M. and Jaattela, M. (2001) J. Cell Biol. 153, 999–1010

13. Nicholson, D.W. (1999) Cell Death Differ. 6, 1028–1042

14. Stoka, V., Turk, B., Schendel, S.L., Kim, T.-H., Cirman, T., Snipast, S.J., Ellerby, L.M., Bredesen, D., Freeze, H., Abrahamson, M. et al. (2001) J. Biol. Chem. 276, 3149–3157

15. Castino, R., Pace, D., Demoz, M., Gargiulo, M., Ariatta, C., Raiteri, E. and Isidoro, C. (2002) Int. J. Cancer 97, 775–779

16. Isahara, K., Ohsawa, Y., Kanamori, S., Shibata, M., Waguri, S., Sato, N., Gotow, T., Watanabe, T., Momoli, T., Urase, K. et al. (1999) Neuroscience 91, 233–249

17. Guicciardi, M.E., Deussing, J., Miyoshi, H., Bronk, S.F., Svingen, P.A., Peters, C., Kauffmann, S.H. and Gores, G.J. (2000) J. Clin. Invest. 106, 1127–1137

18. Mathiasen, I.S., Lademann, U. and Jaattela, M. (1999) Cancer Res. 59, 4848–4856

19. Roberts, L.R., Kurosawa, H., Bronk, S.F., Fesmier, P.J., Agellon, L.B., Leung, W.-Y. and Mao, F. (1997) Gastroenterology 113, 1714–1726

20. Jones, B., Roberts, P.J., Faubion, W.A., Kominami, E. and Gores, G.J. (1998) Am. J. Physiol. 275, G723–G730

21. Jaattela, M. (1999) Exp. Cell Res. 248, 30–43

22. Frosch, B.A., Berquin, I., Emmert-Buck, M.R., Moin, K. and Sloane, B.F. (1999) APMIS 107, 28–37

23. Berquin, I.M. and Sloane, B.F. (1996) Adv. Exp. Med. Biol. 389, 281–294

24. Rhaissi, H., Bechet, D. and Ferrara, M.(1993) Biochimie 75, 899–904

25. Yan, S., Berquin, I.M., Troen, B.R. and Sloane, B.F. (2000) DNA Cell Biol. 19, 79–91

26. Yan, S. and Sloane, B.F. (2003) Biol. Chem., in the press

27. Buttice, G., Duterque-Coquillaud, M., Basuyaux, J.P., Carrere, S., Kurkinen, M. and Stehelin, D. (1996) Oncogene 13, 2297–2306

28. Sasaki, H., Yukiue, H., Moiriyama, S., Kobayashi, Y., Nakashima, Y., Kaji, M., Kiriyama, M., Fukai, I., Yamakawa, Y. and Fujii, Y. (2001) J. Surg. Res. 101, 242–247

29. Davidson, B., Risberg, B., Goldberg, I., Nesland, J.M., Berner, A., Trope, C.G., Kristensen, G.B., Bryne, M. and Reich, R. (2001) Am. J. Surg. Pathol. 25, 1493–1500

30. Sato, Y., Abe, M., Tanaka, K., Iwasaka, C., Oda, N., Kanno, S., Oikawa, M., Nakano, T. and Igarashi, T. (2000) Adv. Exp. Med. Biol. 476, 109–115

31. Teruyama, K., Abe, M., Nakano, T., Iwasaka-Yagi, C., Takahashi, S., Yamada, S. and Sato, Y. (2001) J. Cell. Physiol. 188, 243–252

32. Keppler, D. and Sloane, B.F. (1996) Enzyme Protein 49, 94–105

33. Qian, F., Chan, S.J., Achkar, C., Steiner, D.F. and Frankfater, A. (1994) Biochem. Biophys. Res. Commun. 202, 429–436

34. Berquin, I.M., Yan, S., Katiyar, K., Huang, L., Sloane, B.F. and Troen, B.R. (1999) J. Leukocyte Biol. 66, 609–616

35. Rozhin, J., Gomez, A.P., Ziegler, G.H., Nelson, K.K., Chang, Y.S., Fong, D., Onoda, J.M., Honn, K.V. and Sloane, B.F. (1990) Cancer Res. 50, 6278–6284

36. Yan, Z., Deng, X., Chen, M., Xu, Y., Ahram, M., Sloane, B.F. and Friedman, E. (1997) J. Biol. Chem. 272, 27902–27907

37. Moin, K., Cao, L., Day, N.A., Koblinski, J.E. and Sloane, B.F. (1998) Biol. Chem. 379, 1093–1099

38. Sameni, M., Elliott, E., Ziegler, G., Fortgens, P.H., Dennison, C. and Sloane, B.F. (1995) Pathol. Oncol. Res. 1, 43–53

39. Mai, J., Sameni, M., Mikkelsen, T. and Sloane, B.F. (2002) Biol. Chem. **383**, 1407–1413
40. Khan, A., Krishna, M., Baker, S.P. and Banner, B.F. (1998) Mod. Pathol. **11**, 704–708
41. Mathieu, M., Vignon, F., Capony, F. and Rochefort, H. (1991) Mol. Endocrinol. **5**, 815–822
42. Lorenzo, K., Ton, P., Clark, J.L., Coulibaly, S. and Mach, L. (2000) Cancer Res. **60**, 4070–4076
43. Kuliawat, R., Klumperman, J., Ludwig, T. and Arvan, P.J. (1997) J. Cell Biol. **137**, 595–608
44. Heuser, J. (1989) J. Cell Biol. **108**, 855–864
45. Coopman, P.J., Thomas, D.M., Gehlsen, K.R. and Mueller, S.C. (1996) Mol. Biol. Cell **7**, 1789–1804
46. Guo, M., Mathieu, P.A., Linebaugh, B., Sloane, B.F. and Reiners, Jr, J.J. (2002) J. Biol. Chem. **277**, 14829–14837
47. Demchik, L.L. and Sloane, B.F. (1999) in Proteases: New Perspectives (Turk, V., ed.), pp. 109–124, Birkhauser Publishing Ltd., Basel
48. Hernandez-Barrantes, S., Bernardo, M., Toth, M. and Fridman, R. (2002) Semin. Cancer Biol. **12**, 131–138
49. Mai, J., Finley, Jr, R.L., Waisman, D.M. and Sloane, B.F. (2000) J. Biol. Chem. **275**, 12806–12812
50. Stahl, A. and Mueller, B.M. (1995) J. Cell Biol. **129**, 335–344
51. Smart, E.J., Ying, Y., Donzell, W.C. and Anderson, R.G. (1996) J. Biol. Chem. **271**, 29427–29435
52. Koleske, A., Baltimore, D. and Lisanti, M.P. (1995) Proc. Natl. Acad. Sci. U.S.A. **92**, 1381–1385
53. Mouraviev, V., Li, L., Tahir, S.A., Yang, G., Timme, T.M., Goltsov, A., Ren, C., Satoh, T., Wheeler, T.M., Ittmann, M.M. et al. (2002) J. Urol. **168**, 1589–1596

Biochem. Soc. Symp. **70**, 277–285
(Printed in Great Britain)

23

Proteolytic and non-proteolytic migration of tumour cells and leucocytes

Peter Friedl[1] and Katarina Wolf

Cell Migration Laboratory, Department of Dermatology, University of Würzburg, Josef-Schneider-Str. 2, 97080 Würzburg, Germany

Abstract

The migration of different cell types, such as leucocytes and tumour cells, involves cellular strategies to overcome the physical resistance of three-dimensional tissue networks, including proteolytic degradation of extracellular matrix (ECM) components. High-resolution live-cell imaging techniques have recently provided structural and biochemical insight into the differential use of matrix-degrading enzymes in the migration processes of different cell types within the three-dimensional ECM. Proteolytic migration is achieved by slow-moving cells, such as fibroblasts and mesenchymally moving tumour cells, by engaging matrix metalloproteinases, cathepsins and serine proteases at the cell surface in a focalized manner ('pericellular proteolysis'), while adhesion and migratory traction are provided by integrins. Pericellular breakdown of ECM components generates localized matrix defects and remodelling along migration tracks. In contrast with tumour cells, constitutive non-proteolytic migration is used by rapidly moving T lymphocytes. This migration type does not generate proteolytic matrix remodelling, but rather depends on shape change to allow cells to glide and squeeze through gaps and trails present in connective tissues. In addition, constitutive proteolytic migration can be converted into non-proteolytic movement by protease inhibitors. After the simultaneous inhibition of matrix metalloproteinases, serine/threonine proteases and cysteine proteases in tumour cells undergoing proteolysis-dependent movement, a fundamental adaptation towards amoeboid movement is able to sustain non-proteolytic migration in these tumour cells (the mesenchymal–amoeboid transition). Instead of using proteases for matrix degradation, the tumour cells use leucocyte-like strategies of shape change and squeezing through matrix gaps along tissue scaffolds. The diversity of protease function in cell migration by different cell types highlights

---

[1]To whom correspondence should be addressed (e-mail peter.fr@mail.uni-wuerzburg.de).

response diversity and molecular adaptation of cell migration upon pharmaco-therapeutic protease inhibitor treatment.

Introduction

The extracellular matrix (ECM) in connective tissues represents a structural scaffold as well as a barrier for mobile cells, such as passenger leucocytes or invading tumour cells. The migration of diverse cells on or within the tissues involves adhesive and proteolytic interactions with components of the ECM, coupled with a dynamic actin cytoskeleton and polarized shape change [1]. Migratory cell translocation may result from single-cell migration, but also from dynamic cell chains and collectives [2,3]. While the molecular migration mechanisms of cell chains and collectives remain largely unknown, the principles of adhesion receptor and protease function in single-cell migration are established in detail. The migration of single cells is determined by a set of characteristic cell parameters, including size, morphodynamic adaptability, integrin expression and function, as well as the rate of turnover of cell–matrix interactions [4,5]. Small leucocytes such as T lymphocytes utilize rapid low-affinity interactions with the ECM substrate and retain significant residual migration after abrogation of integrin function, which is reminiscent of the movement of the lower amoeba *Dictyostelium discoideum* ('amoeboid movement') [6,7]. In contrast, stromal cells such as fibroblasts and solid tumour cells are 10–30-fold larger in volume than lymphocytes and develop a slow migration type sustained by an elongated spindle-shaped ('mesenchymal') morphology, a high degree of integrin-mediated adhesion and force generation, and concomitantly a slow turnover of cell–substrate interactions [8].

Upon tumour progression, multiple classes of ECM-degrading enzymes are up-regulated and activated in tumour cells, including matrix metalloproteinases (MMPs), serine proteases and cathepsins [9]. In mobile as well as resident cells, secreted enzymes and proteases expressed at the cell surface mediate the cleavage of many ECM components, such as collagens, fibronectin and laminin [10], thereby increasing the pericellular space of tissue scaffolds in basement membranes and interstitial tissues. Local ECM degradation creates physicochemical trails for migrating cells, and is therefore thought to contribute to tumour invasion and dissemination [10]. We here review the spatial and temporal contributions of proteases to ECM degradation and cell migration, and further discuss recent knowledge of alternative, non-proteolytic cell strategies to overcome physical matrix constraints.

Methodology used to visualize pericellular proteolysis

Several methods have been developed to localize the active subset of proteases expressed and utilized by cells to degrade ECM substrata. *In situ* proteolysis in tissue sections and living cells can be visualized by *in situ* zymography, substrate degradation assays using fluorescent substrata, physical substrate detection analysis, and labelling with cleavage-epitope-specific antibodies. In conjunction, these techniques provide a relatively complete picture of the location and extent of pericellular ECM degradation at the cellular and, in part, subcellular level.

In situ zymography

This technique is a topographic modification of conventional zymography after electrophoresis. Thin substrate-containing gels, such as gelatin or collagen incorporated in a carrier sheet (e.g. acrylamide), are overlaid on to non-denatured frozen tissue sections. The clearance zone of the protein substrate is assessed qualitatively by Coomassie Blue staining or fluorescence detection. Depending on the substrate incorporated into the carrier, a spectrum of gelatinases and proteases can be specifically detected and related to the histological topography and tissue composition of the same or an adjacent tissue section [11]. However, diffusion of proteases may decrease both sensitivity and spatial resolution at the cellular and subcellular level. A further disadvantage of tissue sections is that analysis of live-cell dynamics is not accessible by *in situ* zymography.

Degradation of immobilized fluorescent probes

To increase sensitivity and also to provide access to living samples, the migration template is coated with fluorescently labelled ECM substrate. Pericellular proteolysis is visualized qualitatively as the zone cleared from fluorescence upon migration. Upon detection by fluorescence microscopy, both sensitivity and spatial resolution are high to the subcellular level. Fluorescent probes are used for two-dimensional (2D) migration assays on surfaces coated with fibronectin [12,13], as well as for three-dimensional (3D) invasion assays using fluorescently labelled type IV collagen within reconstituted basement membrane equivalents [14] or within fibrillar type I collagen matrices [15]. If combined with semi-quantitative detection of fluorescence released into the supernatant, the net degradation of ECM substrata by intact cells can additionally be quantified [15]. After injection into live animals, a fluorogenic pseudosubstrate that mimics the natural substrate is cleaved predominantly in proteolytic tumour nodules and metastases, which can be detected by whole-body fluorescence detection and by histology [16].

Confocal backscatter and autofluorescence reconstruction

High-resolution physical detection of 3D ECM scaffolds and structural changes therein is obtained by single-photon confocal backscatter microscopy [17,18] or second harmonic generation imaging using two-photon microscopy [15,19]. Because these techniques detect structural features of ECM networks, such as the backscatter of the laser light and autofluorescence, independent of labelling or fixation, reorganization of the tissue texture within live samples such as collagen matrices *in vitro* as well as connective tissues of living organisms are accessible for real-time imaging [15,19]. The backscatter/autofluorescecnce signal allows the dynamic reconstruction of collagen fibril and bundle structure up to a pixel resolution of 100–200 nm, and includes the visualization of fibre distortion, traction, bundling and clumping, as well as the generation of proteolytic matrix defects by migrating cells [5,18]. As a major disadvantage, phototoxicity induced by the laser light currently hampers the long-term imaging of live cell dynamics (beyond a few hours).

IR micro-spectroscopy

Proteolysis of helical collagens causes unwinding of the triple helix and cleavage of peptide bonds. Both processes result in spectral changes in the absorption characteristics of IR light at wavelengths between 1500 and 1700 nm that is detected by Fourier transform IR micro-spectroscopy [20]. This technique is used to show pericellular and diffuse degradation of reconstituted basement membrane and gelatin substrate in fixed as well as living cells up to the cellular level. As disadvantages, relatively low spatial resolution (3–10 μm), the masking of the ECM signal by the superimposed cell body and demanding hardware requirements may limit this technique to specific applications.

Detection of cleavage-site-specific epitopes

Upon proteolytic cleavage, proteases generate the exposure of previously hidden epitopes near the cleavage site. Cleavage-site-specific neoepitopes can be detected by monoclonal or polyclonal antibodies and represent the cleaved protein in a transition state between initial and complete degradation. Cleavage-site-specific epitopes are exposed in type II collagen upon cartilage degradation by MMPs [21], in the γ2 chain of laminin-5 after proteolysis by MMP-2 [22] and in type I collagen upon remodelling by migrating tumour cells (K. Wolf and P. Friedl, unpublished work).

These above techniques can be combined with biochemical assays and bright-field videomicroscopy to delineate the location and extent of pericellular proteolysis exerted by invading tumour cells *in vitro* and *in vivo*. As we will detail below, these techniques are also useful in dissecting similarities and differences in protease location, function, and related ECM remodelling in other migrating cell types, such as leucocytes.

Proteolytic migration by tumour cells

Fibrillar collagen is the main structural component of the interstitial matrix, and is therefore an important substrate for studies on how migrating cells change the structure, and thereby lower the biomechanical resistance, of connective tissues. Native collagen is degraded by several proteases, including MMPs-1, -2, -8 and -13, membrane-anchored membrane type 1 (MT1)-MMP, MT3-MMP, and cathepsins B, K and L [23–25]. The contribution of these and other proteases to the tumour invasion and dissemination process was shown by studies using inhibitor compounds. Blocking of MMPs, serine proteases and/or cathepsins impairs invasive tumour invasion and migration in several *in vitro* models, including the amnionic membrane, reconstituted basement membrane, and collagen matrices [26–28], as well as in experimental metastasis *in vivo* ([29] and references cited therein).

Where and how proteases support the migration of cells was revealed by *in situ* detection of proteolysis in tumour cells generating dynamic interactions with the ECM. Tumour cells overexpressing MMPs and other proteases generate pericellular proteolytic substrate degradation along their migration tracks on 2D ECM substrata, such as fibronectin, gelatin or Matrigel [12,13,30]. Likewise, in 3D ECM models, tube-like matrix defects are formed that represent the paths of previous migration and which facilitate the migration of neighbouring cells [17]. The subcellular focalization of surface proteases towards the tips of regions interacting with the ECM substrate occurs in conjunction with that of adhesion receptors of the β1 and β3 integrin families on gelatin, fibronectin and vitronectin [15,31], or of CD44 on hyaluronan [32]. Urokinase-type plasminogen activator co-localizes with β1 integrins in tumour cells on fibrin substrate [33]. MT1-MMP is detected co-localized with clustered β1 integrins near the leading edge of migrating tumour cells at traction and bundling sites to collagen fibres [15,31], suggesting that adhesion and proteolysis are focalized towards outward edges and form a functional unit [12]. It is currently unclear, however, by which mechanisms proteases are recruited to focal substrate contacts or how substrate degradation is spatially and temporally regulated without challenging cell attachment and force generation. However, it is clear that highly specific pericellular cleavage of ECM components contributes to the removal of matrix barriers to favour the advancement of the cell body [2,10].

Non-proteolytic migration by T lymphocytes

In contrast with tumour cells that move proteolytically, T lymphocytes utilize a different, more rapid migration strategy that does not involve the development of stringently focalized interactions with ECM substrata. Similar to the lower amoeba *Dictyostelium discoideum*, hallmarks of movement in T lymphocytes are elliptoid yet flexible cell morphology and dynamic polarized pseudopod protrusions and retractions that generate fast low-affinity crawling with velocities of up to 25 μm/min on surfaces and through 3D collagenous scaffolds [34–37]. Migrating T cells lack the focalization of adhesion receptors such as integrins and CD44 towards substrate contacts, and display a diffuse cortical actin cytoskeleton devoid of stress fibres [6,38].

Similar to tumour cells, activated T cells express a spectrum of ECM-degrading proteases at the mRNA level, including MT1- and MT4-MMPs, MMP-9, cathepsin L and urokinase-type plasminogen activator, yet little or no expression of MT1-MMP or other proteases occurs at the cell surface [39]. The 'amoeboid' crawling through the collagen fibre network is driven by flexible morphological adaptation along collagen fibres, followed by squeezing and gliding of the cell through pre-existing matrix gaps [40], independent of collagenase function and fibre degradation [39]. If proteases, including collagenases, are inhibited by broad-spectrum inhibitors targeting MMPs, serine/threonine proteases and cathepsins under non-toxic conditions, T cells do not exhibit changes either in their shape and morphodynamics or in their kinetics of interaction with collagen fibres, resulting in undiminished migration velocities [39]. Amoeboid migration within the interstitial ECM hence embodies the prototype of a non-proteolytic migration mechanism that allows cells to overcome matrix barriers by 'supramolecular' physical strategies, such as shape change. Because of the more diffuse organization of surface integrins and the actin cytoskeleton, it is likely that, in cells with amoeboid-type movement, membrane-anchored proteases do not undergo sufficient focalization to provide proteolysis during the periods of loose and short-lived adhesion to the ECM substrate.

Further studies will be required to investigate whether migration-associated pericellular ECM remodelling is mandatory for the trafficking of other leucocytes, such as neutrophils and monocytes/macrophages [41]. In addition, it will be important to differentiate the tightly regulated process of migration-associated focalized proteolysis that confers relatively minor tissue reorganization from the more extensively damaging 'bystander proteolysis' caused by proteases released from intracellular vesicles by activated leucocytes during the processes of acute and chronic inflammation [41].

Plasticity in protease function in tumour cells (mesenchymal–amoeboid transition)

In addition to the above concept of pericellular ECM degradation, accumulating evidence suggests that, despite their established pro-invasive function, ECM-degrading enzymes may be partly or fully dispensable upon tumour cell motility and dissemination. After blocking of MMPs or serine proteases, significant residual migration of individual cells is observed in different migration and invasion models, including collagen matrices, reconstituted basement membrane and polymerized fibrin [27,28,42,43]. *In vivo*, protease inhibitor-based targeting of MMPs and serine proteases has resulted in unexpectedly small benefit in some animal tumour models and in clinical trials in humans, suggesting that a principal protease-independent dissemination capacity remains intact in tumour cells [29,44–46]. These studies, however, leave unresolved the mechanisms by which cells may maintain migratory dissemination in the absence of ECM-degrading capacity. One possibility is that proteolytic compensation could be provided by enzymes not inhibited in these studies; alternatively, unknown protease-independent compensation strategies could provide cell mobility.

In highly proteolytic and invasive HT-1080 fibrosarcoma and MDA-MB-231 mammary carcinoma cells, abrogation of constitutive proteolysis pathways by inhibition of MMPs, serine proteases and cysteine proteases, including cathepsins, results in an unexpectedly small decrease in migration through 3D collagen matrices. A similar effect is seen in the mouse dermis after intradermal tumour cell injection, as monitored by intravital microscopy [15]. In response to protease inhibitors, a cellular and molecular adaptation reaction allows the tumour cells to sustain migration: the constitutive proteolytic and fibroblast-like ('mesenchymal') migration type is replaced by a non-proteolytic, amoeba-like migration strategy (mesenchymal–amoeboid transition) [15]. After pericellular proteolysis is lost (as detected by zymography, fluorescent substrate degradation and confocal backscatter analyses), the tumour cells are able to maintain high migration rates via lymphocyte-like gliding into matrix gaps, squeezing through pores and concomitant cell compression by outside fibrillar scaffolds conferring cell constriction in narrow regions ('constriction rings' down to 1 μm diameter). Consistent with a newly acquired capacity for amoeboid crawling, changes in cell–matrix contact assembly and turnover include the exclusion of MT1-MMP from fibre binding sites and reduced surface focalization of β1 integrins [15]. Thus, as well as occurring in lymphocytes, non-proteolytic dissemination strategies may be maintained by tumour cells to allow motility and dissemination under adverse conditions.

Conclusions

In summary, the contribution of ECM-degrading proteases to cell motility can be delineated by two opposite conceptual poles that represent a spectrum from highly proteolytic to non-proteolytic (Figure 1). While most, if not all, mesenchymally moving tumour cells utilize ECM-degrading enzymes for the pericellular proteolysis of ECM components, the amoeboid, non-proteolytic migration type employed by T lymphocytes occurs independently of proteolytic matrix degradation and remodelling. Consequently, amoeboid T cell migration is insensitive to treatment with pharmacological protease inhibitors. In proteolytically moving tumour cells, abrogation of the ECM-degrading capacity may either yield reduced migration rates [26–28] or, alternatively, induce an adaptation programme towards non-proteolytic amoeboid crawling at significant migration rates, described here as a mesenchymal–amoeboid transition (Figure 1, double-headed arrow). The mesenchymal–amoeboid transition represents an interesting, novel response pathway in molecular cell dynamics. Stringent, adhesive and proteolytic cell–ECM contacts, as developed by mesenchymally moving cells, can be converted into less well characterized, more diffuse and less proteo-

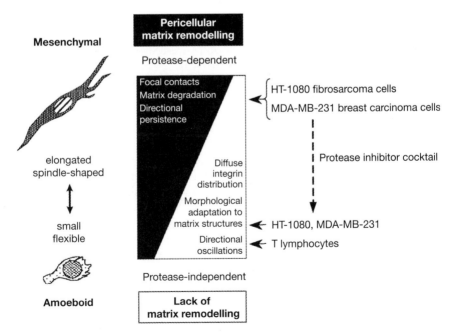

Figure 1 Diversity and adaptation in protease function in different cell types.

lytic interactions with the substrate, reminiscent of the type of cell movement employed by *Dictyostelium* [37].

In conclusion, non-proteolytic amoeboid movement may represent a robust supramolecular physical mechanism employed by cells to overcome matrix barriers, acting as a 'salvage' pathway to sustain cell translocation, not only in lymphocytes but also in tumour cells, independently of matrix protease function. Ultimately, this cellular and molecular plasticity that allows amoeboid-type movement provides an interesting alternative pathway that may contribute to undiminished tumour cell dissemination and disease progression upon MMP inhibitor therapy in cancer patients [29].

This research was supported by the Deutsche Forschungsgemeinschaft (grant no. FR 1155/2-3).

References

1. Lauffenburger, D.A. and Horwitz, A.F. (1996) Cell **84**, 359–369
2. Friedl, P. and Bröcker, E.-B. (2000) Cell. Mol. Life Sci. **57**, 41–64
3. Hegerfeldt, Y., Tusch, M., Brocker, E.B. and Friedl, P. (2002) Cancer Res. **62**, 2125–2130
4. Cox, E.A. and Huttenlocher, A. (1998) Microsc. Res. Tech. **43**, 412–419
5. Friedl, P., Zanker, K.S. and Bröcker, E.-B. (1998) Microsc. Res. Tech. **43**, 369–378
6. Friedl, P., Entschladen, F., Conrad, C., Niggemann, B. and Zanker, K.S. (1998) Eur. J. Immunol. **28**, 2331–2343
7. Brakebusch, C., Fillatreau, S., Potocnik, A.J., Bungartz, G., Wilhelm, P., Svensson, M., Kearney, P., Korner, H., Gray, D. and Fassler, R. (2002) Immunity **16**, 465–477
8. Maaser, K., Wolf, K., Klein, C.E., Niggemann, B., Zanker, K.S., Brocker, E.B. and Friedl, P. (1999) Mol. Biol. Cell **10**, 3067–3079
9. Birkedal-Hansen, H. (1995) Curr. Opin. Cell Biol. **7**, 728–735
10. Murphy, G. and Gavrilovic, J. (1999) Curr. Opin. Cell Biol. **11**, 614–621
11. Kurschat, P., Wickenhauser, C., Groth, W., Krieg, T. and Mauch, C. (2002) J. Pathol. **197**, 179–187
12. Nakahara, H., Howard, L., Thompson, E.W., Sato, H., Seiki, M., Yeh, Y. and Chen, W.T. (1997) Proc. Natl. Acad. Sci. U.S.A. **94**, 7959–7964
13. d'Ortho, M.P., Stanton, H., Butler, M., Atkinson, S.J., Murphy, G. and Hembry, R.M. (1998) FEBS Lett. **421**, 159–164
14. Sameni, M., Moin, K. and Sloane, B.F. (2001) Neoplasia **2**, 496–504
15. Wolf, K., Mazo, I., Leung, H., Engelke, K., von Andrian, U.H., Deryugina, E.I., Strongin, A.Y., Brocker, E.B. and Friedl, P. (2003) J. Cell Biol. **160**, 267–277
16. Bremer, C., Tung, C.H. and Weissleder, R. (2001) Nat. Med. (N.Y.) **7**, 743–748
17. Friedl, P., Maaser, K., Klein, C.E., Niggemann, B., Krohne, G. and Zanker, K.S. (1997) Cancer Res. **57**, 2061–2070
18. Friedl, P. and Brocker, E.B. (2001) in Image Analysis: Methods and Applications, 2nd edn (Hader, D.P., ed.), pp. 9–21, CRC Press, Boca Raton, FL
19. Campagnola, P.J., Millard, A.C., Terasaki, M., Hoppe, P.E., Malone, C.J. and Mohler, W.A. (2002) Biophys. J. **82**, 493–508
20. Federman, S., Miller, L.M. and Sagi, I. (2002) Matrix Biol. **21**, 567–577
21. Wu, W., Billinghurst, R.C., Pidoux, I., Antoniou, J., Zukor, D., Tanzer, M. and Poole, A.R. (2002) Arthritis Rheum. **46**, 2087–2094
22. Gianelli, G., Falk-Marzillier, J., Schiraldi, O., Stetler-Stevenson, W.G. and Quaranta, V. (1997) Science **277**, 225–228

23. Aimes, R.T. and Quigley, J.P. (1995) J. Biol. Chem. **270**, 5872–5876
24. Ohuchi, E., Imai, K., Fujii, Y., Sato, H., Seiki, M. and Okada, Y. (1997) J. Biol. Chem. **272**, 2446–2451
25. Sassi, M.L., Eriksen, H., Risteli, L., Niemi, S., Mansell, J., Gowen, M. and Risteli, J. (2000) Bone **26**, 367–373
26. Mignatti, P., Robbins, E. and Rifkin, D.B. (1986) Cell **47**, 487–498
27. Kurschat, P., Zigrino, P., Nischt, R., Breitkopf, K., Steurer, P., Klein, C.E., Krieg, T. and Mauch, C. (1999) J. Biol. Chem. **274**, 21056–21062
28. Ntayi, C., Lorimier, S., Berthier-Vergnes, O., Hornebeck, W. and Bernard, P. (2001) Exp. Cell Res. **270**, 110–118
29. Coussens, L.M., Fingleton, B. and Matrisian, L.M. (2002) Science **295**, 2387–2392
30. Sameni, M., Dosescu, J. and Sloane, B.F. (2001) Biol. Chem. **382**, 785–788
31. Belkin, A.M., Akimov, S.S., Zaritskaya, L.S., Ratnikov, B.I., Deryugina, E.I. and Strongin, A.Y. (2001) J. Biol. Chem. **276**, 18415–18422
32. Mori, H., Tomari, T., Koshikawa, N., Kajita, M., Itoh, Y., Sato, H., Tojo, H., Yana, I. and Seiki, M. (2002) EMBO J. **21**, 3949–3959
33. Wei, Y., Lukashev, M., Simon, D.I., Bodary, S.C., Rosenberg, S., Doyle, M.V. and Chapman, H.A. (1996) Science **273**, 1551–1555
34. Devreotes, P.N. and Zigmond, S.H. (1988) Annu. Rev. Cell Biol. **4**, 649–686
35. Schor, S.L., Allen, T.D. and Winn, B. (1983) J. Cell Biol. **96**, 1089–1096
36. Sanchez-Madrid, F. and del Pozo, M.A. (1999) EMBO J. **18**, 501–511
37. Friedl, P., Borgmann, S. and Brocker, E.B. (2001) J. Leukocyte Biol. **70**, 491–509
38. Entschladen, F., Niggemann, B., Zänker, K.S. and Friedl, P. (1997) J. Immunol. **159**, 3203–3210
39. Wolf, K., Muller, R., Borgmann, S, Brocker, E. B, and Friedl, P. (2003) Blood, in the press
40. Friedl, P. and Brocker, E.B. (2000) Dev. Immunol. **7**, 249–266
41. Owen, C.A. and Campbell, E.J. (1999) J. Leukocyte Biol. **65**, 137–150
42. Deryugina, E.I., Luo, G.X., Reisfeld, R.A., Bourdon, M.A. and Strongin, A. (1997) Anticancer Res. **17**, 3201–3210
43. Hiraoka, N., Allen, E., Apel, I.J., Gyetko, M.R. and Weiss, S.J. (1998) Cell **95**, 365–377
44. Della, P.P., Soeltl, R., Krell, H.W., Collins, K., O'Donoghue, M., Schmitt, M. and Kruger, A. (1999) Anticancer Res. **19**, 3809–3816
45. Kruger, A., Soeltl, R., Sopov, I., Kopitz, C., Arlt, M., Magdolen, V., Harbeck, N., Gansbacher, B. and Schmitt, M. (2001) Cancer Res. **61**, 1272–1275
46. Zucker, S., Cao, J. and Chen, W.T. (2000) Oncogene **19**, 6642–6650

Subject index